Lecture Notes in Mathematics

Edited by A. Dold and B. Eckmann

835

Heiner Zieschang
Elmar Vogt
Hans-Dieter Coldewey

Surfaces and Planar Discontinuous Groups

Revised and Expanded Translation
Translated from the German by J. Stillwell

Springer-Verlag
Berlin Heidelberg New York 1980

Authors

Heiner Zieschang
Universität Bochum
Institut für Mathematik
Universitätsstr. 150
4630 Bochum 1
Federal Republic of Germany

Elmar Vogt
Freie Universität Berlin
Institut für Mathematik I
Hüttenweg 9
1000 Berlin 33
Federal Republic of Germany

Hans-Dieter Coldewey
Allescherstr. 40b
8000 München 71
Federal Republic of Germany

Revised and expanded translation of:
H. Zieschang/E. Vogt/H.-D. Coldewey,
Flächen und ebene diskontinuierliche Gruppen
(Lecture Notes in Mathematics, vol. 122)
published by Springer-Verlag Berlin-Heidelberg-New York, 1970

AMS Subject Classifications (1980): 20Exx, 20Fxx, 30F35, 32G15, 51M10, 57Mxx

ISBN 3-540-10024-5 Springer-Verlag Berlin Heidelberg New York
ISBN 0-387-10024-5 Springer-Verlag New York Heidelberg Berlin

Printing and binding: Beltz Offsetdruck, Hemsbach/Bergstr.
2141/3140-543210

In memoriam Kurt Reidemeister

Introduction

For the two-dimensional manifolds, the surfaces, the classical topological problems - classification and Hauptvermutung - have long been solved, and more delicate questions can be investigated. However, the most interesting side of surface theory is not the topological, but the analytic. Results of the complex analytic theory, e.g., are often purely topological, but their proofs are not, using deep theorems of function theory. This results from a natural and close connection with discontinuous groups of motions in the non-euclidean or euclidean plane.

The following lectures deal in the first place with combinatorial topological theorems on surfaces and planar discontinuous groups; thus we have adopted the concept of the book "Einführung in die kombinatorische Topologie" by K. Reidemeister. Admittedly, in chapters 1-5 we have not kept strictly to the combinatorial conception, but have changed to another category where this seems convenient. Thus we are far from striving for the purity of the above book, and e.g. group theoretic theorems are sometimes proved geometrically, and vice versa.

In chapter 6 we consider Riemann surfaces and give an elementary foundation of a theory due to O. Teichmüller for the solution of the modular problem. In chapter 7 we prove the triangulation theorem and the Hauptvermutung for surfaces; these theorems help explain why, in 2-dimensional topology, the combinatorial (or PL) theory covers most interesting questions.

In order to read the main parts of the first five chapters only basic knowledge of group theory is necessary; the combinatorial group theory and surface theory is developed from scratch. In some of the later sections in the chapters we use some results from algebraic topology for convenience. However these sections can be omitted at a first reading. For chapters 6 and 7 a basic knowledge of general and algebraic topology, non-euclidean geometry and complex analysis is necessary; we hope that we have described a simple approach to the problems considered. I have not tried to list all the mathematical work that is connected with the combinatorial theory of surfaces and to value all articles appropriately.

This English version of "Flächen und ebene diskontinuierliche Gruppen" contains substantial additions to the original text; among them:

- A closer consideration of the Reidemeister-Schreier method in section 2.1 which leads to an algebraic proof of the Riemann-Hurwitz formula in 4.14.
- Proofs of the Grushko theorem and the Freiheitssatz in 2.9 and 2.11.
- Construction of mappings between surfaces with given degrees in 3.5.
- The theory of intersection numbers for curves on surfaces and Hopf's theory of mappings between surfaces in 3.6 and 3.7.
- The classification of finitely generated discontinuous groups of the plane which do not necessarily have compact fundamental region in 4.11.
- On decompositions of planar groups in 4.12.
- On the action of finite groups on surfaces in 4.15. Generalized versions of the Baer and Nielsen theorems in 5.14-16.
- Proofs of the Schönflies theorem, the triangulation theorem and the Hauptvermutung for surfaces in chapter 7.

In completing the lectures on surfaces I have been lucky to have a big number of helpers. As already said in the German version 1970 my thanks go to A.B. Sossinskij, A.V. Černavskij and A.M. Macbeath for suggestions and criticism. Elmar Vogt made the notes of my lectures in Frankfurt, and Dieter Coldewey worked with me on sections 6.9-11. My thanks go also to E. Gramberg, F. Hillefeld, R. Keller, N. Peczynski, W. Reiwer and B. Zimmermann for their discussions on the different problems. In addition I wish to thank Frau Faber for typing the original German text and Frau Schwarz for typing the English variant, W. Bitter for drawing the figures and G. Krause and B. Wicha for reading the proofs. Considerable help in the completion of the book was given to me by John Stillwell who not only translated the original text, but helped to make it more understandable and to insert new parts.

CONTENTS

INTERDEPENDENCE OF TOPICS

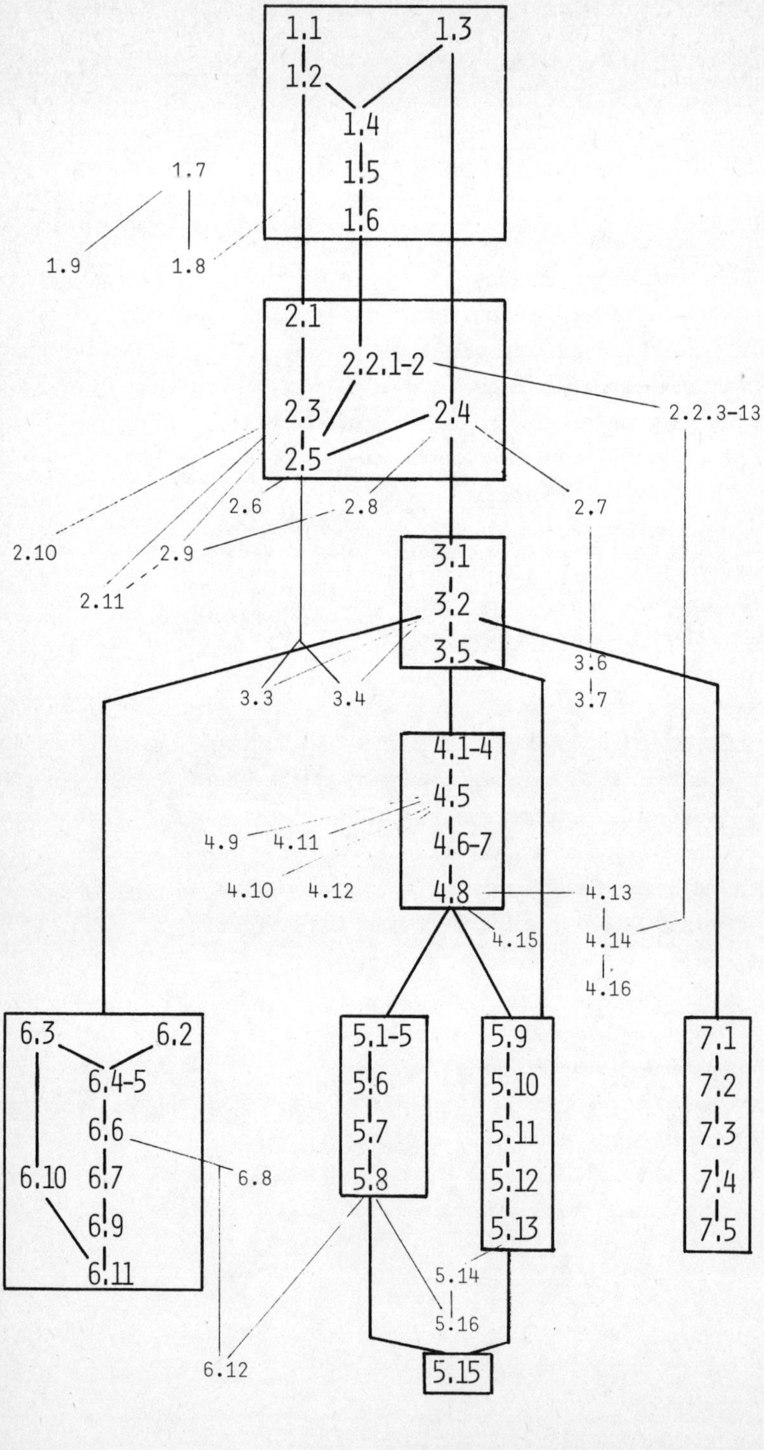

1. FREE GROUPS AND GRAPHS

1.1 FREE GROUPS

We consider 1-dimensional complexes (graphs) and their fundamental groups which have an interesting combinatorial property: they are the so-called free groups. We lead into the problems of combinatorial group theory and develop the methods of Nielsen and Reidemeister-Schreier. Basic for our approach is the close connection of complexes with groups and coverings with subgroups, which is important for later chapters. Here we begin this theory assuming only the definition of a group.

1.1.1 Definition. Given a system of symbols $(S_i)_{i\in I}$ we call each expression of the form

$$W = S_{\alpha_1}^{\varepsilon_1} S_{\alpha_2}^{\varepsilon_2} \dots S_{\alpha_n}^{\varepsilon_n}, \ \varepsilon_i = \pm 1$$

a *word in the* S_i, and the empty word is denoted by "1". The *product of words* is defined by writing the symbols of one after those of the other. We call two words $W = W_1 S_i^{\varepsilon} S_i^{-\varepsilon} W_2$ and $V = W_1 W_2$ elementarily equivalent. Two words W and V are called *equivalent* when there is a finite sequence of words $W = W_1, W_2, \dots, W_m = V$ where the W_i and W_{i+1} are elementarily equivalent to each other. The equivalence class of the word W is denoted by [W]. We define the product of two equivalence classes [W] and [V] by

$$[W] \cdot [V] = [WV].$$

The product is well-defined; for if $W = W_1, W_2, \dots, W_n = W'$ and $V = V_1, \dots, V_m = V'$ are two sequences of elementarily equivalent words then $WV = W_1V_1, W_2V_1, \dots, W_nV_1, W_nV_2, \dots, W_nV_m = W'V'$ is likewise such a sequence. The set of equivalence classes constitutes a group under this multiplication, the identity element of which is the class of the empty word. The inverse element of
$[W] = [S_{\alpha_1}^{\varepsilon_1} \dots S_{\alpha_n}^{\varepsilon_n}]$ is $[W^{-1}] = [S_{\alpha_n}^{-\varepsilon_n} \dots S_{\alpha_1}^{-\varepsilon_1}]$.

1.1.2 Definition. This group S is called the *free group* on the free generators $[S_i]$, $i \in I$, and the cardinal number of the index set is I called its *rank*. We write $S = \langle (S_i)_{i\in I} | \rangle$ and for finite or countable I we often write $S = \langle S_1,\ldots,S_n | \rangle$ or $S = \langle S_1, S_2,\ldots | \rangle$ or similar. We show later that the groups of different rank are not isomorphic (see 1.7, Theorem 1.7.7). We confine ourselves to free groups of at most denumerable rank. The free group $S = \langle (S_i)_{i\in I} | \rangle$ with the free generators $s_i = [S_i]$ has the following property

1.1.3 Proposition. (a) *For each group* G *and elements* $(x_i)_{i\in I}$, $x_i \in G$, *there exists a unique homomorphism* $\varphi: S \to G$ *with* $\varphi(s_i) = x_i$, $i \in I$. *This is often expressed by the diagram*

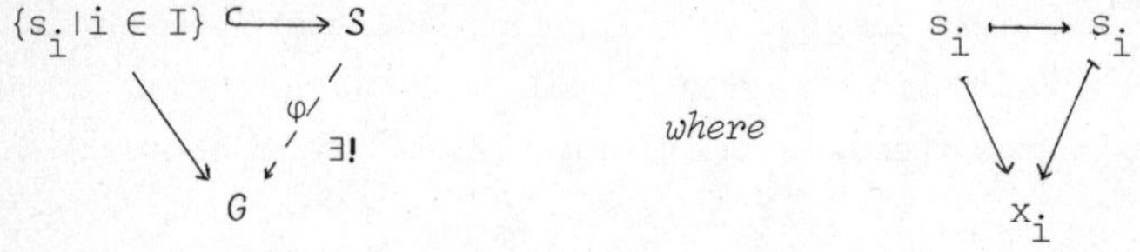

The proof will be evident after the next section 1·2 and exercise.

This property is called a *universal property*. It characterizes S up to isomorphism.

1.1.4 Theorem. (b) *Let* F *be a group and* $(t_i)_{i\in I}$ *elements of* F *such that for any group* G *and any mapping* $(t_i)_{i\in I} \to G$, $t_i \mapsto h_i$, *there is a unique homomorphism* $\varphi: F \to G$ *with* $\varphi(t_i) = h_i$. *Then* $F \cong S$.

Proof. Because of the property (a) we find a homomorphism $\varphi: S \to F$ with $\varphi(s_i) = t_i$. Then by assumption there is a homomorphism $\psi: F \to S$ with $\psi(t_i) = s_i$. Hence $\psi \circ \varphi: S \to S$ and $(\psi \circ \varphi)(s_i) = s_i$. By the uniqueness property (a) it follows that $\psi \circ \varphi = id_S$. The same argument, now using (b), shows that $\varphi \circ \psi = id_F$. It follows that φ and ψ are isomorphisms.

The concept of a free group can also be introduced by the universal property 1.1.4. This is the appropriate way from a categorical point of view, see [Massey 1967, 3.5]. Our elementary approach leads directly to the question of which words define the same element of the group. This so-called *word problem* has a simple solution for free generators of a free group.

Exercise: E 1.1-3.

1.2 WORD AND CONJUGACY PROBLEMS

Given a free generating system for a free group it is easy to decide whether two words represent the same element, or conjugate elements.

1.2.1 Definition. The word $W = S_{\alpha_1}^{\varepsilon_1} \dots S_{\alpha_n}^{\varepsilon_n}$ ($\varepsilon_i = \pm 1$) is called a *reduced word* when we do not have $S_{\alpha_i}^{\varepsilon_i} = S_{\alpha_{i+1}}^{-\varepsilon_{i+1}}$ for any $i = 1,\dots,n-1$.

1.2.2 Construction. The following process converts a given word

$$W = S_{\alpha_1}^{\varepsilon_1} \dots S_{\alpha_n}^{\varepsilon_n}$$

into an equivalent reduced word. We define $W_1 = S_{\alpha_1}^{\varepsilon_1}$.

□

If $W_{i-1} = S_{\beta_1}^{\eta_1} \dots S_{\beta_j}^{\eta_j}$, let $W_i = W_{i-2}$ if $S_{\beta_j}^{\eta_j} = S_{\alpha_i}^{-\varepsilon_i}$; otherwise let $W_i = W_{i-1}S_{\alpha_i}^{\varepsilon_i}$.

W_n is a reduced word equivalent to W.

Now let V be a reduced word equivalent to W, and let $V = V_1, V_2, \dots, V_m = W$ be a sequence of successively elementarily equivalent words. We shall show that the above process converts W into V. It is clear that the process leaves V unchanged.

If $V_i = GH$ and $V_{i+1} = GS_j^{\varepsilon_i}S_j^{-\varepsilon_j}H$ and if K(X) denotes the reduced word associated with X by the above process, then one easily sees that $K(GS_j^{\varepsilon_j}S_j^{-\varepsilon_j}) = K(G)$ and consequently $K(V_i) = K(V_{i+1})$. Thus we have shown:

1.2.3 Proposition. *Two words are equivalent if and only if they have the same reduced word.*

We give the solution of the conjugacy problem as

1.2.4 Proposition. *Two reduced words* W_1 *and* W_2 *represent conjugate elements if and only if they are of the form*

$$W_1 = H^{-1}KJH,\ W_2 = L^{-1}JKL.$$

Here KJ, JK *respectively do not end with the inverse of the initial letter. A reduced word with this property is called* cyclically reduced *and we say that* JK *results from* KJ *by* cyclic interchange.

□

Proof as exercise E 1.4. Other exercise E 1.5,6.

1.3 GRAPHS

Closely related to free groups are 1-dimensional complexes which we introduce now. The connection with groups will be the subject of the next section.

1.3.1 Definition. A *graph* or 1-dimensional complex is an at most denumerable system of points, called *vertices*, and directed line segments, called *edges*, with the following properties:

(a) Each edge contains vertices called its *initial* and *final* point (these may coincide).

(b) For each edge σ there is an inverse edge σ^{-1}. We have $(\sigma^{-1})^{-1} = \sigma$.

(c) The initial point of σ is the final point of σ^{-1} and the final point of σ is the initial point of σ^{-1}.

In future we shall call a pair of edges inverse to each other a *geometric edge*.

1.3.2 Definition. A *path* ω is a finite sequence of edges $\sigma_1, \sigma_2, \ldots, \sigma_n$ such that the final point of σ_i equals the initial point of σ_{i+1}. We write

$$\omega = \sigma_1\sigma_2 \ldots \sigma_n, \text{ or } \omega = \sigma_1 \ldots \sigma_n.$$

A path is called *closed* when the final point of σ_n is the initial point of σ_1, it is called *reduced* if no $\sigma\sigma^{-1}$ *(spur)* appears in it.

A graph is called *connected* when any two vertices occur as initial and final points of a path $\omega = \sigma_1 \ldots \sigma_n$. It is called finite if it contains only finitely many vertices and edges. The *degree* of a point is defined to be the number of edges for which it is the initial point.

Examples of graphs are the street network of a city, the track network of a railway, telegraph lines and labyrinths. A question about graphs, the Königsberg bridge problem, is considered to be the starting point of topology. It was asked whether a *single* path could cross all the bridges in the figure below, without crossing any more than once.

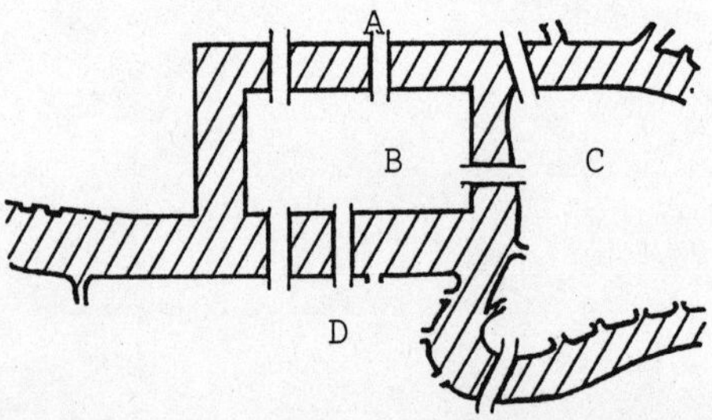

This is equivalent to the question whether one can traverse the graph

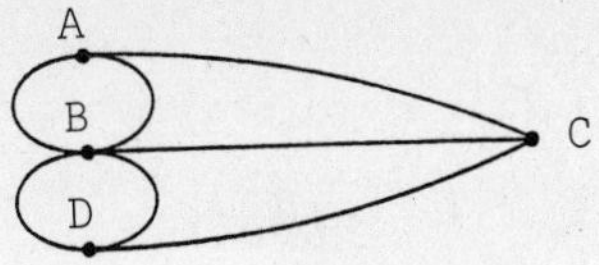

in a single path in which each geometric edge appears exactly once. Euler showed that no path with this property exists, by the following:

1.3.3 Theorem. *If all points in a connected finite graph* C *are of even degree, then* C *may be traversed in a closed path in which each geometric edge appears once. If* C *contains* 2n *vertices of odd degree, then one needs* n *paths* $\omega_1, \omega_2, \ldots, \omega_n$ *in order to traverse each geometric edge of* C *exactly once.*
Proof as exercise E 1.7. □

1.3.4 Definition. A connected graph that contains no reduced closed path other than 1 is called a *tree.*

1.3.5 Theorem. *In a connected graph* C *there is a tree which contains all vertices of* C.

Proof. One takes an arbitrary vertex P_0 of C. The subgraph B_0 consists of P_0. The subgraph B_i of C results from B_{i-1} by taking each of the vertices which are one edge distant from B_{i-1}, together with a connecting edge and its inverse.

$$B = \bigcup_{i=1}^{\infty} B_i \text{ is a tree which contains all vertices of } C.$$

□

A tree which contains all the vertices of a graph C will be said to *span* the graph C. The tree we have constructed possesses a minimal property with respect to P_0. Namely, it connects each vertex of C to P_0 with a path in B as short as the "shortest" such path in C (i.e. containing the minimal number of edges).

Since a tree contains no reduced closed paths and each path determines a unique reduced path, any two vertices in a tree may be connected by a unique reduced path.

Let B be a tree spanning C, σ an edge which is not contained in B, with distinct endpoints P and Q. Let ω be the reduced path from P to Q in B. If we remove an edge σ' of ω from B and insert σ instead, then we obtain a new spanning tree B' which is called *neighbouring* to B. If two spanning trees B and B' differ only with respect to k edges, then one can convert B into B' by k of these operations. For this reason the following definition makes sense:

1.3.6 Definition. *The connectivity number* of a graph is the number of geometric edges which lie outside a spanning tree. If there are only finitely many vertices, it is equal to

$$\alpha_1 - \alpha_0 + 1$$

where α_1 is the number of geometric edges and α_0 is the number of vertices.

1.4 THE FUNDAMENTAL GROUP OF A GRAPH

1.4.1 Construction and Definition. From now on we assume that C is a connected graph and let P be a point of C. By analogy with words in a free group it follows that for each path there is a uniquely determined reduced path. We call two closed paths with initial point P equivalent when they yield the same reduced path. As with words in a free group, this gives an equivalence relation. If we have two paths ω_1 and ω_2 where the final point of ω_1 equals the initial point of ω_2, then we can define the product $\omega = \omega_1\omega_2$ as the path which traverses first ω_1, then ω_2. We multiply equivalence classes of closed paths with initial point P representative-wise. As with free groups one sees that the product is well-defined and that the set of equivalence classes under this multiplication forms a group, the *fundamental group* (sometimes called the *edge group*).

1.4.2 Theorem. *The fundamental group of a (connected) graph is free.*

Proof. Let B be a spanning tree. Each edge which does not lie in B may be associated with a symbol S_i or S_i^{-1}, where inverse edges have inverse symbols and distinct edges have distinct symbols. A closed path with initial point P is then assigned the word $W = S_{\alpha_1}^{\varepsilon_1} \dots S_{\alpha_n}^{\varepsilon_n}$ in the symbols $S_i^{\pm 1}$, when it traverses successively the edges corresponding to the symbols $S_{\alpha_1}^{\varepsilon_1}, \dots, S_{\alpha_n}^{\varepsilon_n}$. Since closed subpaths which lie wholly in B can be reduced to their initial point, paths with equivalent words are themselves equivalent and conversely. In addition, the product of two paths corresponds to the product of the associated words. Thus the fundamental group is isomorphic to the free group on the generators S_i.

□

A path $\sigma_1 \dots \sigma_m \sigma_m^{-1} \dots \sigma_1^{-1}$ is called a *spur* (see 1.3.2). A closed path in a tree consists of spurs, and we have seen in general that the reduced form of an arbitrary path is obtained by successively removing all spurs.

If we take any other point P' as initial point instead of P then we obtain an isomorphic group. To show this take a path ν on the tree connecting P with P' and associate the path ω which begins and ends in P with the path $\nu^{-1}\omega\nu$. This yields a one-one correspondence between the classes of paths.

Examples of spanning trees:

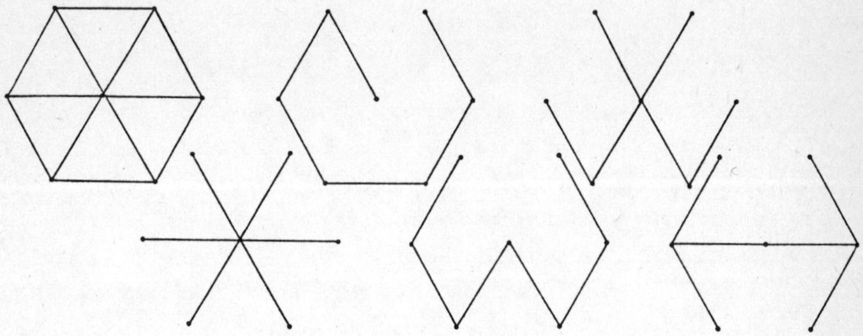

Exercise: E 1.8-9.

1.5 COVERINGS OF GRAPHS

1.5.1 Definition. By a *covering* f: C' → C of a graph C by a graph C' we mean a mapping of the vertices, respectively edges, of C' onto the vertices, respectively edges, of C, with the following properties:

(a) If σ' is an edge with initial point P' and final point Q' then $f(\sigma')$ is an edge with initial point f(P') and final point f(Q').

(b) $(f(\sigma'))^{-1} = f(\sigma'^{-1})$.

(c) If f(P') = P then f maps the set of edges $\sigma'_1, \sigma'_2, \ldots$ with initial point P' one-one onto the set of edges $\sigma_1, \sigma_2, \ldots$ with initial point P.

When $f(\sigma') = \sigma$ we say that σ' *lies over* σ. Let f: C' → C be a covering. Then the image of a spur in C' is again a spur in C, and consequently the images of equivalent paths are equivalent. If now $\sigma'\tau'$ is a path in C' such that $f(\tau') = f(\sigma')^{-1}$ then by (c) we also have $\tau' = \sigma'^{-1}$. Thus a path mapped onto a spur is itself a spur, and paths with the same initial point which are mapped on to equivalent paths are themselves equivalent. Further, it is clear that f carries the product of two paths into the product of their images. Thus f induces a monomorphism of the fundamental group of C' relative to P' into the fundamental group of C relative to P = f(P'):

1.5.2 Proposition. *The fundamental group of a covering graph is isomorphic to a subgroup of the fundamental group of the underlying graph.*

If one takes, instead of P', any other point lying over P as basepoint for the fundamental group of C' then one obtains a subgroup conjugate to U.

Conversely, for any subgroup U of the fundamental group S of C we now construct a covering f: C' → C such that the fundamental group of C' is mapped on to U by the induced monomorphism.

Let $g_1, g_2, \dots$ be representatives of the right cosets U_g of S modulo U. In this connection $g_1 = 1$ will be the representative of U. Let B be a spanning tree of C, and for each coset Ug_i we take a copy of it, B_i. An edge σ of C which does not lie in B corresponds to an element $s \in S$. Let Q and R be the initial and final point of σ respectively, let Q_i and R_i be the corresponding points in B_i. If $Ug_i s = Ug_j$, let $\tau(i,\sigma)$ be an edge with initial point Q_i and final point R_j, and let $\tau(j,\sigma^{-1})$ be the inverse edge. Let C' be the union of the B_i together with the edges $\tau(i,\sigma)$, where $i = 1,2,\dots$ and $\sigma \in C - B$. C' is a connected graph and the mapping f which carries the B_i "identically" into B and $\tau(i,\sigma)$ into σ is a covering.

Let the fundamental group S of C have the base-point P and let P_1 be the vertex over P in B_1. Let S' be the fundamental group of C' relative to P_1. If a path with initial point P_1 lies over a closed path, then it is also closed if it lies wholly in B or else runs through a sequence of edges outside the B_i

$$\tau(i_1,\sigma_1),\ \tau(i_2,\sigma_2),\dots,\tau(i_n,\sigma_n)$$

such that $s_1 s_2 \dots s_n$ is an element of U. Here the s_i are the elements of S which correspond to the σ_i. Thus the monomorphism induced by the covering maps S' into U. Conversely, if ω is a path in C which corresponds to an element of U, and if $\sigma_1, \sigma_2, \dots, \sigma_m$ are the edges of C - B which it traverses in turn, then $s_1 s_2 \dots s_m$ lies in U and ω is equal to $\omega_1\sigma_1\omega_2\sigma_2\dots\omega_m\sigma_m\omega_{m+1}$, where the ω_i are paths in B, and the σ_i are segments outside B. f maps just the following path with initial point P_1 on to ω:

$$\omega_{1,1}\ \tau(1,\sigma_1)\omega_{2,j_2}\ \tau(j_2,\sigma_2)\dots\omega_{m,j_m}\ \tau(j_m,\sigma_m)\omega_{m+1,1}$$

where ω_{i,j_i} is the path in B_{j_i} over ω_i and $\tau(j_i,\sigma_i)$ leads from B_{j_i} to $B_{j_{i+1}}$. It is closed. Thus S' is mapped isomorphically onto U. Hence we have proved

<u>1.5.3 Theorem</u>. *A covering* f: C' → C *induces a monomorphism of the fundamental group* S' *of* C' *onto a subgroup* U *of the fundamental group* S *of* C, *as long as the base point of the former lies over that of the latter. Conversely, for each subgroup* U *of* S *there is a covering* f: C' → C *such that the monomorphism induced by* f *maps*

the fundamental group of C' *isomorphically onto* U.

□

Exercise: E 1.10

1.6 THE REIDEMEISTER-SCHREIER METHOD FOR SUBGROUPS

The topological results of the last section can be used for a *group-theoretic calculation* of generators of subgroups. This method has been introduced by [Reidemeister 1927] and [Schreier 1927]. We will develop this next.

If S is a free group freely generated by $s_1, s_2, \ldots$ and if the graph C has one vertex and an edge $\sigma_i^{\pm 1}$ for each generator $s_i^{\pm 1}$, then the fundamental group of C is isomorphic to S. The spanning tree here consists of a single vertex. If U is a subgroup of S and if C' is the covering complex associated with U, then U is isomorphic to the fundamental group of C', and hence free. This proves the first part of

1.6.1 Theorem. *Subgroups of free groups are free. If the subgroup has finite index* i, *then*

$$\text{Rank } U = (\text{Rank } S - 1) \cdot i + 1.$$

By construction C' consists of $\alpha_0 = i$ points and $\alpha_1 = \text{Rank } S \cdot \alpha_0$ geometric edges. For the construction of the tree we need $(\alpha_0 - 1)$ edges, the connectivity number of C' is therefore Rank $S \cdot i - i + 1$.

□

Let B' be a spanning tree of C'. Each edge of B' corresponds to a generator of S, namely, to the class of paths onto which it is mapped by $f: C' \to C$. Let $P_1', P_2', \ldots$ be the vertices of C' (they all lie over the single point of C) and let P_1' be the base point for the fundamental group of C'. The points correspond uniquely to the right cosets of S modulo U. For each point P_i' there is exactly one reduced path in B' which runs from P_1' to P_i'. The initial subwords of the reduced words for such paths correspond to paths from P_1' to the P_j'. The word which corresponds to P_i' represents a group element w_i from the coset Ug_i (see the construction of the covering for a given subgroup in Theorem 1.5.3). Instead of the g_i we can take the w_i as coset representatives. These satisfy the *"Schreier condition"* with respect to the generators s_i:

1.6.2 Condition. *Each initial subword of a* w_i *(as a word in the* s_i*) is itself a coset representative.*

The w_i are understood to be reduced words in this connection.

Conversely, if a system of representatives $\{g_i\}$ satisfies the Schreier condition, then these representatives correspond to a spanning tree B' of C'. Generators of the subgroup U correspond to paths $\alpha'\sigma'(\overline{\alpha'\sigma'})^{-1}$, where σ' is a segment of $C' - B'$, α' is the unique reduced approach path in B' to the initial point of σ' and $\overline{\alpha'\sigma'}$ is that to the final point of σ'. The elements $as(\overline{as})^{-1}$, generate the subgroup U, where a runs through the Schreier coset representatives and s runs through the generators of S. Here $\overline{as}$ corresponds to the $\overline{\alpha'\sigma'}$, and is thus the coset representatives of Uas. If $\alpha'\sigma'$ is a coset representative as well as α', then $as(\overline{as})^{-1} = 1$. The remaining $as(\overline{as})^{-1}$ will be free generators, since the edge σ' lying over σ does not belong to the tree.

1.6.3 Theorem. *Let S be a free group with free generators $s_1, s_2, \ldots$ and U a subgroup. If $g_1, g_2, \ldots$ constitute a Schreier representative system for U relative to the generators $s_1, s_2, \ldots$ then the elements $g_i s_j (\overline{g_i s_j})^{-1}$ generate the subgroup U, and in fact those different from 1 freely generate the subgroup U. It is $\overline{g_i s_j}$ the representative of $U\, g_i s_j$).*

Exercise: E 1.11-12

1.7 THE NIELSEN METHOD

Now we consider the Nielsen method for handling subgroups of free groups. This important tool of combinatorial group theory was introduced in [Nielsen 1921]; at first glance it seems to have no connection with combinatorial topology. That this is not the case will be shown in the next section.

Let S be a free group on the generators $s_1, s_2, \ldots$. In this section words which represent elements are always reduced.

1.7.1 Definition. The *length* $\ell(w)$ of an element w from S is the length of the reduced word for w. It depends on the system of generators.

1.7.2 Definition. Let $v_1, \ldots, v_n$ be elements of S. If $\ell(v_i^\eta v_j^\varepsilon) < \ell(v_i)$ for $i \neq j, \eta, \varepsilon = \pm 1$ then $v_1', \ldots, v_n'$ with $v_k' = v_k$, $k \neq i$ and $v_i' = (v_i^\eta v_j^\varepsilon)^\eta$ generate the same subgroup U as $v_1, \ldots, v_n$. Such a change of generators for U is called a *Nielsen process of the first kind*. Obviously this represents a reduction in $\sum_{k=1}^{n} \ell(v_k)$. We denote the new elements again by $v_1, \ldots, v_n$.

After finitely many such processes we find generators $v_1, \ldots, v_n$ for which

1.7.3 $\ell(v_i^{\pm 1} v_j^{\pm 1}) \geq \ell(v_i)$ $(i \neq j)$

This inequality says that multiplication by v_i on the right or the left does not remove more than half of any $v_j^{\pm 1}$.

Suppose that multiplication by v_k^{ε} on the left and by v_j^{η} on the right $(\varepsilon, \eta = \pm 1)$ completely cancels v_i, $i \neq j, k$, i.e. v_k^{ε} cancels the left half of v_i and v_j^{η} cancels the right half. (More than half cannot be cancelled because of (1.7.3). Then v_i has even length 2m. Since the number of v_i is finite, only finitely many generators appear in the v_i. We order the elements generated by these s_j, firstly according to length, then among elements of the same length we take any fixed order which does not distinguish between inverse elements. By replacing a v_i of even length by its inverse if necessary we can always arrange that $v_i = uw^{-1}$, where u and w have equal length and u is ahead of w in the ordering. (Here, u does not end with the same symbol as w.) Each element has only finitely many predecessors in the ordering.

In what follows we revert to the first Nielsen process each time (1.7.3) is violated, and then begin afresh with the *Nielsen process of the second kind* which we now describe.

1.7.4 Definition. Let $v_k = uw^{-1}$ be the first element of even length among the v_i. If a v_j has the (reduced) form $v_j = wz$ or $v_j = zw^{-1}$ then we replace it by $v_j' = uz$ or $v_j' = zu^{-1}$ respectively. We call these *changes with respect to* v_k. This will be done in each case, i.e., also when w is only part of the front half of $v_j^{\pm 1}$. After these changes are made no element $v_j^{\pm 1}$ $(j \neq k)$ begins with w or ends with w^{-1}. The lengths of the v_i do not change as a result, since otherwise a Nielsen process of the first kind would be possible. We then apply these changes successively with respect to the elements v_i which have the same length as v_k. Since these v_i possess a half of greatest order, and we only replace halves by earlier halves, this process ends in *finitely* many steps. Then we carry out the process with respect to the next shortest element v_i of even length.

In this process no element v_i shorter than v_k is changed, otherwise we could carry out the first process. Since each process reduces either the length or the order, after finitely many steps we obtain a generating system $w_1, w_2, \ldots, w_n$ for $\mathcal{U}$ to which both the Nielsen processes no longer apply. Generators equal to 1 can also appear, let these be $w_{m+1}, \ldots, w_n$.

1.7.5 Proposition. The elements $w_1, \ldots, w_m$ have the following *Nielsen properties:*

(a) They generate $\mathcal{U}$.

(b) $\ell(w_j^{\varepsilon} w_i^{\eta}) \geq \ell(w_i)$ for $i \neq j$ or $i = j$ and $\varepsilon = \eta$; $\varepsilon, \eta \in \{1, -1\}$.

(c) $\ell(w_k^{\varepsilon} w_i w_j^{\eta}) > \ell(w_k) + \ell(w_j) - \ell(w_i)$ for $i \neq j, k$ or $k = i$, $\varepsilon \neq -1$ or $i = j$, $\eta = -1$

Now let $z = \prod_{i=1}^{k} w_{\alpha_i}^{\varepsilon_i}$ be a reduced word in the w_i, $i = 1,\ldots,m$. It follows from the Nielsen properties that

1.7.6
$$\ell(z) \geq k + \frac{1}{2}\ell(w_{\alpha_1}) + \frac{1}{2}\ell(w_{\alpha_k}) - 2$$
$$\ell(z) \geq \ell(w_{\alpha_i}) + k - 2$$

for no w_i is completely cancelled, and at least half of the initial and final element remains. It follows that $\ell(z) > 0$ for $k \geq 1$.

Thus if $z = 1$, z must be the empty word in the w_i, thus $w_1,\ldots,w_m$ *freely generate* U. Our considerations give a further proof that the finitely generated subgroups of free groups are free. The proof can be extended to show that all subgroups of free groups are free, using a transfinite induction in place of the induction.

1.7.7 Theorem. *Two free groups are isomorphic if and only if they have the same rank.*

Proof. It is trivial that the condition is sufficient.

Let $v_1,\ldots,v_m$ be generators of a free group S which is freely generated by $s_1,\ldots,s_n$. By means of a Nielsen process we obtain from the $v_1,\ldots,v_m$ elements $w_1,\ldots,w_s, 1,\ldots,1$ $(s \leq m)$ where the w_i freely generate S. Thus we can write

$$s_i = \prod_{j=1}^{r} w_{i_j}^{\varepsilon_j}$$

It follows from $\ell(s_i) = 1$ that $r = 1$ and $s_1 = w_{i_1}^{\varepsilon_1}$. For that reason $m \geq n$ and hence $m = n$ by symmetry. Further, it follows immediately that a free group of infinite rank cannot be generated by finitely many elements.

□

The theorem also holds when the rank is not required to be denumerable.

1.7.8 Proposition. *If the generators* $v_1,\ldots,v_m$ *of* U *have the Nielsen property then they are the shortest generators which exist, i.e. if* $w_1,\ldots,w_m$ *are other free generators of* U *and if* $v_1,\ldots,v_m$*, as well as* $w_1,\ldots,w_m$ *are ordered acording to length, then*

$$\ell(w_i) \geq \ell(v_i),\ i = 1,\ldots,m.$$

□

Exercises: E 1.13-17

1.8 GEOMETRIC INTERPRETATION OF THE NIELSEN PROPERTY

In this section we give the geometric interpretation of the Nielsen property for generators of a subgroup from [Reidemeister-Brandis 1959].

Let $\mathcal{S}$ be the fundamental group of a graph C which has only one vertex and let U be the covering complex for the subgroup $\mathcal{U}$. We take the free generators for $\mathcal{S}$ to be those belonging to C. By a minimal spanning tree with respect to P' of U we mean a spanning tree which, for each point Q' of U, contains a path of minimal length connecting P' and Q'. As we saw in the proof of Theorem 1.3.3 each graph possesses a minimal spanning tree. The proof of Theorem 1.3.5 shows how we find free generators for $\mathcal{U}$. If one determines the generators from a minimal spanning tree, these have the Nielsen property with respect to the generators of $\mathcal{S}$ chosen from C.

Conversely,

1.8.1 Theorem. If a finite system of generators of the subgroup $\mathcal{U}$ of $\mathcal{S}$ has the Nielsen property, then it is obtained from a minimal spanning tree.

Proof. Let U be the covering graph for $\mathcal{U}$ and let P' be the basepoint for the fundamental group of U. Let $v_1,\dots,v_m$ be a Nielsen generating system and let $v_1,\dots,v_r$ be the generators of odd length. For a given initial point in U, each generator of $\mathcal{S}$ determines an edge of U, and each v_i, $i = 1,\dots,m$ therefore determines a reduced closed path in U. If we take all paths which correspond to the $v_1,\dots,v_r$ and remove their middle segments then we obtain a tree B_1 of U. If B_1 had a reduced closed path $\neq 1$ there would be initial or final segments v_i', v_j' of v_i, v_j respectively, or their inverses, such that $v_i'v_j'$ is a closed path, $v_i'v_j'$ is a word in the Nielsen generators, but because of the property 1.7.5 c) $v_i'v_j'$ itself is not a Nielsen generator, since by suitable multiplication with $v_i^{\pm 1}$ and $v_j^{\pm 1}$, $v_i'v_j'$ is completely cancelled.

Thus $v_i'v_j'$ is a product $v_{\alpha_1}^{\varepsilon_1}v_{\alpha_2}^{\varepsilon_2}\dots v_{\alpha_r}^{\varepsilon_r}$. Let $v_{\alpha_1}^{\varepsilon_1} = w_1w_1'$, $v_{\alpha_2}^{\varepsilon_2} = w_1'^{-1}w_2$ where w_1w_2 is reduced. If w_1 is not merely an initial subword of $v_i'v_j'$, but even one of v_i', then $v_{\alpha_1}^{\varepsilon_1}$ is completely cancelled by multiplication with v_i or v_i^{-1} on the left and $v_{\alpha_2}^{\varepsilon_2}$ on the right, in contradiction with 1.7.5 (c). If w_1 extends into v_j', then we have the same argument for $v_{\alpha_r}^{\varepsilon_r}$.

B_1 must therefore be a tree, which we shall now extend. Because of 1.7.5 (b) and 1.7.5 (c), at most one half of a generator $v_{r+1},\dots,v_m$ of even length is contained in B_1. If a half is present in its entirety, then we add to B_1 all of the other half except the segment which meets the half already in B_1. We do this for all the generators v_i, $i > r$ which have a half in the tree already constructed. If at this stage one of the generators still has neither half in the tree constructed so far, we add one half to the tree and proceed as before. As above, one may convince oneself that each step results in a tree, so that we finally arrive at a spanning tree B which contains each v_i, $i = 1,\dots,m$ except for one segment. If we choose the right direction, the generating system relative to B for the fundamental group of U with basepoint P' corresponds to the Nielsen generators $v_1,\dots,v_m$ and B is a minimal tree.

□

However, if one starts from a minimal spanning tree, takes generators, and again constructs a minimal spanning tree, then one obtains a possibly different tree.

1.9 AUTOMORPHISMS OF A FREE GROUP OF FINITE RANK

Let S be a free group of finite rank with free generators $s_1,s_2,\dots,s_n$. If α is an automorphism of S, then $\alpha s_1,\dots,\alpha s_n$ is again a system of generators of S. Conversely, if $v_1,\dots,v_n$ is a system of generators of S, then $s_i \mapsto v_i$ defines an automorphism. Now we can convert the s_i into the v_i by Nielsen processes and obtain:

<u>1.9.1 Theorem</u>. *Each automorphism of a free group S with free generators* $s_1.s_2.\dots,s_n$ *is a product of the following automorphisms* [Nielsen 1919, 1924]:

(a) $s_1 \mapsto s_1s_2$, $s_i \mapsto s_i$ $\quad i > 1$

(b) $s_1 \mapsto s_1^{-1}$, $s_i \mapsto s_i$ $\quad i > 1$

(c) Permutation of generators.

□

J.H.C. Whitehead has given a constructive process for the solution of the following problem: Given elements w_1, $w_2 \in S$, is there an automorphism of S which sends w_1 to w_2? J.H.C. Whitehead uses the following system of automorphisms:

1.9.2
$$\begin{array}{lll} s_{\alpha_j} \mapsto s_{\alpha_j} & & 1 \le j \le n_1 \\ s_{\alpha_j} \mapsto s_{\alpha_j} s_{\alpha_1}^{-\varepsilon} & & n_1 < j \le n_2 \\ s_{\alpha_j} \mapsto s_{\alpha_1}^{\varepsilon} s_{\alpha_j} & & n_2 < j \le n_3 \\ s_{\alpha_j} \mapsto s_{\alpha_1}^{\varepsilon} s_{\alpha_j} s_{\alpha_1}^{-\varepsilon} & & n_3 < j \le n \end{array}$$

where $\begin{pmatrix} 1 & \dots & n \\ \alpha_1 & \dots & \alpha_n \end{pmatrix}$ is a permutation and $\varepsilon = \pm 1$, and $1 \le n_1 \le n_2 \le n_3 \le n$ are natural numbers.

□

[Whitehead 1936, 1936'] obtained his results by topological considerations which seem difficult to understand. Later [Rapaport 1958] and [Higgins-Lyondon 1974] gave combinatorial group theoretic proofs. See also [Lyndon-Schupp 1977, pp. 31-38]. In a refinement [McCool 1974] gave length conditions for the systems appearing in the application of the Whitehead method, cf. [Lyndon-Schupp 1977, pp. 38-41].

1.9.3 Theorem. *If* w_1 *is equivalent to a word of smaller length then there is an automorphism 1.9.2, which sends* w_1 *to an element of smaller length. If* w_1 *and* w_2 *both have minimal length and they are equivalent then there is a chain* $A_1,\dots,A_r$ *of automorphisms 1.9.2 such that* $A_r\dots A_1(w_1) = w_2$ *and* $\ell(A_j\dots A_1 w_1) = \ell(w_1)$, $1 \le j < r$. *Since there are only finitely many automorphisms 1.9.2 and finitely many words of a given length, the process is finite. (Two elements are called equivalent here if one can be carried into the other by automorphisms.)*

□

1.9.4 Example. Take S_2 = free group on s,t and $w_1 = s^a t^b$, $w_2 = s^c t^d$ with $a,b,c,d \ge 2$. We consider Whitehead equivalence and length. It suffices to look at just the following Whitehead automorphisms, since all others result from them by conjugation:

$$(s,t) \mapsto (s,ts^{\varepsilon}),\ (s,t) \mapsto (st^{\varepsilon},t) \text{ where } \varepsilon \in \{-1,1\}.$$

In the first case

$$s^q t^b \mapsto s^q (ts^{\varepsilon})^b \text{ with } \ell(s^q(ts^{\varepsilon})^b) = \begin{cases} a+2b & \text{for } \varepsilon = 1 \\ a+2b-2 & \text{for } \varepsilon = -1 \end{cases}$$

so $\ell(s^a t^b) \le \ell(s^a(ts^{\varepsilon})^b)$, with equality only for $b = 2$. The second case is analogous. In both cases it follows that $s^a t^b$ has minimal length among all elements Whitehead equivalent to it. Thus $s^a t^b$ is not Whitehead equivalent to a free generator of S_2, and a fortiori $(s^a t^b, s^c t^d)$ is not Whitehead equivalent to a free generator system. This is particularly interesting for the case where $\begin{vmatrix} a & b \\ c & d \end{vmatrix} = \pm 1$, because $s^a t^b$, $s^c t^d$ then generate the abelianized group.

EXERCISES

E 1.1 Prove that $\mathbb{Z}$ is a free group of rank 1

E 1.2 Prove that $\mathbb{Z} \oplus \mathbb{Z}$ is not free

E 1.3 Prove that a free group of rank ≥ 1 is infinite.

E 1.4 Prove Proposition 1.2.4

E 1.5 Let F be a free group and $x, y \in F$ such that $xy = yx$. Prove that there is an element $z \in F$ such that x and y are both powers of Z.

E 1.6 Deduce that if a non-trivial abelian group is free then it is isomorphic to $\mathbb{Z}$.

E 1.7 Prove Theorem 1.3.3.

E 1.8 Find a spanning tree for the following graph and hence find the rank of its fundamental group.

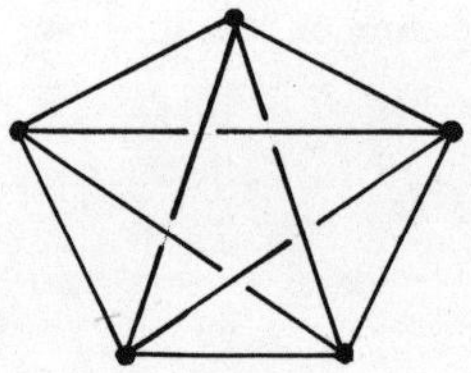

E 1.9 A change in the spanning tree corresponds to a change in the generators of the fundamental group. Describe the effect on the generators when we pass to a neighbouring tree.

E 1.10 Using spanning trees of the following graphs (a), (b), (c),

(a)

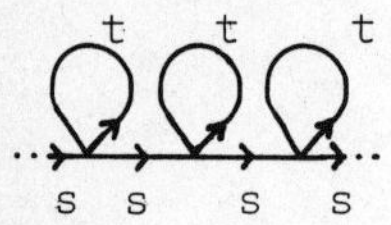

(b)

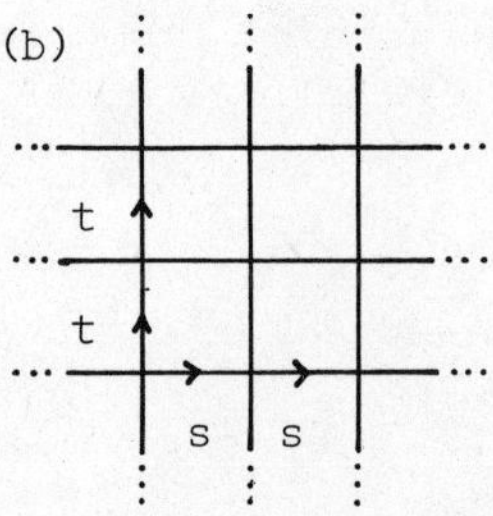

(c)

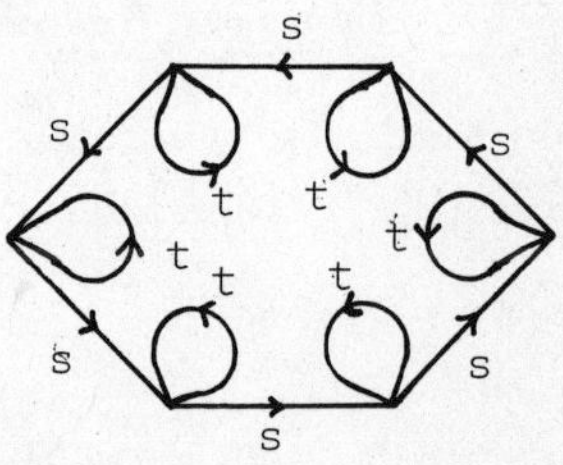

find generating systems for the subgroups of $\langle s,t \mid \rangle$ determined by (a), (b), (c) as coverings of 

and find the rank and index of each of these subgroups.

E 1.11 Show algebraically that the Schreier system of elements given in Theorem 1.6.3 generates U freely.

E 1.12 Repeat exercise 1.5.1 using the Reidemeister-Schreier method.

E 1.13 Prove Proposition 1.7.8.

E 1.14 Prove that the subgroup U of $G = \langle s,t \mid \rangle$ generated by s^2t^3 and s^3t^4 has infinite index, but that after "abelianizing" (i.e. allowing the generators to commute) the factor group G/U is trivial.

E 1.15 Let $G = \langle s,t \mid \rangle$ and let U be the subgroup generated by s^2t^3, $sts^{-1}t^{-1}$, $s^4t^5s^4t^5$. Determine all generating systems for U that have the Nielsen property.

E 1.16 Prove that if elements $x_1,\ldots,x_n$ of a free group have the Nielsen property and if $w \neq 1$ lies in the subgroup they generate, then

$$\ell(wx_i^{\varepsilon}) < \ell(w)$$

for some x_i^{ε}.

E 1.17 Let φ be a homomorphism from a finitely generated free group F onto a free group G. Then F has a free generating set $Z = Z_1 \cup Z_2$ such that φ maps $\langle Z_1 \mid \rangle$ isomorphically onto G, and $\langle Z_2 \mid \rangle$ onto 1.

E 1.18 Prove that the elements

$$R_{(7,3)} = vuvu^{-1}v^{-1}uvu^{-1}v^{-1}u^{-1}vuv^{-1}u^{-1}$$
$$R_{(7,5)} = vuv^{-1}uvu^{-1}vu^{-1}v^{-1}uv^{-1}u^{-1}vu^{-1}$$

cannot be transformed into each other by an automorphism of the free group with free generators u,v. (The reason for interest in this example, and an explanation of the names for the elements, will appear in Chapter 2, see 2.1.10 (b) and 2.11.3.)

E 1.19 Let S be the free group with free generators s,t and let $2 \le a \le b$, $2 \le c \le d$. Show that if there is an automorphism of S that maps s^at^b to s^ct^d then $a = c$ and $b = d$.

E 1.20 Let S be the free group with free generators $s_1,\ldots,s_n$, $t_1,\ldots,t_m$ and let w be a word in $s_1,\ldots,s_n$ which is of minimal length with respect to the automorphism of the subgroup U generated by $s_1,\ldots,s_n$. Show that there is no automorphism α of S such that $\ell(\alpha(w)) < \ell(w)$.

E 1.21* a) Show that an arbitrary automorphism α of the free group S with two free generators s,t satisfies

$$\alpha(sts^{-1}t^{-1}) = u(sts^{-1}t^{-1})^{\pm 1}u^{-1},$$

where u is a suitable element of S.

b) If an endomorphism β of S satisfies the equation

$$\beta(sts^{-1}t^{-1}) = sts^{-1}t^{-1},$$

then β is an automorphism [Nielsen 1918], [Chang 1960], [Mal'cev 1962].

E 1.22* a) If S is as above and $(a,b) = 1$, $|a|, |b| \geq 2$, show that an endomorphism β with $\beta(s^a t^b) = s^a t^b$ is an automorphism.

b) If an automorphism satisfies $\alpha(s^a t^b) = s^{a'} t^{b'}$ then $a' = \pm a$, $b' = \pm b$ or $a' = \pm b$, $b' = \pm a$. (This exercise is difficult, cf. [Lyndon 1959], [Zieschang 1962].)

2. 2-DIMENSIONAL COMPLEXES AND COMBINATORIAL PRESENTATIONS OF GROUPS

Each group is a factor group of a free group, so in addition to a system of generators we need a description of the kernel. This leads to the notion of a relation and presentation of a group. In the following we consider group presentations and their connection with 2-dimensional complexes. We generalize the methods of the first chapter to prove some subgroup theorems using coverings of these complexes.

2.1 TIETZE'S THEOREM

2.1.1 Definition. Let $(S_i)_{i \in I}$ be a system of symbols and let $(R_j(S))_{j \in J}$ be a system of words in the S_i and S_i^{-1}. We say that a group G has the *(combinatorial) presentation* $\langle (S_i)_{i \in I} \mid (R_j(S))_{j \in J} \rangle$ when

(a) there are elements $s_i \in G$, corresponding to the S_i, which generate G,

(b) for all j the relations $R_j(s) = 1$ are satisfied in G and

(c) for each system $(v_i)_{i \in I}$ of elements of a group H with $R_j(v) = 1$, $s_i \mapsto v_i$ defines a homomorphism $G \to H$.

We then write $G = \langle (S_i)_{i \in I} \mid (R_j)_{j \in J} \rangle$ and say that G is given by the *generators* $(S_i)_{i \in I}$ and the *defining relations* $(R_j)_{j \in J}$.
Mostly we consider finite or countable systems and write
$\langle S_1, \ldots, S_n \mid R_1(S), \ldots, R_m(S) \rangle$ or $\langle S_1, \ldots \mid R_1(S), \ldots \rangle$ or similar. For the free group with generators $S_1, \ldots$ we write $\langle S_1, \ldots \mid \rangle$.

This manner of speaking is not strictly correct, since the elements $S_1, S_2 \ldots$ are not elements of G, but it is often convenient, in order to avoid continual changes of notation, to treat symbols, words or group elements interchangeably. In all cases it will be clear from the context which meaning is intended. Later we shall also write

$$G = \langle s_1, s_2, \ldots \mid R_1(s), R_2(s) \ldots \rangle$$

where the s_i are elements of G. We obviously have the

2.1.2 Uniqueness theorem. *Two groups having the same presentation are isomorphic.*

Let S be the free group on the generators $\bar{s}_1, \bar{s}_2, \ldots$ and R the smallest normal subgroup of S which contains the elements $R_1(\bar{s})$, $R_2(\bar{s}),\ldots$. The group S/R is then generated by the elements $s_i = R\overline{s_i}$, and $R_i(s) = 1$ for all i.

If a group H contains elements $v_1, v_2, \ldots$ such that $R_i(v) = 1$ for all i, then $\overline{s_i} \mapsto v_i$ defines a homomorphism $\bar{\alpha}\colon S \to H$ which maps R onto the 1 of H. Thus $\bar{\alpha}$ induces a homomorphism $\alpha\colon S/R \to H$, which maps $s_i = R\overline{s_i}$ onto v_i:

2.1.3 Existence theorem. *For each system* $(S_1,S_2,\ldots;\ R_1(S),\ R_2(S),\ \ldots)$ *there is a group* $G = \langle s_1,s_2,\ldots \mid R_1(s),R_2(s),\ldots\rangle$.

□

2.1.4 Remark. We see that any further relation of G, regarded as an element $R(\bar{s})$ of S, lies in R. We call all words associated with elements of R *consequence relations* of the $R_i(S)$. They are products of conjugates of the $R_i(S)$ and their inverses.

It is clear that a group G can have different presentations. We now give processes (2.1.5) - (2.1.8) by means of which one can pass from one presentation to all others. Let $G = \langle (S_i)_{i\in I} \mid (R_j(S))_{j\in J}\rangle$.

2.1.5 Addition of new generators U_i and as many new relations $R_{U_i}(S,U_i) = U_i W_i^{-1}$ where W_i is a word in the S_j.

2.1.6 The operation inverse to 2.1.5 .

2.1.7 Addition of consequence relations. Among these we also admit trivial relations, i.e. words equal to 1 in the free group.

2.1.8 The operation inverse to 2.1.7 .

These operations are named after Tietze, who also proves

2.1.9 Theorem. *Two presentations* $\langle (S_i)_{i\in I} \mid (R_j(S))_{j\in J}\rangle$ *and* $\langle (T_i)_{i\in I'} \mid (P_j(T))_{j\in J'}\rangle$ *define isomorphic groups if and only if presentation* $\langle (S_i)_{i\in I} \mid (R_j(S))_{j\in J}\rangle$ *may be converted into* $\langle (T_i)_{i\in I'} \mid (P_j(T))_{j\in J'}\rangle$ *by finitely many Tietze transformations.*

Proof: It remains only to show that the condition is necessary. Assuming that both presentations define the same group, let S be the free group on the generators $(\bar{s}_i)_{i\in I}$ and T the free group on the generators $(\bar{t}_i)_{i\in I'}$. By hypothesis there is a homomorphism $\varphi\colon S \to G$ (onto) with kernel R, so that $S/R \cong G$, and a homomorphism $\psi\colon T \to G$ (onto) with kernel K, so that $T/K \cong G$. Let $X_1, X_2, \ldots$ be words for the elements $x_1, x_2, \ldots$ of T such that $\psi(x_i) = \varphi(\bar{s}_i)$, $Y_1, Y_2, \ldots$ words for the elements $y_1, y_2, \ldots$ of S with $\varphi(y_i) = \psi(\bar{t}_i)$. Then the operation 2.1.5 converts

$<(S_i)_{i\in I} \mid (R_j(S))_{j\in J}>$ into

$<(S_i)_{i\in I} \cup (T_i)_{i\in I'} \mid (R_j(S))_{j\in J} \cup (T_iY_i^{-1})_{i\in I'}>$ and by 2.1.7 we obtain

$<(S_i)_{i\in I} \cup (T_i)_{i\in I'} \mid (R_j(S))_{j\in J} \cup (P_j(T))_{j\in J'} \cup (T_iY_i^{-1})_{i\in I'} \cup (S_iX_i^{-1})_{i\in I}>$, by 2.1.8,

$<(S_i)_{i\in I} \cup (T_i)_{i\in I'} \mid (P_j(T)_{j\in J'} \cup (S_iX_i^{-1})_{i\in I}>$ and finally, by 2.1.6,

$<(T_i)_{i\in I'} \mid (P_j(T))_{j\in J'}>$.

□

If we confine ourselves to presentations which contain only finitely many generators, then 2.1.5 and 2.1.7 can be replaced by

2.1.5' Addition of a new generator U and a relation $R_U(S,U) = UW^{-1}$, where W is a word in the S_i.

2.1.7' Addition of finitely many consequence relations.

Theorem 2.1.9 obviously holds for presentations in which only finitely many generators and defining relations appear when the operations are 2.1.5' , 2.1.6 , 2.1.7' and 2.1.8 .

2.1.10 Examples.

(a) The group $<s,t \mid s^2t^3, s^3t^4>$

$\cong <s,t \mid s^2t^3, s^3t^4, st>$ as $s^3t^4 = s(s^2t^3)t$

$\cong <s,t \mid s^2t^3, st>$ for same reason

$\cong <s \mid s^2s^{-3}>$ as $t = s^{-1}$

$\cong 1$

(b) ([Funcke 1975]) We consider the group K with generators

$$a_i,\ b_i,\ x_i,\ y_i,\quad 0 \le i \le 6$$

and the defining relations

$b_0x_6^{-1}$	$a_0y_6^{-1}$
$b_1y_1^{-1}b_0^{-1}y_2$	$a_1x_1^{-1}a_0^{-1}x_2$
$b_2x_3b_1^{-1}x_2^{-1}$	$a_2y_3a_1^{-1}y_2^{-1}$
$b_3y_5^{-1}b_2^{-1}y_6$	$a_3x_5^{-1}a_2^{-1}x_6$
$b_4x_0^{-1}b_3^{-1}x_1$	$a_4y_0^{-1}a_3^{-1}y_1$
$b_5y_4b_4^{-1}y_3^{-1}$	$a_5x_4a_4^{-1}x_3^{-1}$
$b_6x_4^{-1}b_5^{-1}x_5$	$a_6y_4^{-1}a_5^{-1}y_5$

$$y_0b_6^{-1} \qquad x_0a_6^{-1}$$

$$b_0 = b_1 = \ldots = b_6 \qquad a_0 = a_1 = \ldots = a_6$$

(This is the group of the 2-bridge knot (7,3), but with more generators than usual, namely one for each segment in the projection, see figure below.)

If we put $u: = x_6 = b_0 = \ldots = b_6 = y_0$ and $v: = y_6 = a_0 = \ldots = a_6 = x_0$ then we obtain the following relations:

$$\begin{array}{ll} y_1 = vuv^{-1} & x_1 = uvu^{-1} \\ y_1 = uy_1u^{-1} & x_2 = vx_1v^{-1} \\ y_3 = v^{-1}y_2v & x_3 = u^{-1}x_2u \\ y_4 = u^{-1}y_3u & x_4 = v^{-1}x_3v \\ y_5 = vy_4v^{-1} & x_5 = ux_4u^{-1} \\ y_6 = uy_5u^{-1} & x_6 = vx_5v^{-1} \\ y_6 = v & x_6 = u \end{array} .$$

Now we drop all the generators except u,v by Tietze transformations, and obtain the defining relations

$$uvu^{-1}v^{-1}uvuv^{-1}u^{-1}vuv^{-1}u^{-1}v^{-1}$$
$$vuv^{-1}u^{-1}vuvu^{-1}v^{-1}uvu^{-1}v^{-1}u^{-1} .$$

The first word is conjugate to the inverse of the second; so it can be dropped, and conjugation of the latter gives the presentation

2.1.11 $K = \langle u,v \mid vuvu^{-1}v^{-1}uvu^{-1}v^{-1}u^{-1}vuv^{-1}u^{-1}\rangle$.

We let $R_{(7,3)}$ denote this relation (cf. E. 1.18). Now we add new generators a,b by

$$a = vuvu^{-1}v^{-1}, \quad b = uvuv^{-1}u^{-1} .$$

Then

$$R_{(7,3)} = ab^{-1}a^{-1}v$$

and

$$vuv^{-1}u^{-1}R_{(7,3)}^{-1}uvu^{-1}v^{-1} = ba^{-1}b^{-1}u.$$

Hence by Tietze transformations:

$$\begin{aligned} K &= \langle u,v,a,b \mid R_{(7,3)},\ a^{-1}vuvu^{-1}v^{-1},\ b^{-1}uvuv^{-1}u^{-1}\rangle \\ &= \langle u,v,a,b \mid ab^{-1}a^{-1}v, ba^{-1}b^{-1}u, a^{-1}vuvu^{-1}v^{-1}, b^{-1}uvuv^{-1}u^{-1}\rangle \\ &= \langle a,b \mid a^{-1}aba^{-1}bab^{-1}aba^{-1}ba^{-1}b^{-1}ab^{-1}a^{-1}, b^{-1}bab^{-1}aba^{-1}bab^{-1}ab^{-1}a^{-1}ba^{-1}b^{-1}\rangle \\ &= \langle a,b \mid ba^{-1}bab^{-1}aba^{-1}ba^{-1}b^{-1}ab^{-1}a^{-1}\rangle \end{aligned}$$

where we first dropped $R_{(7,3)}$ because $R_{(7,3)} = ab^{-1}a^{-1}v$, next expressed u,v in terms of a,b, and finally could drop the second relation because it is conjugate to the inverse of the first. The new relation is conjugate to the relation $R_{(7,5)}$ from E 1.8 where we found that $R_{(7,3)}$ and $R_{(7,5)}$ were not equivalent under an automorphism of the free group $\langle u,v \mid \ \rangle$. We come back to this example in 2.11.

(The two presentations of K reflect the fact that the 2-bridge knots (7,3) and (7,5) are the same. The (7,3) and (7,5) projections give the $R_{(7,3)}$ and $R_{(7,5)}$ presentations resp.. Thus the groups must be the same for geometric reasons.)

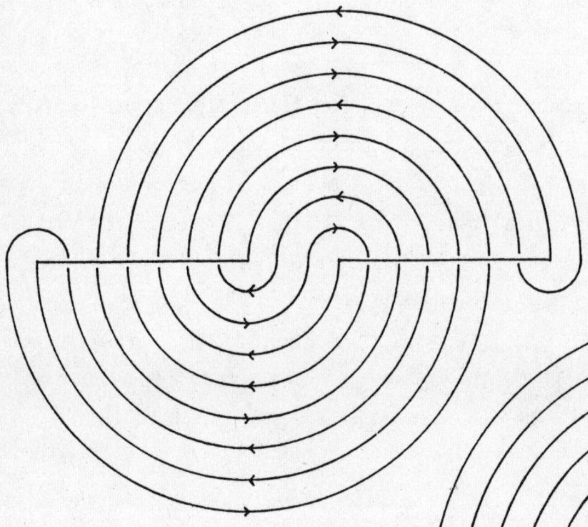

(7,3)

(7,5)

Those familiar with the method for obtaining presentations of knot groups may note that the relations in the two columns of the original presentation come from the vertices of the bridges. To get $R_{(7,3)}$ directly one adds $b_o = \ldots = b_6$, $a_o = \ldots = a_6$, and to get $R_{(7,5)}$ one adds $x_o = \ldots = x_6$, $y_o = \ldots = y_6$.)

2.1.12 Definition. The *rank* of a group G is the minimum number of generators required for a presentation of G.

This generalizes the notion of rank of a free group (1.7.7) and will be used later, see e.g. 2.9.2 .

Exercises: E 2.1, 2.15.

2.2 THE REIDEMEISTER-SCHREIER METHOD

Let U be a subgroup of $G = \langle (S_i)_{i \in I} \mid (R_j(S))_{j \in J} \rangle$ and let S be the free group with the free generators $(S_i)_{i \in I}$. Let φ be the canonical epimorphism of S onto G, and U' the pre-image $\varphi^{-1}U$.

If $(G_k)_{k \in K}$ is a Schreier representative system for the right cosets of S modulo U', then the $G_kS_i\overline{G_kS_i}^{-1}$ freely generate the subgroup U' when we omit the generators for which $G_kS_i = \overline{G_kS_i}^{+1}$, see 1.6 . Let X_{ik} be a system of symbols which correspond uniquely to the non-trivial (free) generators $G_kS_i\overline{G_kS_i}^{-1}$ of U'. The normal subgroup R of S generated by $(R_j(S))_{j \in J}$ is contained in U', and U'/R is isomorphic to U. Thus, R considered as subgroup is generated by the $XR_j(S)X^{-1}$, $X \in S$, $j \in J$, and hence by $UG_kR_j(S)G_k^{-1}U^{-1}$, $U \in U'$, $j \in J$, $k \in K$. Consequently, R is the normal subgroup of U' defined by the $G_kR_j(S)G_k^{-1}$. Therefore if one writes $G_lR_j(S)G_l^{-1}$ as a word $R_{jl}(X_{ik})$ in the X_{ik}, $\langle X_{ik} \mid R_{jl}(X_{ik}) \rangle$ is a presentation of U:

2.2.1 Theorem. *Let U be a subgroup of $G = \langle (S_i)_{i \in I} \mid (R_j(S))_{j \in J} \rangle$ and let the elements $g_1, g_2, \ldots$, written as words in the S_i, be a Schreier representative system for the right cosets modulo U. Then the elements $g_ks_i \cdot \overline{g_ks_i}^{-1}$ generate the subgroup U. If we omit trivial generators $g_ks_i\overline{g_ks_i}^{-1}$, then the $g_lR_j(s)g_l^{-1}$, written in the non-trivial generators, constitute the defining relations.*

□

2.2.2 Example. Let $G = \langle v_1, \ldots, v_n \mid v_1^2 \ldots v_n^2 \rangle$ and let $\alpha: G \to \mathbb{Z}_2$ be the homomorphism which maps all v_i to -1. We seek the kernel of α.

Representatives are 1, v_n and generators are then the elements

$$x_i = 1 \cdot v_i(\overline{v_i})^{-1} = v_i v_n^{-1} \qquad i = 1,\dots,n-1$$

$$y_i = v_n v_i(\overline{v_n v_i})^{-1} = v_n v_i \qquad i = 1,\dots,n-1$$

and $z = v_n^2$.

We have two relations $v_1^2\dots v_n^2 = 1$ and $v_n v_1^2 \dots v_n^2 v_n^{-1} = 1$. The first may be written in the form

$$v_1 v_n^{-1} v_n v_1 \cdot v_2 v_n^{-1} v_n v_2 \cdots v_{n-1} v_n^{-1} v_n v_{n-1} v_n^2$$

and thus is the same as

$$x_1 y_1 x_2 y_2 \cdots x_{n-1} y_{n-1} z = 1.$$

Then we can execute the Tietze transformation 2.1.6 replacing the generator z by $(x_1 y_1 \dots x_{n-1} y_{n-1})^{-1}$. The second relation may be written in the form

$$v_n v_1 v_1 v_n^{-1} \cdot v_n v_2 v_2 v_n^{-1} \cdots v_n v_{n-1} v_{n-1} v_n^{-1} \cdot v_n^2 = 1$$

and thus is the same as

$$y_1 x_1 y_2 x_2 \cdots y_{n-1} x_{n-1} z = 1$$

which becomes

$$y_1 x_1 y_2 x_2 \cdots y_{n-1} x_{n-1} y_{n-1}^{-1} x_{n-1}^{-1} \cdots y_1^{-1} x_1^{-1} = 1$$

after substituting for z.

Thus the kernel of α has the presentation

$$\langle X_1,\dots,X_{n-1},Y_1,\dots,Y_{n-1} \mid Y_1 X_1 Y_2 X_2 \dots Y_{n-1} X_{n-1} Y_{n-1}^{-1} X_{n-1}^{-1} \dots Y_1^{-1} X_1^{-1} \rangle.$$

This presentation may be converted into

$$\langle A_1, B_1, \dots, A_{n-1}, B_{n-1} \mid A_1 B_1 A_1^{-1} B_1^{-1} \dots A_{n-1} B_{n-1} A_{n-1}^{-1} B_{n-1}^{-1} \rangle$$

by Tietze transformations. We commonly use the abbreviation $[x,y]$ for the commutator $xyx^{-1}y^{-1}$. Then we have the presentation $\langle A_1,B_1,\dots,A_{n-1},B_{n-1} \mid \prod_{i=1}^{n-1}[A_i,B_i] \rangle$.

Theorem 2.2.1 of Schreier can be strengthened [Gramberg-Zieschang 1979]. Namely it often turns out that some of the relations $g_l R_j(s) g_l^{-1}$, expressed in the generators $g_k s_i \overline{g_k s_i}^{-1}$, are obvious consequences of others. We give criteria which permit the number of relations to be diminished. (What follows is used in this text only in 4.14 and can be skipped for the first reading.)

2.2.3 Illustrative example. We consider the homomorphism $\varphi\colon G = \langle s,t \mid s^3, t^3, (st)^3\rangle \to \mathbb{Z}_3 = \{0,1,2\}$ where $s,t \mapsto 1$, and we will determine the kernel. As coset representatives we choose $1, s, s^2$. Then the Reidemeister-Schreier generators are

$$x = s^3,\ y_0 = ts^{-1},\ y_1 = sts^{-2},\ y_2 = s^2t$$

For the defining relations we have to express the nine relations

$$s^i s^3 s^{-i},\ s^i t^3 s^{-i},\ s^i (st)^3 s^{-i} \qquad (i = 0, 1, 2)$$

as words in the new generators. We see that the first three relations are all equal to $s^3 = x$. In the second case we get

$$\begin{aligned} t^3 &= ts^{-1} \cdot sts^{-2} \cdot s^2t = y_0 y_1 y_2 \\ st^3s^{-1} &= sts^{-2} \cdot s^2t \cdot ts^{-1} = y_1 y_2 y_0 \\ s^2t^3s^{-2} &= s^2t \cdot ts^{-1} \cdot sts^{-2} = y_2 y_0 y_1 \end{aligned}$$

and these are the same up to a cyclic permutation, i.e. they lead to only one defining relation. In the third case we obtain

$$\begin{aligned} (st)^3 &= sts^{-2} \cdot s^3 \cdot ts^{-2} \cdot s^2t = y_1 x y_0 y_2 \\ s(st)^3s^{-1} &= s^2t \cdot sts^{-2} \cdot s^3 \cdot ts^{-1} = y_2 y_1 x y_0 \\ s^2(st)^3s^{-2} &= s^3 \cdot ts^{-1} \cdot s^2t \cdot sts^{-2} = x y_0 y_2 y_1 \end{aligned}$$

and these three relations are the "same". Hence

$$\begin{aligned} \text{Kernel } \varphi &= \langle x, y_0, y_1, y_2 \mid x, y_0 y_1 y_2,\ x y_0 y_2 y_1\rangle \\ &= \langle y_0, y_1, y_2 \mid y_0 y_1 y_2,\ y_0 y_2 y_1\rangle \\ &= \langle y_1, y_2 \mid y_2^{-1} y_1^{-1} y_2 y_1\rangle \\ &\cong \mathbb{Z} \oplus \mathbb{Z} \end{aligned}$$

We will show that these simplifications are of a general nature.

2.2.4 Assumptions and definitions.

Let $G = \langle (S_i)_{i\in I} \mid (R_j)_{j\in J}\rangle$ and let J be partially ordered by $<$. We assume that the order is transitive and that $j_1 < j_2$ implies $j_1 \neq j_2$. We abbreviate G to $\langle S \mid R\rangle$. Similarly, let $\hat{G} = \langle S \mid\ \rangle$ be the free group on $(S_i)_{i\in I}$, and let $\varphi\colon \hat{G} \to G$ be the canonical homomorphism. Moreover, let

$$G_R = \langle S \mid \{R' \in R \mid R' < R\}\rangle$$

for each $R = R_j \in R$, and let $\varphi_R\colon \hat{G} \to G_R$ be the canonical epimorphism.

Let stab_R denote the stabilizer of $\{\varphi_R(R), \varphi_R(R^{-1})\}$ in G_R and $\mathrm{Stab}_R = \varphi_R^{-1}(\mathrm{stab}_R) \subset \hat{G}$.

Let $U \subset G$ be a subgroup and K a Schreier representative system for the cosets $\varphi^{-1}(U)$ in $\hat{G}$. Representatives $K, K' \in K$ are called R-equivalent iff there is an $X \in \hat{G}$ with $XRX^{-1}R^{\varepsilon} \in \mathrm{Kern}\,\varphi_R$ such that $K'XK^{-1} \in \varphi^{-1}(U)$. We write $K \sim_R K'$.

Let $\lambda(R,K) = |\{K' \in K \mid K' \sim_R K\}|$ be the number of representatives which are R-equivalent to K.

The R-equivalence is obviously an equivalence relation and $\lambda(R,K)$ is the cardinality of the R-equivalence class of K. The following interpretation of $\lambda(R,K)$ is sometimes convenient:

2.2.5 Lemma. $\lambda(R,K) = [\varphi(\mathrm{Stab}_R) : (\varphi(K)^{-1}U\varphi(K) \cap \varphi(\mathrm{Stab}_R))]$.

Proof. By definition,

$$\begin{aligned}\lambda(R,K) &= |\{K' \in K \mid K\,\mathrm{Stab}_R\,K'^{-1} \cap \varphi^{-1}(U) \neq \emptyset\}|\\ &= |\{K' \in K \mid \varphi(\mathrm{Stab}_R) \cap \varphi(K^{-1})U\varphi(K') \neq \emptyset\}|.\end{aligned}$$

As the products $K^{-1}K'$, $K' \in K$, form a system of coset representatives for the subgroup $\varphi(K^{-1})U\varphi(K)$ it follows that

$$\{\varphi(\mathrm{Stab}_R) \cap \varphi(K^{-1})U\varphi(K') \mid K' \in K\}$$

is the decomposition of $\varphi(\mathrm{Stab}_R)$ into cosets with respect to the subgroup $\varphi(\mathrm{Stab}_R) \cap \varphi(K^{-1})U\varphi(K)$. This implies 2.2.5.

□

An immediate consequence is the following statement:

2.2.6 Corollary. $\lambda(R,K)$ *does not depend on* K *in one of the following cases:*

(a) U *is a normal subgroup of* G.

(b) U *is torsionfree and* stab_R *is finite.*

□

2.2.7 A construction. For each $R \in R$ we choose a *choice function* $\Theta_R\colon K \to K$ which maps each R-equivalence class of representatives to one of its members, i.e.,

$\Theta_R(K) \sim_R K$ and $\Theta_R(K) = \Theta_R(K')$ iff $K \sim_R K'$.

We now state the basic result:

2.2.8 Theorem. *Let $G = \langle S \mid \mathcal{R}\rangle$ and $\mathcal{U}$ be a subgroup of G with a Schreier system $\mathcal{K}$ of coset representatives and let $(\Theta_R)_{R \in \mathcal{R}}$ be a system of choice functions. Then*

2.2.9 $\quad \mathcal{U} = \langle S(\mathcal{U},\mathcal{K}) \mid \{(\Theta_R(K)R\ \Theta_R(K)^{-1})_\tau \mid R \in \mathcal{R},\ K \in \mathcal{K}\}\rangle$

where τ is the Reidemeister-Schreier rewriting process corresponding to $\mathcal{K}$ and $S(\mathcal{U},\mathcal{K})$ denotes as before the Reidemeister-Schreier generators corresponding to $\mathcal{K}$, see the beginning of this chapter.

Proof. Let $K', K'' \in \mathcal{K}$, $K' \sim_R K''$ and $K''XK'^{-1} \in \varphi^{-1}(\mathcal{U})$ for some $X \in \mathrm{Stab}_R$. Then $K'' \mathrel{\hat{=}} UK'X^{-1}$ for some $U \in \varphi^{-1}(\mathcal{U})$, where $\hat{=}$ denotes equality in the free group. The equations

$$K''RK''^{-1} \mathrel{\hat{=}} UK'X^{-1}RXK'^{-1}U^{-1} =_R UK'R^{\varepsilon}K'^{-1}U^{-1} \text{ for a suitable } \varepsilon \in \{1,-1\}$$

imply that $(K''RK''^{-1})_\tau$ is derivable in $\mathcal{U}$ from $(K'RK'^{-1})_\tau$ and the relations $\{\Theta_{R'}(K)R'\ \Theta_{R'}(K)^{-1} \mid K \in \mathcal{K},\ \mathcal{R} \ni R' < R\}$. (Here $=_R$ denotes equality in G_R.)

□

2.2.10 Definition. The presentation 2.2.9 is called an *order reduced Reidemeister-Schreier presentation of $\mathcal{U}$.*

By a simple calculation we obtain the following two corollaries:

2.2.11 Corollary. *Let $G = \langle S \mid \mathcal{R}\rangle$ where $|\mathcal{R}| < \infty$ and let $\mathcal{U}$ be a subgroup of finite index which is either normal in G or which is torsionfree and* stab_R *is finite for every $R \in \mathcal{R}$. Then the order reduced Reidemeister-Schreier presentation of $\mathcal{U}$ has $[G:\mathcal{U}] \cdot \sum_{R \in \mathcal{R}} \frac{1}{\lambda(R,1)}$ relations.*

□

2.2.12 Corollary. *Let $G = \langle S \mid \mathcal{R}\rangle$ be a group with one defining relation* $R(S) \mathrel{\hat{=}} W(S)^k$, $k \geq 1$. *Let $\psi\colon G \to \mathbb{Z}_k$ be an epimorphism such that $W(S)$ is mapped to a generator of $\mathbb{Z}_k$. Then the order reduced Reidemeister-Schreier presentation of the kernel of ψ has one defining relation.*

□

In Chapter 4 we give a further application of theorem 2.2.8 and prove that subgroups of Fuchsian groups are again Fuchsian. The following example concerns presentations which will be of interest later, the so-called "planar" presentations.

<u>2.2.13 Example</u>. Let $G = \langle S'' \mid R' \cup R'' \rangle$ where

$S'' = \{y_1,\ldots,y_m\}$, $R'' = \{y_1^2,\ldots,y_m^2\}$ and $R' = \{(y_1y_2)^{h_1},\ldots,(y_{m-1}y_m)^{h_{m-1}}, (y_my_1)^{h_m}\}$.

We write $R_j = (Y_jY_{j+1}^{-1})^{h_j}$ for $1 \le j \le m$ where $j + 1$ is considered modulo m. Let $f\colon G \to \mathbb{Z}_2, y_i \mapsto -1$ $(1 \le i \le m)$, and U be the kernel of f. We take $K = \{1,y_m\}$. Let $y_i^2 < R_j$ for all $i, j \in \{1,\ldots,m\}$. Now $G_{y_i^2} = \hat{G} = \langle Y_1,\ldots,Y_m \mid \rangle$, hence,

$$\mathrm{stab}_{y_i^2} = \mathrm{Stab}_{y_i^2} = \{Y_i^\ell \mid \ell \in \mathbb{Z}\}.$$

As $y_my_i \in U$ for $1 \le i \le m$ it follows that $Y_m \in \mathrm{Stab}_{y_i^2}\, \varphi^{-1}(U)$, hence $y_i \sim_{y_i^2} 1$. We define $\Theta_{y_i^2}(1) = \Theta_{y_i^2}(y_m) = 1$. Now $Y_iR_jY_i^{-1} \mathrel{\hat{=}} R_j^{-1}$, i.e. $Y_i \in \mathrm{Stab}_{R_j}$, and $y_my_i \in U$, hence, we obtain in addition that $y_m \sim_{R_i} 1$, and we define $\Theta_{R_i}(1) = \Theta_{R_i}(y_m) = 1$.

The Reidemeister-Schreier generators form the system $S'(U,K) = \{z_i := y_iy_m^{-1}, \bar{z}_i = y_my_i$ for $1 \le i \le m-1$, $\bar{z}_m = y_m^2\}$. (The (here unmotivated) notation S'' and $S'(U,K)$ will be explained in 4.13, where is also given a geometric interpretation.)

$$\Theta_{y_i^2}(K)y_i^2\Theta_{y_i^2}(K)^{-1} = y_i^2 = z_i\bar{z}_i \text{ for } 1 \le i \le m-1,$$

$$y_m^2 = \bar{z}_m,$$

$$\Theta_{R_i}(K)R_i\Theta_{R_i}(K)^{-1} = R_i = (y_iy_{i+1})^{h_i} = (y_iy_m^{-1}y_my_i)^{h_i} = (z_i\bar{z}_{i+1})^{h_i}$$

for $1 \le i \le m-2$,

$$R_{m-1} = (y_{m-1}y_m)^{h_{m-1}} = (y_{m-1}y_m^{-1}\cdot y_m^2)^{h_{m-1}} = (z_{m-1}\bar{z}_m)^{h_{m-1}}$$

and

$$R_m = (y_my_1)^{h_m} = \bar{z}_1^{h_m}$$

we obtain $R'(U,K)$ as the set of all these relations. The presentation $\langle S''(U,K) \mid R'(U,K) \rangle$ of U is planar as follows from theorem 4.13.7 or can easily be checked. (The notion *planar* in defined in 4.13.6, see also Theorem 4.14.1.)

If we drop $\bar{z}_m$ as generator as well as relation and replace each $\bar{z}_i$ by

$z_i^{-1}(1 \le i \le m-1)$ and drop the relations $z_i\bar{z}_i$ we obtain another planar presentation of $\mathcal{U}$:

$$\mathcal{U} = \langle z_1,\ldots,z_{m-1} \mid (z_1 z_2^{-1})^{h_1},\ldots,(z_{m-2} z_{m-1}^{-1})^{h_{m-2}},\ z_{m-1}^{h_{m-2}},\ z_1^{-h_m}\rangle.$$

If we define $\hat{z}_{m-1} = z_{m-1}$ and $\hat{z}_i = z_i z_{i+1}$ for $1 \le i \le m-2$ we get $z_i = \hat{z}_i \ldots \hat{z}_{m-1} (1 \le i \le m-1)$ and $z_i z_{i-1}^{-1} = \hat{z}_i$ $(1 \le i \le m-2)$. Thus $\mathcal{U}$ has the (almost canonical) presentation.

$$\langle \hat{z}_1,\ldots,\hat{z}_{m-1} \mid \hat{z}_1^{h_1},\ldots,\hat{z}_{m-2}^{h_{m-2}},\ \hat{z}_{m-1}^{h_{m-1}},\ (\hat{z}_1\ldots\hat{z}_{m-1})^{-h_m}\rangle.$$

If we define $\hat{z}_m = (\hat{z}_1\ldots\hat{z}_{m-1})^{-1}$ we obtain the canonical presentation

$$\mathcal{U} = \langle \hat{z}_1,\ldots,\hat{z}_m \mid \hat{z}_1^{h_1},\ldots,\hat{z}_m^{h_m},\ \hat{z}_1\cdot\ldots\cdot\hat{z}_m\rangle.$$

The steps to simplify the presentation are the elementary Tietze processes. A similar example is considered in 2.2.2.

The Reidemeister-Schreier method was first introduced by Reidemeister in [Reidemeister 1927] using an arbitrary system of coset representatives, i.e. without postulating the Schreier condition 1.6.2. Shortly after that, Schreier in [Schreier 1927] introduced the restriction on the representatives and obtained the subgroup presentation of 2.2.1. Without the restriction one gets more generators and defining relations, see [Reidemeister 1932, pp. 69-81], though these presentations may be useful sometimes. The modification of the Reidemeister-Schreier method arose from the observation that relations are often redundant. An interesting application will be considered in 4.13 . Both are from [Gramberg-Zieschang 1979]. The modification applies very well to relations that are powers, and commutator relations can also be treated efficiently.

Exercises: E 2.2-5, 22-23.

2.3 FREE PRODUCTS WITH AMALGAMATION

Now we consider some constructions to obtain "bigger" groups from given groups. These processes correspond to the direct sum in the category of abelian groups.

2.3.1 Definition. Given groups $G = \langle (S_i)_{i \in I} \mid (R_j)_{j \in J} \rangle$ and $H = \langle (T)_{i \in I'} \mid (U_j)_{j \in J'} \rangle$, we call the group $G * H = \langle (S_i)_{i \in I} \cup (T_i)_{i \in I'} \mid (R_j(S))_{j \in J} \cup (U_j(T))_{j \in J'} \rangle$ the *free product* of G and H.

This group is welldefined up to isomorphism. If $\langle X_1, X_2, \ldots, \mid P_1(X), P_2(X), \ldots \rangle$ and $\langle Y_1, Y_2, \ldots \mid Q_1(Y), Q_2(Y), \ldots \rangle$ are two other presentations of G and H respectively then there is a finite sequence of Tietze transformations $T_1, T_2, \ldots, T_n$ which convert $\langle X_1, \ldots \mid P_1(X), \ldots \rangle$ into $\langle S_1, \ldots \mid R_1(S), \ldots \rangle$ and likewise Tietze transformations $T_1', \ldots, T_m'$ which convert the second presentations of H into the first. Then $T_1, T_2, \ldots, T_n, T_1', \ldots, T_m'$ convert $\langle X_1, \ldots, Y_1, \ldots \mid P_1(X), \ldots, Q_1(Y), \ldots \rangle$ into $\langle S_1, \ldots, T_1, \ldots \mid R_1(S), \ldots, U_1(T), \ldots \rangle$ since the T_i concern neither the T_j nor the U_k and the T_i' concern neither the S_j nor the R_ℓ.

2.3.2 Definition. Let U and V be subgroups of G and H respectively and let $\varphi: U \to V$ be an isomorphism. Further, let $(v_k)_{k \in K}$ be a system that generates U, V_k the word in the S_i corresponding to v_k, and $\varphi(V_k)$ a word in the T_i corresponding to $\varphi(v_k)$. Then we define

$$G * H_{U=V} = \langle (S_i)_{i \in I} \cup (T_i)_{i \in I'} \mid (R_j(S))_{j \in J} \cup (U_j(T))_{j \in J'} \cup (V_k \varphi(V_k)^{-1})_{k \in K} \rangle$$

The latter relations $V_k \varphi(V_k)^{-1}$ naturally suggest that we regard $G * H_{U=V}$ as the free product of G and H with the subgroups U and V identified. This identification can be made at the outset, so that $A = G \cap H \cong U$. One then writes $G *_A H$ and calls it the *free product of G and H with amalgamated subgroup A*.

This definition can obviously be generalized to an arbitrary number of factors. In this case identifications are always made with the same subgroup. The free product of groups G_i, $i \in I$, is denoted $\mathop{\ast}\limits_{i \in I} G_i$, or similar.

In order to solve the word problem in free products with amalgamation we give a normal form for the group elements. We regard the amalgamated subgroup A as a subgroup of G and H and choose representatives $g_1, g_2, \ldots$ of the right cosets of G modulo A and $h_1, h_2, \ldots$ of the right cosets classes of H modulo A. In this connection, the coset A will always be represented by 1.

Each element $w \in G *_A H$ may be written as a product

$$w = x_1 y_1 x_2 y_2 \ldots x_n y_n$$

where $x_i \in G$ and the $y_i \in H$. First of all we convert the product $x_1y_1\ldots x_ny_n$ into one in which none of the x_i, y_i apart from x_1 or y_n lies in A. If $y_n \in A$, then it shall equal 1. There is an obvious process, working backwards from the right, which brings an arbitrary product into this form:

Let $y_n = a_1\bar{y}_n$ with $a_1 \in A$ and let $\bar{y}_n$ be the representative of the coset of Ay_n, $x_na_1 = a_2\overline{x_na_1}$ etc. Continuing this process eventually brings w into the form

$$w = ag_{i_1}h_{j_1}g_{i_2}h_{j_2} \ldots g_{i_\ell}h_{j_\ell}$$

where $a \in A$ and at most one of g_{i_1} or h_{j_ℓ} can equal 1. We call this the *normal form* of w. We have just describes a process for bringing any representation $w = x_1y_1\ldots x_ny_n$ into the normal form.

2.3.3 Lemma. *The normal form of an element is unique.*

We follow the proof of [van der Waerden 1948]. Let Θ be the set of normal forms and $z \in G$ or H. Given $ax_1y_1\ldots x_ny_n \in \Theta$, let $(ax_1y_1\ldots x_ny_n)\Pi_z$ be the normal form which one obtains when the above reduction process is applied to $ax_1y_1\ldots x_ny_nz$. This defines a mapping $\Pi_z : \Theta \to \Theta$. If z and z' are both from G or H, then $\Pi_{zz'} = \Pi_z\Pi_{z'}$. The proof of this assertion is not difficult, however it is lengthy and involves many case distinctions.

Π_1 is the identity, so $\Pi_{z^{-1}}\Pi_z = \Pi_z\Pi_{z^{-1}} = id$. Thus we can regard Π_z as an element of the group K_Θ of all permutations of Θ, and $z \mapsto \Pi_z$ defines a homomorphism $\alpha: G \to K_\Theta$ (or $\beta: H \to K_\Theta$). Let K be the smallest subgroup of K_Θ which contains $\alpha(G)$ and $\beta(H)$. Then K satisfies the conditions 2.1.1 (a) and (b) for the presentation of $G *_A H$ and hence there is a homomorphism

$$\Pi : G *_A H \to K$$

with $\Pi_{ax_1y_1\ldots x_ny_n} = ax_1y_1\ldots x_ny_n$.

Π is therefore a monomorphism, and different normal forms are mapped to different permutations. In particular, different normal forms correspond to different elements. □

The following properties may be easily derived from these facts:

2.3.4 Proposition. *G and H can be regarded as subgroups of $G *_A H$.* □

2.3.5 Proposition. *If $\varphi: G \to X$ and $\psi: H \to X$ are homomorphisms with $\varphi(a) = \psi(a)$ for* *all* $a \in A$*, then there exists a uniquely determined homomorphism:* Φ*:* $G *_A H \to X$ *which agrees with* φ *on* G *and with* ψ *on* H*.*

□

<u>2.3.6 Proposition</u>. *If* $[G *_A H : G] < \infty$ *then either* $A = G$ *or* $A = H$ *(and* $G *_A H$ *equals* G *or* H *respectively).*

□

The uniqueness of $G *_A H$ up to isomorphism also follows from 2.3.5, as the free product of two groups G and H with amalgamated subgroup A has the following universal property:

<u>2.3.7 Property</u>. *Let* K *be a group and let* $\gamma: G \to K$*,* $\eta: H \to K$ *be homomorphisms such that* $\gamma(a) = \eta(a)$ *for* $a \in A$*. Let* X *be some other group, and let* $\varphi: G \to X$*,* $\psi: H \to X$ *be homomorphisms such that* $\varphi(a) = \psi(a)$ *for* $a \in A$*. Then there is a unique homomorphism* $\chi: K \to X$ *such that* $\varphi = \chi \circ \gamma$ *and* $\psi = \chi \circ \eta$*, i.e. such that the following diagrams are commutative.*

□

<u>2.3.8 Proposition</u>. (a) *By 2.3.7,* K *is determined up to isomorphism.*

(b) γ *and* η *are monomorphisms.*

(c) $K \cong G *_A H$.

Proof. Let K and K' have the universal property; mappings which involve K' are denoted by a prime. Since $\gamma': G \to K'$ and $\eta': H \to K'$ are homomorphisms it follows that there is a homomorphism $\chi: K \to K'$ such that

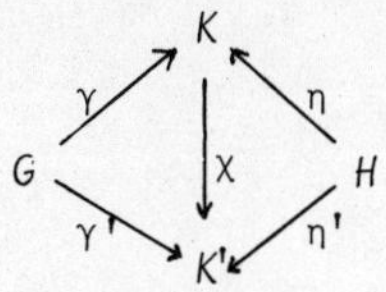

is commutative. Since K' also has the universal property there is a homomorphism $\chi': K' \to K$ making

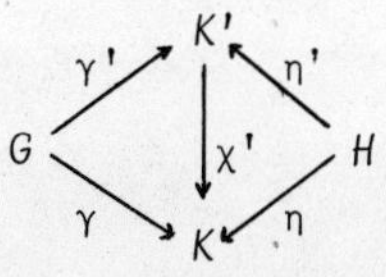

commutative. We now have the commutative diagrams

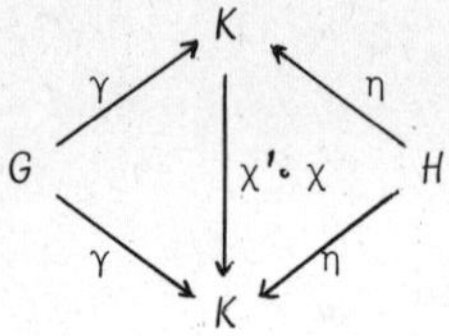

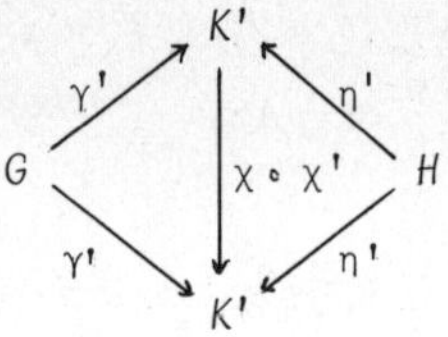

The identity mappings $K \to K$ and $K' \to K'$ resp. would solve the same problems, hence by the uniqueness property of χ in 2.3.7 it follows that

$$\chi' \circ \chi = \mathrm{id}_K \text{ and } \chi \circ \chi' = \mathrm{id}_{K'}.$$

Thus χ and χ' are isomorphisms.

Now (b) follows from

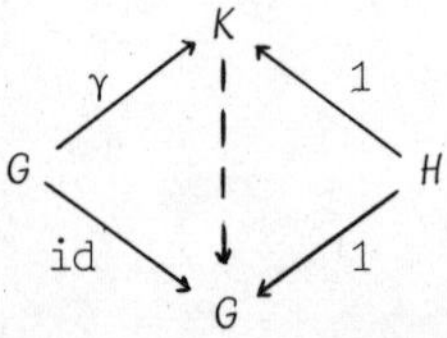

where 1 is the trivial homomorphism, and it follows from 2.3.5 that $G *_A H$ has the property required for (c). □

2.3.9 Proposition. *An element* z *belongs to the center of* $G *_A H$ *if and only if* z *is in* A *and in the centers of* G *and* H $(G \neq A \neq H)$.

Proof. The condition is obviously sufficient. Conversely, if z is in the center of $G *_A H$ and has the normal form

$$z = ax_1y_1 \ldots x_ny_n, \; y_n \neq 1$$

then zx and xz have distinct normal forms for $x \in G$, $x \notin A$. Similarly for $y_n = 1$ and $x_n \neq 1$. Thus z must belong to A and the centers of G and H. □

2.3.10 Proposition. (a) *An element of finite order is conjugate to an element of* G *or* H.

(b) *Each finite subgroup of* $G *_A H$ *is conjugate to a subgroup of* G *or* H.

(c) *If* $1 \neq x \in G$ *is a proper power of an element from* $G *_A H$ *then* x *is already a proper power of an element of* G.

Proof: Let $z \in G *_A H$ and $z^m = 1$, $z = ax_1y_1 \ldots x_ny_n$. If $n = 1$ and x_1 or y_n equals 1 we have nothing to prove. If $x_1 = 1$ then $y_nzy_n^{-1}$ has a shorter normal form than z, likewise $x_nzx_n^{-1}$ for $y_n = 1$. These shortenings are only impossible for $x_1 \neq 1$ and $y_n \neq 1$. But then if we write $ax_1y_1 \ldots x_ny_n$ m times in succession and convert into normal form no factors are lost, so z^m cannot lie in A, let alone equal 1.

The proofs of (b) and (e) are exercises.

□

<u>2.3.11 Proposition</u>. *Assume that* $x \in H$, $x \notin A$, $y \in G *_A H$ *and* $yxy^{-1} \in H$. *Then either* $y \in H$ *or* x *is conjugate to an element of* A.

□

<u>2.3.12 Proposition</u>. *Let systems of coset representatives modulo* A *in* G *and* H *be given as before. Then each element of* $G *_A H$ *has a unique normal form*

$$w = r_1 \ldots r_m \, k \, \bar{r}_m^{-1} \ldots \bar{r}_1^{-1}$$

where the r_i, $\bar{r}_i$ *are coset representatives different from* 1, *consecutive* r_i *(resp.* $\bar{r}_i$*) are from different groups and* k *is either from* A, *in which case* r_m *and* $\bar{r}_m^{-1}$ *are from different groups, or* k *is from* G *(resp.* H*), in which case* r_m *and* $\bar{r}_m^{-1}$ *belong to* H (resp. G). ***The length of*** w *is* $2m$ *if* $k \in A$ *and* $2m+1$ *if* $k \notin A$.

□

A subgroup theorem analogous to that for free groups holds for free products:

<u>2.3.13 Theorem</u> (A.G. Kurosh). *Subgroups of free products are free products. More precisely,* $U \subset G * H$ *may be written as a free product*

$$U = \mathop{*}_i U_i * S$$

where S *is a free group and the* U_i *are conjugate to subgroups of* G *or* H.

An interesting application of the Kurosh theorem is the following:

<u>2.3.14 Theorem</u>. *Let* $G = G_1 * \ldots * G_n = H_1 * \ldots * H_m$ *where the* G_i *and* H_i *are non-trivial and indecomposable into proper free products. Then* $n = m$ *and there is a permutation* $\begin{pmatrix} 1 & \ldots & n \\ k_1 & \ldots & k_n \end{pmatrix}$ *such that* H_i *is isomorphic to* G_{k_i}. *If* H_i *is not an infinite cyclic group then* H_i *is conjugate (in* G*) to* G_{k_i}.

Proof. Consider the subgroup H_m of G. By 2.3.13, H is a free product of groups conjugate to subgroups of the factors G_i of G, and a free group. Since H_m is indecomposable we have either

(a) H_m is conjugate to a subgroup of a factor G_{k_m}

or (b) $H_m \cong \mathbb{Z}$ (the only indecomposable free group).

In case (a) we consider G_{k_m}. By the same reasoning, it is conjugate to a subgroup of one of the factors H_j or else $\cong \mathbb{Z}$. In the first case it follows that H_m is conjugate to a subgroup of H_j, hence by 2.3.11 $m = j$ and G_{k_m} and H_m are conjugate. In the second case $H_m \cong \mathbb{Z}$. Since H_m is a free factor of G the generator of H_m cannot be a proper power of an element of G (2.3.10 (c)), hence H_m and G_{k_m} are conjugate. The normal subgroup N generated by H_m is the same as that for G_{k_m}, and

$$G/N \cong H_1 * \ldots * H_{m-1} \cong G_1 * \ldots * G_{k_m-1} * G_{k_m+1} * \ldots * G_n$$

Now we can apply induction.

For case (b) we consider $H_{m-1}, H_{m-2}, \ldots, H_1$ in turn to try and secure situation (a) and apply induction. If this fails then $H_i \cong \mathbb{Z}$ for $1 \le i \le m$, hence G is the free group of rank m. Therefore all G_i are indecomposable free groups, i.e. $\cong \mathbb{Z}$, and $m = n$ is proved in 1. 7.7.

□

We carry out the proof of 2.3.13 in 2.6. At the end of the proof there are some remarks on strengthenings of the theorem. For free products with non-trivial amalgamation only a much more complicated and weaker-looking theorem is true, see [Karrass-Solitar 1970]. The subgroups are built by several "free product with amalgamation" constructions when the amalgamated subgroups vary.

Free products (with amalgamation) first appeared in the study of discontinuous groups of the plane in [Fricke-Klein 1897], where they were called "group compositions". The general concept is developed explicitly in [Schreier 1927] where there is an existence proof based on the solution of the word problem. Nowadays it is more usual to use the universal property 2.3.7 as a starting point.

Exercises: E 2.6,7

2.4 2-DIMENSIONAL COMPLEXES

The presentation of a group by generators and relations can also be reflected geometrically, but now we have to add faces (for the relations) to the graph that corresponds to the generators.

2.4.1 Definition. A *2-dimensional complex or 2-complex* is a system of vertices, edges and (oriented) faces with the following properties:

(a) The vertices and edges constitute a graph C.

(b) For each face ϕ there is a closed path ω in C. The set of paths ω' which result from ω by cyclic interchange is called the class of *positive boundary paths* of ϕ. The path ω^{-1} bounds ϕ negatively. The positive boundary path of ϕ is often denoted by $\partial\phi$.

(c) For each face ϕ there is an oppositely oriented face ϕ^{-1}, the positive boundary paths of which are the negative boundary paths of ϕ. A pair $\{\phi,\phi^{-1}\}$ is called a *geometric face.*

2.4.2 Definition. Two paths in the 1-dimensional subcomplex of F are called *equivalent* when one may be converted into the other by a finite sequence of the following modifications:

(a) Removal or insertion of spurs (the old equivalence in C).

(b) If $\omega = \omega_1\omega_2\omega_3$ and if $\omega_2^{-1}\omega_2'$ is the boundary path of a face, then ω may be replaced $\bar{\omega} = \omega_1\omega_2'\omega_3$.

2.4.3 Proposition and Definition. *The equivalence classes of closed paths with fixed initial point* P *constitute a group, the* fundamental group *of* F *relative to* P. *This group is also called* first homotopy group *and denoted* by $\pi_1(F)$.

□

2.4.4 Presentation of the fundamental group. For the sake of simplicity let C be connected. We obtain a presentation of the fundamental group $\mathcal{W}$ of F in the following way: let B be a spanning tree of C. As in 1.4, generators of $\mathcal{W}$ correspond to segments which do not lie in C. We obtain relations, and indeed "definitively many", by circuits around the faces, i.e. words which correspond to paths of the form $\xi\partial\phi\xi^{-1}$ where ϕ is a face and ξ is an approach path from P to the beginning of ϕ. Since one may always alter relations by conjugation without altering the group, we can always choose ξ in B so that the approach paths no longer appear in the relations.

2.4.5 Definition. Two 2-complexes are called *related* if one may be converted into the other by a finite sequence of the following *elementary transformations (processes):*

(a) Let σ be an edge which goes from P_1 to P_2. The modified complex consists of all pieces of the old complex except σ and σ^{-1}, a new vertex P, and the new edges

σ_1, σ_2 and their inverses, where σ_1 goes from P_1 to P and σ_2 from P to P_2.

In boundary paths, σ is replaced by $\sigma_1\sigma_2$.

$$P_1 \xrightarrow{\ \sigma\ } P_2 \;\rightarrow\; P_1 \xrightarrow{\ \sigma_1\ P\ \sigma_2\ } P_2$$

This process is called *subdivision of an edge.*

(b) Let ϕ be a face of the complex and $\partial\phi = \omega_1\omega_2$. The new complex consists of all parts of the old one except ϕ and ϕ^{-1}. In addition it contains a new segment σ which runs from the initial point of ω_1^{-1} to the final point of ω_1^{-1}, and its inverse σ^{-1}, and faces ϕ_1 and ϕ_2 with $\partial\phi_1 = \omega_1\sigma$ and $\partial\phi_2 = \omega_2\sigma^{-1}$ together with their inverses. (Subdivision of a face.)

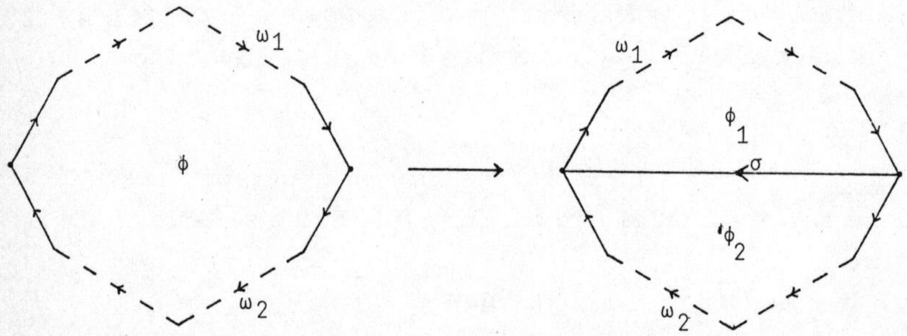

(c) The processes inverse to (a), (b), as long as they are possible.

It should be noted that removal of a geometric edge from the complex is not always possible

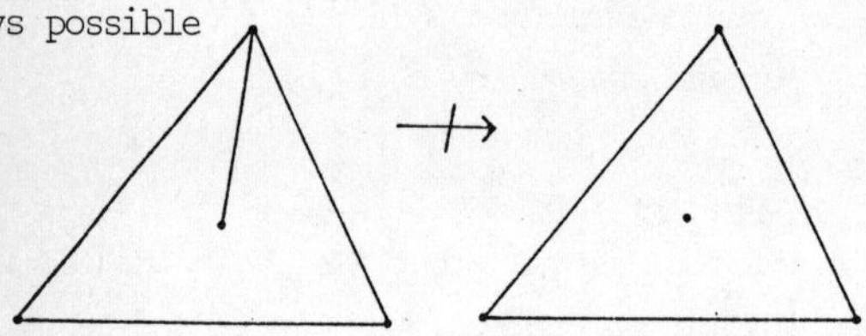

or

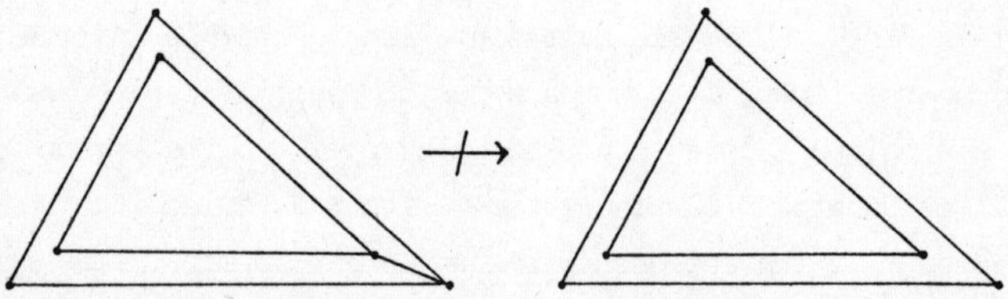

Similarly, one must be careful with vertices, for

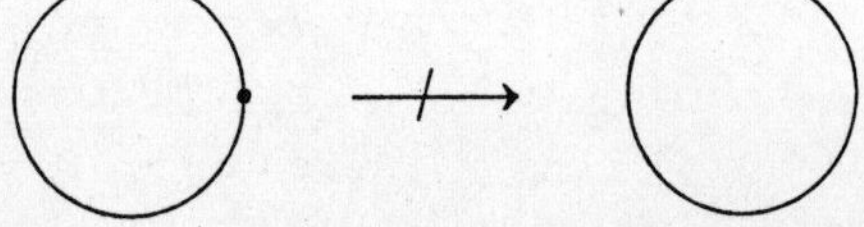

2.4.6 Proposition. *The number* $\alpha_o - \alpha_1 + \alpha_2$ *where* α_o *and* α_1, *as before, are the numbers of vertices and geometric edges, and* α_2 *is the number of geometric faces, remains invariant under these processes. We call* $\alpha_o - \alpha_1 + \alpha_2$ *the* Euler characteristic *of the 2-complex.*

2.4.7 Proposition. *Related 2-dimensional complexes have isomorphic fundamental groups.*

Proof. If for example we carry out the first process and if σ lies in the tree we have nothing to prove; if σ does not lie in the tree, then we must add σ_1 or σ_2 to the old tree in order to obtain a spanning tree. The edge not added gives the generator corresponding to σ. In the second process we introduce a new generator s and a new relation sw_1, where w_1 is a word in the old generators. This is the Tietze operation 2.1.5'.

Exercise: E 2.8 □

2.5 COVERINGS

Covering complexes will now be defined analogously to coverings of graphs.

2.5.1 Definition. A mapping $f: F' \to F$ is called a *covering* when (a), (b) and (c) hold.

(a) The restriction of f to the 1-dimensional subcomplex gives a covering $f|C': C' \to C$ of graphs.

(b) Faces are mapped to faces and $f(\phi^{-1}) = (f(\phi))^{-1}$. The image of the positive boundary path of a face ϕ is the positive boundary path of $f(\phi)$ (traversed once).

(c) If ω appears k times as positive boundary of faces from F then each path ω' in F' which covers ω also appears k times as positive boundary of faces of F'. (Here are two cases: 1) The boundary ω of a face ϕ is the m-th power of a path ω_o where $m \geq 1$ is maximal, then ω is counted m times. 2) ω is the boundary of different faces. See the figure.)

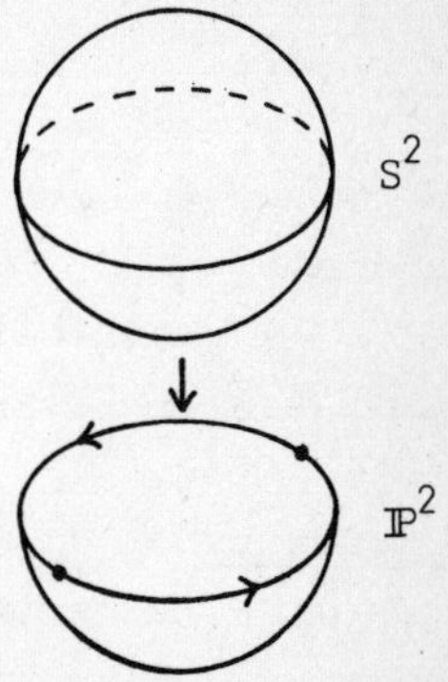

For some applications it is necessary to strengthen (c) to a local homeomorphism condition, analogous to condition 1.5.1 (c) for 1-dimensional coverings. Since this will not play an important role in this book (see however exercise E 3.7) we leave the reader to formulate a suitable statement of local homeomorphism for 2-complexes.

Let f: $F' \to F$ be a covering, let F be connected and let P' be a point that lies over $P \in F$. As with coverings of graphs one sees that f induces a monomorphism of the fundamental group $\mathcal{U}$ of F' relative to P' into the fundamental group $\mathcal{W}$ of F relative to P. Indeed, condition (c) says that each path mapped onto a path equivalent to the identity is itself equivalent to the identity.

Conversely, each subgroup $\mathcal{U}$ of $\mathcal{W}$ corresponds to a covering:

Let C be the graph of F and let $\mathcal{W}^*$ be the fundamental group of C relative to the basepoint P, computed with the help of a tree B. One obtains the free group $\mathcal{W}^*$ when one drops all relations from the presentation of $\mathcal{W}$ obtained from the tree B. Then there is a canonical homomorphism $\alpha: \mathcal{W}^* \to \mathcal{W}$. Let $\mathcal{U}^* = \alpha^{-1}(\mathcal{U})$ and let $f_1: C' \to C$ be the covering associated with $\mathcal{U}^*$. We obtain the desired covering f: $F' \to F$ by spanning each path of C' which maps to the boundary of a face of F by a face. Such a path is closed, for its image is equivalent to the identity path in F, and that says that it corresponds to an element of $\mathcal{U}^*$ in C. Now f is the natural extension of f_1 to F'.

Thus we can carry over Theorem 1.5.3 word for word if we interpret the C and C' there as 2-complexes. (The theorem also remains true for complexes of arbitrary dimension.) Namely

2.5.2 Theorem. *A covering* f: $C' \to C$ *induces a monomorphism of the fundamental group* $\mathcal{W}'$ *of* C' *onto a subgroup* $\mathcal{U}$ *of the fundamental group* $\mathcal{W}$ *of* C, *as long as the basepoint of the former lies over that of the latter. Conversely, for each subgroup* $\mathcal{U}$ *of* $\mathcal{W}$ *there is a covering* f: $C' \to C$ *such that the monomorphism induced by* f *maps the fundamental group of* C' *isomorphically onto* $\mathcal{U}$.

□

2.5.3 Definition. An *isomorphism* h between two 2-complexes F_1 and F_2 is a correspondence which is one-to-one and maps the faces, edges and vertices of F_1 to the faces, edges and vertices, respectively, of F_2 in such a way that inverse elements are mapped to inverse elements and the boundary relations are preserved. We mean by the latter that if $\sigma_1 \ldots \sigma_n$ is the positive boundary path of ϕ in F_1 then $h(\sigma_1) \ldots h(\sigma_n)$ is the boundary of $h(\phi)$; if σ is an edge from P_1 to P_2 in F_1 then $h(\sigma)$ is an edge from $h(P_1)$ to $h(P_2)$.

2.5.4 Definition. If f: $F' \to F$ is a covering, then an isomorphism h: $F' \to F'$ induces a mapping of F when parts of F' which lie over the same part of F are also sent to such parts by the isomorphism. An isomorphism h which induces the identity in this way (i.e. fh = f) is called a *covering transformation.* The covering transformations constitute a group under composition. We shall determine this group. A covering transformation can never map an edge or face to its inverse. In particular, it follows from the property 1.5.1 (c) of coverings that

2.5.5 Lemma. *If P' is a point over P and ω is a path in F beginning at P, then there is exactly one path lying over ω which begins at P'.*

□

Likewise, it follows from 1.5.1 (c) together with 2.5.1 (b) that

2.5.6 Lemma. *A covering transformation which has a fixed point, or which maps an edge or a face to itself is the identity.*

□

Let P be the basepoint for the fundamental group G of F, $P'_1, P'_2, \dots$ the vertices lying over P, and U the subgroup of G upon which the fundamental group of F' relative to the basepoint P'_1 is mapped by the homomorphism induced by f. The representatives $g_1 = 1, g_2, \dots$ of the cosets of G modulo U are chosen so that the path with initial point P'_1 lying over a curve which represents g_i ends in P'_i. (By Lemma 2.5.5 the path which lies over the curve representing g_i is uniquely determined, and in 1.1.5 we have seen that two paths in F' with initial point P'_1 which lie over two curves of the same coset lead to the same endpoint over P. If the two curves are in different cosets, then the endpoints are different vertices lying over P.)

A covering transformation which maps P'_i to P'_j defines a one-to-one correspondence between the closed paths beginning at P'_i and P'_j. Any two corresponding paths lie over the same underlying path. Thus the homomorphisms, induced by f, of the fundamental groups of F' with basepoint P'_i and the fundamental group of F' with basepoint P'_j are onto the same subgroup of G i.e. $g_i^{-1}\, U\, g_i = g_j^{-1}\, U\, g_j$. Thus we have proved half of the following theorem.

2.5.7 Theorem. *Let U, P, P'_1, P'_2, ..., g_1, g_2, ... be as above. Then there is a covering transformation which maps P'_i to P'_j if and only if $g_i^{-1}\, U\, g_i = g_j^{-1}\, U\, g_j$.*

Conversely, if $g_i^{-1}\, U\, g_i = g_j^{-1}\, U\, g_j$ then the two paths which lie over the same closed path of F and which begin at P'_i and P'_j are either both closed or both open. For this reason we can define the following mapping:

If Q is a vertex of F', ω' a path from P'_i to Q which lies over ω, and ω" the unique path over ω beginning at P'_j, then h maps Q to the final point of ω". h is well-defined on the vertices of F', for if $\bar{\omega}'$ is another path from P'_i to Q over the underlying path $\bar{\omega}$ then $\omega'\bar{\omega}'^{-1}$ is closed. But then the path $\omega''\bar{\omega}''^{-1}$ with initial point P'_j over the underlying path $\omega\bar{\omega}^{-1}$ is also closed, and $\bar{\omega}''$ is the path beginning at P_j which lies over $\bar{\omega}$. It is now obvious how h can be extended to a covering transformation.

□

The covering transformation h constructed in the proof of Theorem 2.5.7 in particular carries P_1' to the point P_ℓ' when $\mathcal{U}\, g_i^{-1} g_j = \mathcal{U}\, g_\ell$. Since a covering transformation is uniquely determined by the image of P_1' (Lemma 2.5.6), each representative g_j, and hence each $\mathcal{U}g_j$, with $g_j^{-1}\, \mathcal{U}\, g_j = \mathcal{U}$ corresponds uniquely to a covering transformation. If g_i, g_j are both from $N = \{x\colon x^{-1}\, \mathcal{U}\, x = \mathcal{U}\}$, the normalizer of $\mathcal{U}$, then the product of the covering transformations associated with $\mathcal{U}g_i$ and $\mathcal{U}g_j$ is the covering transformation associated with $\mathcal{U}g_i g_j$. Thus we have proved:

2.5.8 Theorem. *The covering transformation group is isomorphic to* $N/\mathcal{U}$. *If* $g_{j_1} = g_1 = 1, g_{j_2}, g_{j_3}, \ldots$ *are the representatives of the cosets modulo* $\mathcal{U}$ *which lie in* N, *then for any two points* P_{j_k}' *and* P_{j_ℓ}' *there is a transformation of the cover which sends* P_{j_k}' *to* P_{j_ℓ}'.

□

If $\mathcal{U}$ is normal, so that $N = G$, then the covering transformation group is transitive on the $P_1', P_2', \ldots$. In particular, if $\mathcal{U} = 1$ then G operates as the covering transformation group on F'.

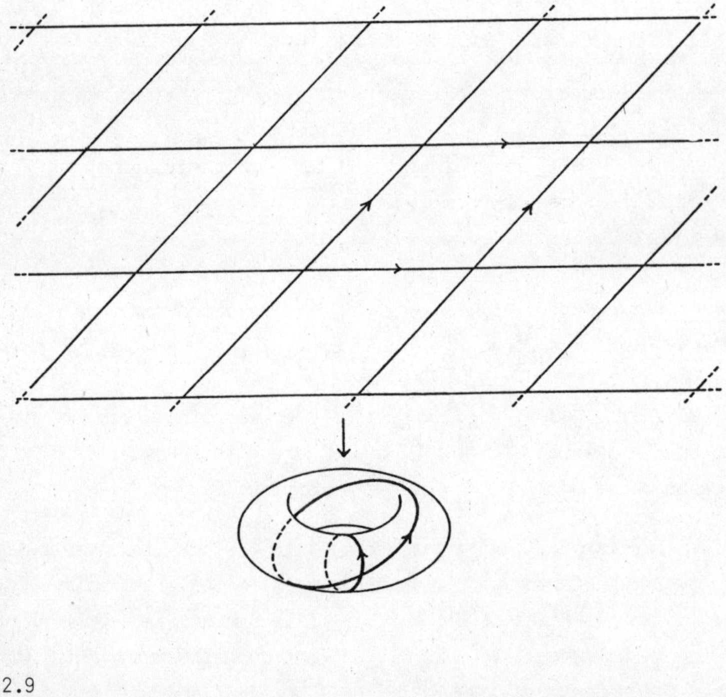

Exercise: E 2.9

2.6 PROOF OF THE KUROSH THEOREM

Let F be a 2-complex. We shall modify F without altering its fundamental group. Divide the complex F into various parts, countries, so that a face and its boundary edges (and their inverses), an edge and its endpoints, a boundary edge and the faces it bounds, belong to the same country. A segment or a face may only belong to one country. We also allow a *no-man's-land* which contains no faces and the edges of which belong to no country. If a vertex P lies in a country L, but also in others, we introduce a new vertex Q and an edge τ from P to Q, let all the edges of the country L emanating from P begin at Q and let the remainder begin at P. The new complex has an isomorphic fundamental group, since we can insert the added edges in the old spanning tree to obtain a spanning tree for the new complex. We carry over the old countries to the new complex and assign τ to the no-man's-land.

After several of these steps we obtain a 2-complex F in which no vertex belongs to different countries. We assume that each country is connected, choose spanning trees for the countries L_1, L_2, ... and then extend them to a spanning tree for F. To present the fundamental group of F we take generators s_{ij} resulting from the edges in L_i, as well as further generators $s_1, s_2, \ldots$ from the no-man's-land. Since the boundary path of a face belongs entirely to one country, it yields a relation for the generators s_{ij} with fixed i. Thus the fundamental group of F has the form $\mathop{\ast}_i G(L_i) * S$, where $G(L_i)$ is the group of paths in country L_i provided with a suitable approach path and S is the free group in the generators from the no-man's-land. The product is taken over all countries.

Let G and H be two groups and let presentations for them both be given. We construct 2-complexes G,H for these with edges from vertices P_1, P_2 for the generators and faces for the relations. We also take a point P and two edges τ_1, τ_2 from P to P_1 and P_2 respectively. The new 2-complex F has $G * H$ as fundamental group relative to the basepoint P.

Corresponding to a subgroup $U \subset G * H$, we take a covering complex F' with basepoint P' over P. In it we choose a spanning tree which contains all edges over τ_1 and τ_2. Also, we define countries $L_1, L_2, \ldots$ in F' by transitive application of: "Two edges belong to the same country of P' when they appear in the boundary of a face". The edges not covered by this are assigned to the no-man's-land. Each country is connected, and we choose an approach path in the tree to each country, beginning at P'. Then the fundamental group is the free product $\mathop{\ast}_i G(L_i) * S$. But since a path in an L_i lies wholly over G or H, the groups $G(L_i)$ are conjugate to subgroups of G or H and the conjugation factor corresponds to the approach path to L_i.

□

Naturally it is possible to make sharper statements when one examines the execution of this construction more precisely. The conjugation factors can be prescribed coset representatives of $G * H$ modulo U which satisfy the Schreier condition, and we can assume that each representative appears at most once with a subgroup of G or H respectively. It is also interesting to compare several product representations of a group, see [R. Baer and F. Levi 1936]. Other proofs are found in [Kuhn 1952], [MacLane 1958], [Weir 1956], [Lyndon-Schupp 1977], [Kurosh 1934, 1967].

2.7 HOMOLOGY GROUPS

A disadvantage of fundamental groups is that they can be non-abelian, and this makes computations and isomorphism decisions difficult or even impossible. Better known topological invariants are the homology groups, which are a major subject of algebraic topology. Although we do not use them very often we will briefly introduce the notions of homology theory that are needed for 2-complexes.

2.7.1 We associate pairs of symbols (f,-f), (s,-s) and a symbol p with each geometric face $\{\phi, \phi^{-1}\}$, each geometric edge $\{\sigma,\sigma^{-1}\}$ and vertex P respectively of a 2-complex F. The free abelian groups $C_0(F)$, $C_1(F)$, $C_2(F)$ in the generators p (where P ranges over all vertices of F), s (where σ ranges over all geometric edges) and f (where ϕ ranges over all geometric faces) are called the *chain groups* of F in dimensions 0,1 and 2 respectively. An element of a chain group is called a *chain*. If ϕ is a face with positive boundary path $\partial\phi = \sigma_1\sigma_2\ldots\sigma_n$ and if f and $s_1 + s_2 + \ldots s_n$ are the chains corresponding to ϕ and $\partial\phi$ respectively then the mapping $f \mapsto s_1 + s_2 + \ldots + s_n$ as ϕ ranges over all surface piece pairs defines a homomorphism $\partial_2\colon C_2(F) \to C_1(F)$ called the boundary homomorphism in dimension 2. We obtain a boundary homomorphism $\partial_1\colon C_1(F) \to C_0(F)$ analogously in dimension 1: here the image of a generator s equals q - p when the segment σ corresponding to s runs from P to Q. Finally, ∂_0 denotes the mapping $C_0 \to 0$. The elements of kernel ∂_i are called *i-cycles* and the elements of the image of ∂_{i+1} are called *i-boundaries*. If s is a boundary then s is often called *null homologous*, and we write $s \sim 0$.

2.7.2 Lemma. $\partial_1\partial_2$ *is the null homomorphism.*

The proof is simple. It suffices to show that $\partial_1\partial_2$ maps the generators of $C_2(F)$ to 0. But this is clear since the boundary path of a surface piece is closed.

□

2.7.3 Definition. The 0,1 and 2-*dimensional homology groups* of F are respectively defined by

$$H_o(F) = {}^{C_o(F)}/_{\text{Image } \partial_1}$$

$$H_1(F) = {}^{\text{Kernel } \partial_1}/_{\text{Image } \partial_2}$$

$$H_2(F) = \text{Kernel } \partial_2$$

Lemma 2.7.2 guarantees that the definition of $H_1(F)$ makes sense. Two elements of $C_o(F)$ or Kernel ∂_1 are called *homologous* when they belong to the same coset modulo Image ∂_1 or Image ∂_2 respectively. Two elements of $C_2(F)$ are called homologous when they are equal. Two homologous elements belong to the same *homology class*.

2.7.4 Theorem. *The homology groups are invariant under the processes 2.4.5 (a) - (c) (i.e. related 2-complexes have isomorphic homology groups in all dimensions).*

The *proof* is not difficult, in fact for some of the cases to be treated it is trivial, and we set it as Exercise E 2.10.

□

2.7.5 Definition. By a *mapping of a 2-complex* F' *into a 2-complex* F we mean a correspondence $f: F' \to F$ which sends faces of F' to faces, edges or vertices of F; edges of F' to edges or vertices of F; and vertices of F' finally to vertices of F, in such a way that boundary relations are preserved. More precisely, this means:

(a) The boundary objects of an element are carried into the image element or its boundary objects. Inverse elements are carried to inverse elements or else to the same vertex.

(b) If the face ϕ' is mapped to ϕ, then after removal of spurs the image of $\partial\phi'$ is a positive power of $\partial\phi$.

(c) If the image of ϕ' is the edge σ, then boundary edges and vertices of ϕ' are associated with the edges σ and σ^{-1} or their endpoints. As a result $f(\partial\phi')$ will be a closed path.

(d) If a face or an edge is mapped to a vertex, so are all its boundary objects.

As examples of mappings between 2-complexes we mention coverings defined in 2.5.

2.7.6 Construction. A mapping $g: F' \to F$ induces homomorphisms

$$g_i: C_i(F') \to C_i(F), \quad i = 0,1,2$$

in the following way.

If g maps the element α' of F' corresponding to a generator a' of $C_i(F')$, $i = 0,1$, onto α then let $g_i(a') = a$ when the chain associated with α belongs to the ith chain complex; otherwise $g_i(\alpha') = 0$. Now g_i extends linearly to the whole of $C_i(F')$.

If the image of a face ϕ' is a face ϕ and $g_1(\partial_2' f') = n\partial_2 f$ holds for the boundary chains, then let $g_2 f' = nf$. If the image of ϕ' has lower dimension, then let $g_2 f' = 0$. Then one obtains:

2.7.7 Proposition. *A mapping* g *between 2-complexes* F' *and* F *induces homomorphisms* $g_{i*}: H_i(F') \to H_i(F)$ *in the following way: if* $\Sigma x_j a_j'$ *is a chain from* $C_i(F')$ *with* $\partial_i(\Sigma x_j a_j') = 0$ *and if* $[\Sigma x_j a_j']$ *denotes the homology class of* $\Sigma x_j a_j'$ *then*

$$g_{i*}[\Sigma x_j a_j'] = [g_i(\Sigma x_j a_j')] = [\Sigma x_j g_i(a_j')]$$

The *proof* is Exercise E 2.11 .

□

2.7.8 Theorem. *Let* F *be a connected 2-complex. Then* $H_1(F)$ *is isomorphic to the abelianized fundamental group* G *of* F *(with arbitrary basepoint).*

Proof. We construct an epimorphism $G \to H_1(F)$, the kernel of which is the commutator subgroup of G.

If the path ω represents the element $\{\omega\} \in G$ and $w \in C_1(F)$ is the chain corresponding to ω, then $\partial_1 w = 0$ since ω is closed. Thus w represents a homology class $[w]$ of $H_1(F)$. One sees immediately that $\{w\} \leftrightarrow [w]$ defines a homomorphism $\varphi: G \to H_1(F)$. Now if $[u]$ is an arbitrary element of $H_1(F)$, then since $\partial_1 u = 0$, u may be decomposed into a sum $w_1 + w_2 + \ldots + w_n$ where the corresponding paths $\omega_1, \omega_2, \ldots, \omega_n$ are closed. By insertion of spurs we can connect each ω_i with the basepoint of the fundamental group. Since spurs are mapped to $0 \in H_1(F)$ by φ, the product of the ω_i together with the approach paths constitutes a closed curve ω with $\varphi\{\omega\} = [u]$. Thus φ is an epimorphism. Since $H_1(F)$ is abelian, the kernel of φ contains the commutator subgroup of G. If on the other hand $\{\omega\}$ is in the kernel of φ and w is the chain of $C_1(F)$ corresponding to ω, then there is a chain $f = \Sigma a_i f_i$ of $C_2(F)$ with $\partial_2 f = \Sigma a_i \partial_2 f_i = w$. The f_i correspond to faces of F. Since the $\partial_2 f_i$ are boundary paths of faces, ω must become equal to a product of boundary paths of faces when its component segments are allowed to commute. Thus ω is homotopic to a path consisting of *commutators*.

□

Exercises: E 2.10 - 12.

2.8* THE FIRST HOMOTOPY GROUP AND THE FUNDAMENTAL GROUP. SEIFERT-VAN KAMPEN THEOREM. PASTING COMPLEXES TOGETHER

In this paragraph complexes are the usual cell-complexes or CW-complexes of arbitrary dimension. The properties of the first homotopy group π_1, expecially as regards coverings, are assumed known. They can be found in textbooks on algebraic topology, see specially [Massey 1967].

If K^i denotes the i-dimensional skeleton of a complex K, then we define the fundamental group G of K to be the fundamental group of the 2-complex K^2.

Let K be a complex with a connected K^1, and let the basepoint * for $\pi_1(K)$ be a point of K^0. Then each closed path emanating from * may be continuously deformed to one in K^1 and each closed surface to one in K^2. Consequently $\pi_1(K) = \pi_1(K^2)$. To each closed path with initial point *, in our sense, there corresponds a continuous mapping $(S^1, *) \to (K^2, *)$ (where S^1 is the 1-sphere with basepoint *), and consequently it represents an element of $\pi_1(K)$. Combinatorially equivalent paths are trivially homotopic, and so we obtain a homomorphism ψ: $G \to \pi_1(K^2)$. Since one may deform a closed path of K^2 with initial point * into a path which lies wholly in K^1, and which fully traverses each segment it enters, ψ is an epimorphism. Moreover, when G acts as the covering transformation on a covering $\tilde{K}^2$ of K^2 (or also on a covering $\tilde{K}$ of K) it is fixed point free. But the coverings defined in 2.5 can also be regarded topologically as coverings. Two different elements of G correspond to different covering transformations, therefore ψ is also a monomorphism.

2.8.1 Proposition. *The fundamental group and the first homotopy group of a complex coincide.*

□

The following theorem gives a process for computing the fundamental group of a complex $K = K_1 \cup K_2$ from the fundamental groups of K_1 and K_2.

2.8.2 Seifert-Van Kampen Theorem. *Let* K *be a connected complex of arbitrary dimension,* K_1, K_2 *connected subcomplexes with connected intersection* $K_1 \cap K_2$. *In the intersection, we choose the same basepoint, for all fundamental groups. Let the fundamental group of* K_i *have presentation* $\langle S_1^{(i)}, S_2^{(i)}, \ldots \mid R_1^{(i)}(S^{(i)}), \ldots \rangle$ *and let that of* $K_1 \cap K_2$ *have generators* $V_1, V_2, \ldots$. *Let* $V_j^{(i)}$ *denote the element of the fundamental group of* K_i *corresponding to* V_j. *Then the fundamental group of* K *obviously has the presentation*

---

*This paragraph is not needed for the understanding of the rest of the text.

$$<S_1^{(1)}, S_2^{(1)}, \ldots, S_1^{(2)}, S_2^{(2)}, \ldots | R_1^{(1)}(S^{(1)}), \ldots,$$

$$R_1^{(2)}(S^{(2)}), \ldots, V_1^{(1)} V_1^{(2)-1}, \ldots>.$$

□

2.8.3 In this section we compute the fundamental group of a complex which results from *pasting together* two isomorphic subcomplexes. Let K be a connected complex, L_1, L_2 connected subcomplexes with empty intersection. We choose the basepoint * for the fundamental group outside $L_1 \cup L_2$. Let $\iota: L_1 \to L_2$ be an isomorphism and K^ι the complex which results from identifying L_1 and L_2 according to ι. Let P and $\iota(P)$ be vertices of L_1 and L_2 respectively, σ_1, σ_2 paths from * to P and $\iota(P)$ respectively, let τ denote the path $\sigma_1\sigma_2^{-1}$ and t its homotopy class. Then one obtains a presentation of the fundamental group of K^ι from that of K by adding the generator t and the relations $\{\sigma_1\lambda\sigma_1^{-1}\} = t\{\sigma_2\iota(\lambda)\sigma_2^{-1}\}t^{-1}$, where λ ranges over the generating paths of the fundamental group of L_1.

For applications of these two results see the exercises E 2.13, 2.8.

2.9 GRUSHKO'S THEOREM

As another application of topological methods to group theory we present a proof of Grushko's theorem [Grushko 1940], due to [Stallings 1965]. The theorem concerns epimorphisms of free groups onto free products, and we state it only for free groups of finite rank, though Stallings' argument in fact extends into the transfinite to deal with the general case.

2.9.1 Theorem. *Let $\psi: S \to \ast_i G_i$ be an epimorphism of a free group S of finite rank onto an arbitrary free product of groups. Then there is a decomposition of S as a free product*

$$S = \ast_i S_i$$

such that $\psi(S_i) \subset G_i$ for each i.

If we denote the restriction of ψ to S_i by ψ_i, then each ψ_i must be an epimorphism onto G_i (otherwise ψ will not be onto $\ast_i G_i$), so the theorem can be interpreted as saying that an epimorphism $S \to \ast_i G_i$ is a "free product" of epimorphisms from free factors of S onto the individual G_i's.

2.9.2 Corollary. *The rank of a free product is the sum of the ranks of the factors.*

The proof is E 2.14.

□

The proof of 2.9.1 is an application of the Seifert-Van Kampen theorem 2.8.2 and the following method of adding faces to a 2-complex without changing the fundamental group.

2.9.3 Construction and Proposition. Let F be a 2-complex and ω a path in F between vertices P_1 and P_2. Then if we introduce a new edge τ from P_1 to P_2 and a new face ϕ with $\partial\phi = \omega\tau^{-1}$ *the new complex* F' *has the same fundamental group as* F. It follows easily from the definition of fundamental group for a 2-complex, 2.4.3, and the construction of generators from a spanning tree, 1.5.2, that τ and ϕ simply add a new generator t and a new relation t = w, where w is the element corresponding to ω. In particular *if* $\pi_1(F)$ *is a free group then* $\pi_1(F')$ *is a free group with the same generating set as* $\pi_1(F)$.

Proof of 2.9.1. We realize S as the fundamental group of a graph F with a single vertex P and an edge σ for each free generator s of S. If $a_1 \dots a_n$ is the normal form of $\psi(s)$ in $\ast_i G_i$ we divide σ into n subedges $\sigma_1, \dots, \sigma_n$, and define a map Ψ from the subdivided F into G by $\sigma_i \mapsto a_i$. In a natural way, Ψ induces the map $\psi: \pi_1(F) = S \to G$.

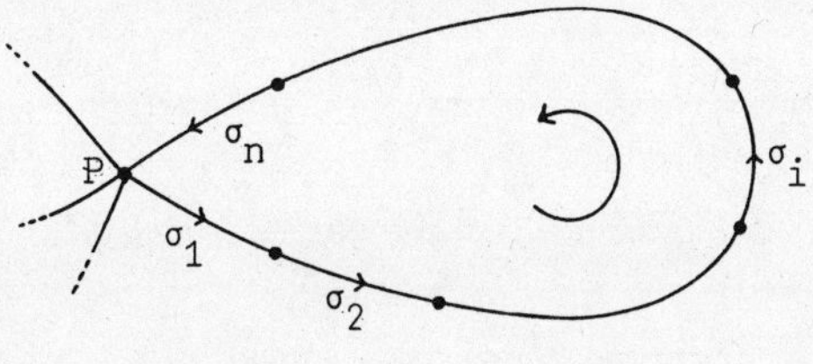

We continue to call the new graph F and let

$$F_i = \{\text{edges of F mapped into } G_i\} \cup \{\text{vertices of F}\}.$$

Then $\cap F_i$ = {vertices of F}, which is in general not connected, but it is (trivially) a disjoint union of trees.

In the course of future modifications, F will always denote the complex most recently constructed, and F_i will be defined as above. The strategy will be to add new edges and faces to F so as to make $\bigcap_i F_i$ "more connected", while preserving $\pi_1(F)$ and the fact that $\bigcap_i F_i$ is a disjoint union of trees. More precisely, we will preserve the following properties:

(1) $\pi_1(F) = S$, with the original generators s.

(2) $\bigcap_i F_i$ is a disjoint union of trees, and $F_{i_1} \cap F_{i_2} = \bigcap_i F_i$ for any $i_1 \neq i_2$.

2.9.4 Definition. A *binding tie for* F is a path ω in F such that
(a) ω connects vertices P_1, P_2 which meet different components of $\bigcap_i F_i$,
(b) $\omega \subset F_i$ for some i,
(c) $\hat{\psi}(\omega) = 1$

Given a binding tie ω for F, we introduce a new edge τ from P_1 to P_2 and a new face ϕ with $\partial\phi = \omega\tau^{-1}$, and set $\hat{\psi}(\tau) = 1$. Then the new F continues to enjoy properties (1) (by 2.9.3) and (2), but $\bigcap_i F_i$ has one less component, since two components of the old $\bigcap_i F_i$ have been joined by the edge τ.

If binding ties exist as long as $\bigcap_i F_i$ is disconnected, we can make $\bigcap_i F_i$ connected (hence a tree) in a finite number of steps. Note that each $F_i \supset \bigcap_i F_i \supset$ {vertices of F} so at this stage each F_i is also connected and (1), (2) and the Seifert-Van Kampen theorem 2.8.2 gives

$$S = \pi_1(F) = \mathop{\ast}_i \pi_1(F_i) = \mathop{\ast}_i S_i,$$

because $\pi_1(\bigcap_i F_i) = \pi_1(\text{tree}) = 1$. The desired epimorphisms $\psi_i: S_i \to G_i$ are precisely those induced by $\hat{\psi}$ on the F_i.

Thus to prove 2.9.1 it remains to prove the following:

2.9.5 Lemma. *If* $\bigcap_i F_i$ *is not connected, there is a binding tie for* F.

Proof. Since $\bigcap_i F_i$ is not connected there is a vertex Q not met by any of the components of $\bigcap_i F_i$ which meet the basepoint P for $\pi_1(F)$. Choose any path μ from P to Q. Since ψ maps $\pi_1(F)$ onto G, there is a closed path ν from P such that $\hat{\psi}(\nu) = \hat{\psi}(\mu)$, and hence if $\alpha = \nu^{-1}\mu$ we have (3) $\hat{\psi}(\alpha) = 1$, and since α runs from P to Q also

(4) $\alpha \not\subset \bigcap_i F_i$, which implies

(5) there is a decomposition $\alpha = \alpha_1\alpha_2\ldots\alpha_t$ of more than one factor, in which successive factors belong to different F_i. In particular

$$1 = \hat{\psi}(\alpha) = \hat{\psi}(\alpha_1)\hat{\psi}(\alpha_2)\ldots\hat{\psi}(\alpha_t)$$

and successive $\hat{\psi}(\alpha_i)$'s lie in different G_i's. Since $\bigcap_i G_i = \{1\}$, this implies that some $\hat{\psi}(\alpha_i) = 1$. If the endpoints of this α_i meet different components of $\bigcap_i F_i$, then

α_i is a binding tie. If not, we construct a new α with the same properties (3), (4), (5), but a smaller number of factors, as follows:

Choose a "basepoint" in each component of $\bigcap_i F_i$, taking P and Q among them, and construct the new product

$$\beta = \alpha_1\rho_1 \; \rho_1^{-1}\alpha_2\rho_2 \; \cdots \; \rho_{t-1}^{-1}\alpha_t$$

where ρ_i is a path in $\bigcap_i F_i$ from the final point of α_i to the basepoint of the component of $\bigcap_\lambda F_\lambda$ which contains it:

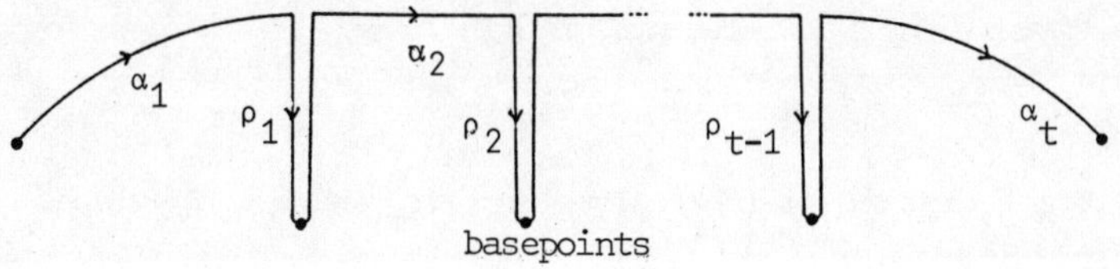

Let $\beta_1 = \alpha_1\rho_1, \ldots, \beta_t = \rho_{t-1}^{-1}\alpha_t$ denote the factors of β. We obviously have $\Phi(\beta) = \Phi(\alpha) = 1$, whence (3) for β, and (4), (5) hold with the β_i in place of the α_i; thus by the same argument we find a β_j such that $\Phi(\beta_j) = 1$. But now, if β_j is not a binding tie it begins and ends at the same "basepoint", and hence can be dropped entirely. By dropping all such β_j's and consolidating any adjacent β_i's from the same K_i which then occur we obtain a new α satisfying (3), (4), (5) but with fewer factors.

When the number of factors can be reduced no further, one of them is a binding tie.

□

Exercise: E 2.14.

2.10 ON THE NIELSEN METHOD IN FREE PRODUCTS

In this section we give without proof a theorem about the Nielsen cancellation method in free products with amalgamation. The proof of the main theorem 2.10.2 is cumbersome, but straightforward if one looks for the obstruction to a direct generalization of the Nielsen method from 1.7.

<u>2.10.1 Construction of an ordering in $G = G_1 *_A G_2$.</u> In the free product we use the normal forms for elements as described in 2.3.12: In each group G_i we choose coset representatives for the cosets Ag, where A itself is represented by 1. Then an ar-

bitrary element x has a uniquely determined presentation

$$g = r_1 \cdots r_m k \bar{r}_m^{-1} \cdots \bar{r}_1^{-1},$$

where the r_j and $\bar{r}_j$ are non-trivial representatives, consecutive from different factors G_i, and k is either from $G_i \setminus A$ for some i and $r_m, \bar{r}_m$ are not in G_i, or $k \in A$ and r_m and $\bar{r}_m$ are from different factors. We define $\ell(g) = \begin{cases} 2m+1 & k \notin A \\ 2m & k \in A \end{cases}$.

Now we well-order the accessible products of coset representatives: first by the length m, secondly those of equal length by some lexicographical order where we first order the set of groups G_i and then the cosets within one group.

Next we order pairs (g,g^{-1}) of elements of G follows:
$(g,g^{-1}) < (g',g'^{-1}) :\Longleftrightarrow g < g' : \Longleftrightarrow$

either $\ell(g) < \ell(g')$

or if $\ell(g) = \ell(g')$ then one of the 'halves' $r_1 \cdots r_m$ and $\bar{r}_1 \cdots \bar{r}_m$ stands before both halves of g',

or if the earlier halves of g and g' coincide then the other half of g stands before that of g'.

If $g < g'$ and $g' < g$ then g and g' (or g'^{-1}) coincide except in the central element.

We are especially interested in orders where, within one factor G_i, each representative is preceded by only finitely many others.

2.10.2 Theorem. *Let* $G = G_1 *_A G_2$ *and let the length* l *and order* $<$ *be defined as in* 1.7 *and* 2.10.1. *Let* $(g_i)_{i \in I}$ *be a system of elements from* G, *and* U *the subgroup generated by the* g_i. *Then one of the following cases occurs:*

2.10.3 *Each element* $w \in U$ *has a representation*

$$w = \prod_{j=1}^{n} g_{i_j}^{\varepsilon_j}, \quad i_j \in I, \quad \varepsilon_j = \pm 1,$$

where $\ell(g_{i_j}) \le \ell(w)$.

2.10.4 *There is a Nielsen process from* $(g_i)_{i \in I}$ *to a shorter system* $(g_i')_{i \in I}$, *i.e.* $g_i' < g_i$ *for all* $i \in I$, *but at least for one* $i \in I$, $g_i \nless g_i'$.

2.10.5 *There is a product* $a = \prod_{j=1}^{h} g_{i_j}^{\eta_j}$ *where* $a \neq 1$ *and* $1 \neq g_{i_j}^{\eta_j} \in A$ *for* $j = 1,\ldots,h$ *and an element* $x \in G_i \setminus A$ *for some* i *such that* $xax^{-1} \in A$.

2.10.6 *There is a product* $\prod_{j=1}^{h} g_{i_j}^{\eta_j}$ $(i_j \in I)$ *and a factor* G_k *such that:*

(a) All g_{i_j} *are in a subgroup which is conjugate to* G_k *and either they all have the same length or their length is* ≤ 1.

(b) At least one g_{i_j} *is not in* A.

(c) $\prod_{j=1}^{h} g_{i_j}^{n_j}$ *is conjugate to a non-trivial element of* A.

(d) All elements g_{i_j} *have the same length* ≥ 1 *and* $\ell(\prod_{j=1}^{h} g_{i_j}^{n_j}) < \ell(g_{i_1})$, *or all* g_{i_j} *are in the same factor* G_i, *some may be in* A, *and* $\prod_{j=1}^{k} g_{i_j}^{n_j}$ *is conjugate to some element from* A *where the conjugation factor is from* G_i.

For a *proof* see [Zieschang 1970].

□

2.10.4 suggests some induction step which corresponds to that from 1.7 and one finally comes, by Nielsen processes, to a system $(g_i')_{i \in I}$ where only the possibilities 2.10.3,5,6 occur and possibly some of the g_i' are trivial. The following example shows that no better result is possible.

2.10.7 Example ([Rosenberger 1974]). Let $G = G_1 *_A G_2$ where

$$G_1 = \langle s_1,\ldots,s_{2n-2} \mid s_1^2,\ldots,s_{2n-2}^2 \rangle \cong \mathbb{Z}_2 * \ldots * \mathbb{Z}_2$$

$$G_2 = \langle s_{2n-1}, s_{2n} \mid s_{2n-1}^2,\ s_{2n}^{2k+1} \rangle \cong \mathbb{Z}_2 * \mathbb{Z}_{2k+1}$$

$$A = \langle s_1 \ldots s_{2n-2} \rangle = \langle (s_{2n-1}\ s_{2n})^{-1} \rangle \cong \mathbb{Z}$$

and $n \geq 2$, $k \geq 1$. Then $x_1 = s_1s_2$, $x_2 = s_1s_3,\ldots,x_{2n-2} = s_1s_{2n-1}$ generate G. In proving this, the following is the important equation:

$$s_{2n}^{-2} = x_1x_2^{-1}x_3x_4^{-1}\ldots x_{2n-2}^{-1}x_1^{-1}x_2x_3^{-1}\ldots x_{2n-3}^{-1}x_{2n-2}.$$

For $n = 2$, situation 2.10.5 occurs, since $x_1 = s_1s_2 \in A$ and $s_1x_1s_1^{-1} = s_2s_1 = x_1^{-1} \in A$, and there is no situation 2.10.6 nor 2.10.4. For $n = 3$ there is a Nielsen process from (x_1,x_2,x_3,x_4) to (x_1,x_2,y_3,x_4) where $y_3 = x_1x_2^{-1}x_3 = s_1s_2s_3s_4 \in A$. Now the situation 2.10.6 occurs, namely G_2 contains

$$s_6^{-2} = x_1x_2^{-1}x_3x_4^{-1}x_1^{-1}x_2x_3^{-1}x_4 = y_3x_4^{-1}x_1^{-1}x_2y_3^{-1}x_1x_2^{-1}x_4$$

$$= (s_1s_2s_3s_4)\underbrace{(s_5s_1)\underbrace{\underset{G_1}{\underset{\cap}{(s_2s_1)}}\underset{G_1}{\underset{\cap}{(s_1s_3)}}\underset{G_1}{\underset{\cap}{(s_4s_3s_2s_1)}}\underset{G_1}{\underset{\cap}{(s_1s_2)}}\underset{G_1}{\underset{\cap}{(s_3s_1)}}}_{= s_2s_3s_4s_1 \in G_1}(s_1s_5)}_{= s_1s_2s_3s_4 \in A}$$

Here there is no situation 2.10.4 or 5.

2.10.8 Remark. Grushko's theorem 2.9.1 is a direct consequence of Theorem 2.10.2 because in a free product the obstructions to the application of Nielsen processes do not occur.

Theorem 2.10.2 was proved in [Zieschang 1970] and was used there to determine the rank of planar discontinuous groups, see 4.15.1 . The proof is rather unpleasant, but to my knowledge there is no simpler discussion. The theorem can be generalized to HNN-groups, see [Peczynski-Reiwer 1978] and groups with a length function, see [Hoare, preprint].

2.11 ON GROUPS WITH ONE DEFINING RELATION

Next to free groups, those with one defining relation seem to be the most accessible. In fact, quite a lot of results are known for them and we will mention some. The most famous is the

2.11.1 Theorem (Freiheitssatz). *Let* $G = \langle t,x_1,\ldots,x_n \mid R(t,x_i)\rangle$ *where* R *is a cyclically reduced word. If* $R(t,x_i)$ *contains* t *then the group generated* by $x_1,\ldots,x_n$ *is a free group of rank* n, *i.e. there are no proper relations between the* x_i.

This theorem was proposed by Dehn and first proved in [Magnus 1930]. Most proofs of it follow the line of the original one, perhaps with a little more sophistication. We take the exposition from [Lyndon-Schupp 1977]. Going back to the free group in the generators the Freiheitssatz takes the following form:

2.11.2 Proposition. *Let* $F = \langle X \mid \;\rangle$ *be the free group with basis* X *and let* R *be a cyclically reduced element from* F. *Then every non-trivial element of the normal closure* $N(R) \lhd F$ *(the smallest normal subgroup containing R) contains all generators from* X *that appear in* R.

Magnus proves this result by showing a more general looking proposition. To formulate it we introduce a convenient notation:

2.11.3 Definition. (a) A *presentation* $G = \langle X | R\rangle$ is called *staggered* if the following conditions are fulfilled: (a) Let $I = \mathbb{Z}$ or $I = \{1,\ldots,n\}$ and let Y be a subset of X that is the disjoint union of sets Y_i, $i \in I$. Let $R = \{r_j | j \in J\}$, where J is a linearly ordered set and each r_j is cyclically reduced and contains some generator from Y.
(b) For $j \in J$ we denote by $a(j)$ the least $i \in I$ such that r_j contains a generator from Y_i, and by $b(j)$ the greatest such i.
(c) For a staggered presentation we require that $j < k$ implies $a(j) < a(k)$ and $b(j) < b(k)$.

Such situations often appear when, starting from a presentation of a group H, the Reidemeister-Schreier method is applied to a normal subgroup U with cyclic factor group H/U, see 2.11.6. We will prove:

2.11.4 Proposition. *Let* $\langle X | R\rangle$ *be a staggered presentation, in the notation from 2.11.3. Suppose that a consequence* w *of* R (i.e. $w \in N(R)$) *contains generators* y *in* Y_i *only for* i *in a certain interval* $a \le i \le b$. *Then* w *is a consequence of those* r_j *that contain generators* y *in* Y_i *only for* $a \le i \le b$. *(We use the usual notation* $[a,b]$ *for the interval* $a \le i \le b$ *in this proof.)*

We prove now 2.11.2 and 2.11.4 by some induction arguments. The proof will be finished in 2.11.6.

2.11.5 Lemma. *The statements* 2.11.2 *for* $\ell(R) \le K$ *and 2.11.4 for* $\ell(R_j) \le K$, $j \in J$, *are equivalent.*

Proof. 2.11.2 restricted to relations with $\ell(R) \le K$ is part of 2.11.4 with the corresponding restriction, hence it remains to prove that the first implies the second. Since w is already a consequence of a finite subset of the relations, it suffices to consider the case that I is finite: $I = \{1,\ldots,n\}$. We argue by induction on n and start with the trivial case $n = 0$. We may suppose that $R = \{R_1,\ldots,R_m\}$ and that $a(1) = 1$ and $b(m) = n$. Let $X' = X\setminus Y_n$, $Y' = Y\setminus Y_n$, $R' = R\setminus\{R_m\}$. Then the presentation $G' = \langle X' | R'\rangle$ satisfies the hypothesis of 2.11.4 with $n' = n-1$. By the induction hypothesis, the normal closure N' of R' in $F' = \langle X' | \rangle$ contains no nontrivial element which is a word in the generators $Z = X\setminus Y$. Thus the image H' of $U := \langle Z | \rangle$ in G' is free with the image of Z as a system of free generators, and Z is embedded into G'.

Similarly, let $X'' = X \setminus (Y_1 \cup \ldots \cup Y_{n-1})$, let $Y'' = Y_n$, and let $R'' = \{R_m\}$. Then the presentation $G'' = \langle X'' | R'' \rangle$ satisfies the hypothesis of 2.11.2, hence the image H'' of U in G'' is free on the image of Z as basis, again Z is embedded into G''. This proves that G is the free product of G' and G'' with their subgroups H' and H'' amalgamated, see 2.3.2.

If, in the notation from 2.11.4, the interval $[a,b]$ does not contain n, then $w \in N(R) \cap F' = N'$, where the last equation is a consequence of Schreier's result that the factors of a free product with amalgamation are embedded into the product (2.3.4). Now we can apply the induction hypothesis to $G' = \langle X' | R' \rangle$. The case that $[a,b]$ does not contain 1 follows by symmetry. In the remaining case we have $[a,b] = [1,n]$ and the conclusion holds vacuously.

□

2.11.6 *Proof of* 2.11.2. Now we will prove 2.11.2 by induction on the length of R. We may assume that all generators of X appear in R and that X contains at least two elements. We consider first the case that some generator, say $t \in X$, has exponent sum 0 in R. We denote the other generators by x_i. The relation R now lies in the normal closure F_1 of the x_i in F, hence $N(R) \triangleleft F_1$ and so $w \in F_1$. The subgroup F_1 is generated by the elements $X_1 = \{x_{ik} = t^{-k}x_it^k : x_i \in X\setminus\{t\},\ k \in \mathbb{Z}\}$. Moreover, it follows that $N(R)$ is the normal closure of $R_1 = \{R_k = t^{-k}Rt^k : k \in \mathbb{Z}\}$ in F_1, where we express these relations by the x_{ik}.

Notice that R_j is obtained from R_0 by increasing all second subscripts by j; thus we have a staggered presentation. Moreover, all the R_k have the same length; this is evidently the total number of occurrences in R of letters other than t or t^{-1}. As t occurs in R, the length of R_j as a word in the generators x_{ik} is smaller than the length of R in the generators X.

We must show that R contains t and all other $x_i \in X$. For the first case we take $Y_k = \{x_{ik} : x_i \in X,\ x_i \neq t\}$. Now R contains t and some part $x_i^e t^l x_{i'}^{e'}$, for $e, e' \in \{1, -1\}$, $l \neq 0$, and $x_i, x_{i'} \in X\setminus\{t\}$. Then R_0 as word over X_1 will contain $x_{i,h+l}$ and $x_{i',h}$ for some h. Thus $a(0) < b(0)$; for the definition of these numbers see 2.11.3 (b). It is clear that $a(k) < b(k)$, and that $k < l$ implies $a(k) < a(l)$ and $b(k) < b(l)$. Hence no R_j contains only generators from Y_0 alone. From the induction assumption and lemma 2.11.5 it follows that we may apply 2.11.4 to $\langle F_1 | R_1 \rangle$. Hence, as a word over X_1, w cannot contain letters from Y_0 only; but then w, as a word over X, must contain t.

In order to show that R contains $x_i \in X\setminus\{t\}$, we define $Y_k := \{x_{ik}\}$. Since R contains x_i, R_0 must contain some x_{ik}. Again applying 2.11.4 for relations of smaller length then $\ell(R)$, we conclude that w, as a word over X_1, must contain some x_{ik}, and therefore, as a word over X, must contain x_i.

It remains to treat the case in which no generator occurs in R with exponent sum 0. We have assumed that X contains at least two elements t and u. It will suffice to show that w contains, say, u. Let τ be the exponent sum of t in R and μ that of u. We embed F into the free group F' with basis $\{x\} \cup (X\setminus\{t\})$ such that $x^{\mu} = t$. Since w is a consequence of R in F, it is also in F'. Now we take for F' the basis $X^* = (X\setminus\{t,u\}) \cup \{x,u^*\}$ where $u^* = ux^{\tau}$. As a word over this basis, R has exponent sum 0 in x. We pass as before to the subgroup F_1' of F' with basis $X' = \{u_k = x^{-k}ux^k,\ x_{ik} = x^{-k}x_ix^k : k \in \mathbb{Z},\ x_i \in X\setminus\{t\}\}$. The length of R relative to this basis is evidently the total number of occurences of letters other then t and t^{-1} in R, as a word over X, and is thus less than the length of R as a word over X. We may thus apply the inductive hypothesis as before to conclude that w, as a word over X', contains some u_k, and therefore, as a word over X, contains u. This completes the proof of the Freiheitssatz.

□

Using the construction from the last proof we will show the following

2.11.7 Proposition. *If two elements* R *and* Q *of a free group* F *have the same normal closure* N *in* F, *then* R *is conjugate to* Q *or* Q^{-1}.

Proof. We may assume that R and Q are non-trivial and cyclically reduced. By the Freiheitssatz, since each is a consequence of the other, they contain exactly the same generators. We may suppose that $F = \langle X \mid \rangle$ and that both R and Q contain all elements of X. The case of a single generator is trivial and we put it aside. We now prove the statement by induction on the length of R. As before, by embedding F into a larger free group F' if necessary, we can suppose that some $t \in X$ has exponent sum 0 in R. As before, we pass to a free group F_1 with a basis $X_1 = \{x_{ik} = t^{-k}x_it^k : x_i \in X\setminus\{t\},\ k \in \mathbb{Z}\}$ and conclude that $N \triangleleft F_1$ is the normal closure of $R_1 = \{R_j = t^{-j}Rt^j : j \in \mathbb{Z}\}$ in F_1 as well as the normal closure of $Q_1 = \{Q_j = t^{-j}Qt^j : j \in \mathbb{Z}\}$.

Let $x_i \in X\setminus\{t\}$. Some R_p will contain x_{io} as x_{ik} of lowest second subscript, and some x_{im} with highest index. Similarly, some Q_ℓ contains x_{io} as x_{ik} of lowest second index, and some x_{in} with highest second subscript. Each R_j with $j \neq p$ contains generators x_{ik} with $k \notin \{0,\dots,m\}$. Now if $n < m$, the Freiheitssatz in the version 2.11.4 applied to the staggered presentation $\langle X_1 \mid R_1\rangle$ contradicts the assumption that Q_ℓ is a consequence of R_1. Thus $n \geq m$, and by symmetry we conclude that $n = m$. But now 2.11.4 implies that Q_ℓ is a consequence of R_p, and, symmetrically, R_p is a consequence of Q_ℓ. This means that R_p and Q_ℓ have the same normal closure in F_1. Since R_p as a word in X_1 is shorter than R as a word over X, we can apply the induction hypothesis to conclude that R_p is conjugate in F_1 to Q_ℓ or Q_ℓ^{-1}. But then R_p is conjugate in F to Q_ℓ or Q_ℓ^{-1}; since R_p is conjugate to R and Q_ℓ to Q, it now follows that R is conjugate to Q or Q^{-1}.

□

Among the various consequences and strengthenings of the Freiheitssatz we will mention only the following 2.11.8. A large number of others can be found in [Lyndon-Schupp 1977, II. 5] where there is also a detailed list of literature.

2.11.8 Proposition. *If the free group G has the presentation* $\langle X \mid R \rangle$ *with one defining relation* R, *then either* R = 1 *or* R *is a member of some basis for* $F = \langle X \mid \rangle$.

Proof. We may suppose that F has finite rank $n \geq 1$ and that $R \neq 1$. By 2.11.1 rank $G \leq n-1$, while by abelianizing we see that rank $G \geq n-1$. From the Grushko theorem 2.9.1 or from E 1.17 it follows that F has a basis $y_1,\dots,y_n$ such that the images of $y_1,\dots,y_{n-1}$ form a basis for G. Since R and y_{n-1} have the same normal closure in F, we conclude from 2.11.7 that R is conjugate to y_n or y_n^{-1}, and hence is an element of a basis of F.

□

2.11.9 *On the Magnus conjecture*. Proposition 2.11.7 suggests a simple way to decide whether two groups with one defining relation are isomorphic. This is the case if one word can be brought into the other or its inverse by a number of Nielsen processes. That this is the only way, is sometimes called the Magnus conjecture [Magnus-Karrass-Solitar 1966, p. 401]. If the Magnus conjecture is true for a group G then it can be decided by the Whitehead method whether a given presentation describes G. Examples where the Magnus conjecture has a positive answer are the fundamental groups of surfaces, see [Zieschang 1970], [Peczynski 1972], [Peczynski, Rosenberger, Zieschang 1975]. The proof is rather unpleasant and is based on the theorem 2.10.2 about the Nielsen method in free products with amalgamation. Other examples for groups with a positive answer to the Magnus conjecture are in papers of Rosenberger.

A counterexample to the Magnus conjecture was given in [Zieschang 1970]. The same example was found independently [McCool-Pietrowski 1971]. For generalizations see [Zieschang 1977, 6.3] . We will describe the example in 2.11.10. It turns out that for each of these (torus knot) groups there are only finitely many non Nielsen-equivalent presentations with one relation [Collins 1978]. [Brunner 1976] gives examples of groups which have infinitely many inequivalent one-relator presentations. Other counterexamples to the Magnus conjecture are obtained from the two Wirtinger presentations of two-bridge-knots, if we take once the overcrossing arcs as generators and the other time the undercrossing ones, see [Funcke 1975]; the simplest of these examples has been treated in 2.1.10 and E 1.16.

2.11.10 Example. Consider $T = \langle s,t \mid s^p t^{-q} \rangle$, $p,q \geq 2$. We take c,r such that $p = 1-cr$. Then by Tietze transformations :

$$\begin{aligned}\langle s,t \mid s^p t^{-q}\rangle &= \langle s,t,u \mid us^{-r}, s^{1-cr}t^{-q}\rangle \\ &= \langle s,t,u \mid us^{-r}, su^{-c}t^{-q}\rangle \\ &= \langle t,u \mid u(t^q u^c)^{-r}\rangle\end{aligned}$$

If q, c, $r \geq 3$ then the word $u(t^q u^c)^{-r}$ has minimal length among all that are obtained from it by Nielsen transformations. The same is true for $s^p t^{-q}$. For the proof, apply the Whitehead automorphisms 1.9.2.

As for the lengths:

$$p+q < |rc| + q + |rq|$$

so it follows from 1.9.3 that there is no isomorphism of the free group on s,t to that on t,u that maps $s^p t^{-q}$ to $u(t^q u^c)^{-r}$.

2.12 REMARKS

Combinatorial group theory, in its origins, was closely connected with combinatorial topology, as should be apparent from the use already made of 1- and 2-dimensional complexes in proving theorems about groups. Indeed, the main lines of research were already laid out in the topological papers [Tietze 1908] and [Dehn 1910]. From his theorem on Tietze transformations, Tietze was led to pose the general problem of deciding when two presentations define isomorphic groups (the *isomorphism problem*); he even suggested that this problem might be unsolvable. Dehn posed the *word problem* and *conjugacy problem* for arbitrary finitely presented groups. We have seen how to solve these problems for free groups in 1.2. Another solution we shall see in 4.9 is for surface groups, obtained by Dehn in 1912.

The word problem has been solved only for a few narrow classes of groups, for example groups with one defining relation [Magnus 1932], less is known about the conjugacy problem (not yet solved for 1-relator groups), and less still about the isomorphism problem. This state of affairs has become some what understandable since the remarkable result of [Novikow 1955] that *no algorithm exists* for the word and conjugacy problems in certain finitely presented groups. Likewise, it was proved in [Rabin 1958] that there is no algorithm for the isomorphism problem.

Specific examples of finitely presented groups with unsolvable word problem are rather complex, but they can be constructed from groups with solvable word problem using free products with amalgamation [Britton 1963].

Topological methods have been applied to free products with amalgamation and the related HNN-*construction* [Higman-Neumann-Neumann 1949]. In [Serre 1977], Serre considers groups acting on trees and their quotient spaces in much the same way that we consider groups acting on the plane in this book (see chapter 4). It turns out that free groups are those which act in such a way that no non-trivial element has a fixed point, free products with amalgamation are those with an edge as quotient, and HNN groups those with a loop as quotient. (In general, an HNN group is one obtained from cyclic groups by iterated amalgamations and HNN constructions.) This yields a powerful method for proving subgroup theorems and for finding presentations of specific groups.

The amalgamation and HNN constructions can also be interpreted in terms of fundamental groups, cf. 2.8. Amalgamation corresponds to pasting two complexes together, and HNN corresponds to pasting a complex to itself ("adding a handle").

Just as it is difficult to extract information about a group from its presentation, so it is difficult to obtain a presentation of a group described in some other way, for example as a group of automorphisms. The first results were those obtained by Poincaré and Klein in the 1880's for automorphisms of planar tessellations. We give a more general form of these results in Chapter 4. Finite presentations can also be obtained for the automorphism group of a free group of finite rank, [Nielsen 1924], and for mapping class groups of surfaces, [McCool 1975], cf. 5.15.

EXERCISES

E 2.1 Prove $\langle S,T \mid S^aT^b, S^cT^d\rangle \cong \mathbb{Z}_D$ if $D = |ad-bc| \neq 0$.
What happens for $D = 0$?

E 2.2 Let $G = \langle S_1,S_2 \mid S_1^4, S_2^4, (S_1S_2)^2\rangle$ and let $\varphi: G \to \mathbb{Z}_4 = \{0,1,2,3\}$ be defined by $S_i \mapsto 1$. Show that the kernel of φ is isomorphic to $\mathbb{Z} \oplus \mathbb{Z}$.

E 2.3 Prove that $G = \langle T,U,X \mid T^{-1}U^{-1}TU, X^{-1}TXT, X^{-1}UXU, X^2\rangle$ is isomorphic to $\langle A,B,C \mid A^2,B^2,C^2,(ABC)^2\rangle$ and contains a subgroup T of index 2 which is isomorphic to $\mathbb{Z}^2$. Show that T is uniquely determined by these properties.

E 2.4 Let $G = \langle S,T \mid S^aT^b\rangle$, $(a,b) = 1$. In this group the factor group by the commutator subgroup, $G/[G,G]$, is isomorphic to $\mathbb{Z}$. Show that the kernel of the canonical mapping $G \to G/[G\ G] = \mathbb{Z} \to \mathbb{Z}/s\mathbb{Z} = \mathbb{Z}_s$ is isomorphic to G when $(s,ab) = 1$. (Prove this first for $a = 2$, $b = 3$ and $s = 5$.)

E 2.5* Determine the subgroups of index 8 of $\langle S,T \mid S^8,T^2,(ST)^8\rangle$ which are isomorphic to $\langle A_1,B_1,A_2,B_2 \mid [A_1,B_1][A_2,B_2] = 1\rangle$ and show that there is exactly one normal subgroup among them [Sanatani 1967].

E 2.6 Let $\varphi: H \to G = G_1 *_A G_2$ be an epimorphism. Prove that $H = \varphi^{-1}(G_1) *_{\varphi^{-1}(A)} \varphi^{-1}(G_2)$. In particular, if G is properly decomposable as a free product with amalgamation, then so is H (see [Stallings 1965], [Zieschang 1976]).

E 2.7 Prove that in a free product with amalgamation, two elements x,y commute only when there is an element z such that zxz^{-1} and zyz^{-1} are in the same factor (cf. E 1.5).

E 2.8 By pasting together, calculate the fundamental groups of the annulus,

the projective plane

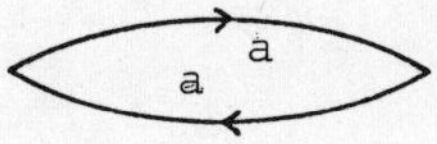

the torus

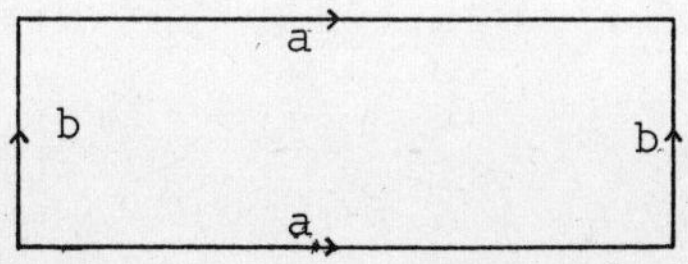

the Klein bottle

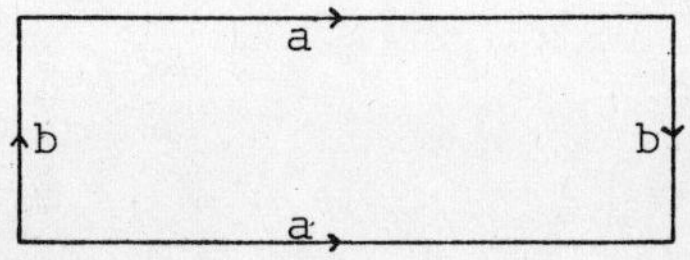

and the "Dunce's hat"

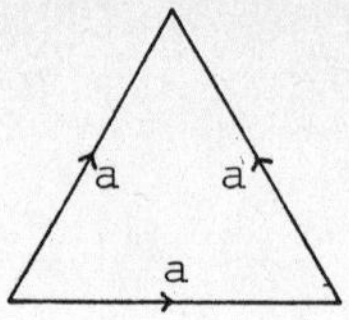

E 2.9 Prove that each finitely generated non-trivial normal subgroup of a finitely generated free group is of finite index. (Hint: For a geometric proof use properties of the covering transformation group.)

E 2.10 Prove Theorem 2.7.4.

E 2.11 Prove Proposition 2.7.7.

E 2.12 Determine H_i(graph) and H_i(torus).

E 2.13 Use Seifert-Van Kampen to determine the fundamental group of

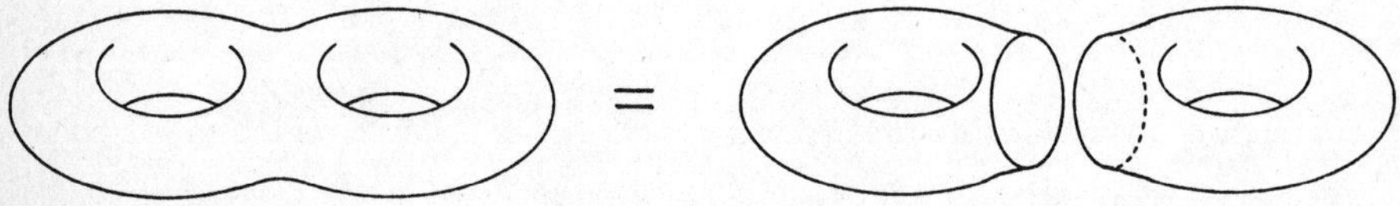

E 2.14 Prove Corollary 2.9.2.

E 2.15 Prove that $<S_1,S_2,S_3,S_4 | S_1^2,S_2^2,S_3^2,S_4^{2m+1},S_1S_2S_3S_4>$ is generated by S_1S_2,S_1S_3.

E 2.16 (a) Prove that Z^n cannot be generated by less than n elements.
    (b) Prove that $Z^n \oplus Z_{a_1} \oplus \ldots \oplus Z_{a_m}$ with $2 \le a_1 | \ldots | a_m$ cannot be generated by less than n+m elements.

E 2.17 Classify the finitely generated abelian groups up to isomorphism.

E 2.18 For g,k > 0 the groups $<T_1,U_1,\ldots,T_g,U_g | \prod_{i=1}^{g} [T_i,U_i]>$ and $<V_1,\ldots,V_k | V_1^2 \ldots V_k^2>$ are not free. Give several proofs.

E 2.19 Determine the elements of finite order in $<S,T,U | S^n, S[T,U]>$, $n \ge 2$.

E 2.20 Classify the groups $<S,T | S^a T^{-b}>$ up to isomorphism ([Schreier 1924]).

E 2.21 Let $G = \langle S,T \mid S^aT^b\rangle$ with $a,b \geq 2$, and let $\alpha, \beta \in \mathbb{Z}$ be such that $(a,\alpha) = (b,\beta) = (\alpha,\beta) = 1$. Prove:

(a) G is generated by S^α, T^β.

(b) If α or β equals 1 then only one defining relation in the generators S^α, T^β is needed.

(c) By a sequence of Nielsen processes (1.7) the system S^α, T^β can be transformed into a system $S^{\alpha'}$, $T^{\beta'}$ with $0 < \alpha' \leq \frac{\beta'\alpha}{2}$ and $0 < \beta' \leq \frac{\alpha'b}{2}$. (Equality occurs only if a or b equals 2.)

Moreover (d) holds, though this is difficult to prove.

(d) For $a \neq b$ the systems (S^α, T^β) and $(S^{\alpha'}, T^{\beta'})$ with $0 < \alpha, \alpha' \leq a/2$ and and $0 < \beta, \beta' \leq b/2$ can be transformed into each other by Nielsen processes iff $\alpha = \alpha'$ and $\beta = \beta'$.

(In fact an even more general result is true, namely that (S^α, T^β) with $0 < \alpha \leq \frac{\beta a}{2}$, $0 < \beta \leq \frac{\alpha b}{2}$ can be transformed into $(S^{\alpha'}, T^{\beta'})$ where $0 < \alpha' \leq \frac{\beta' a}{2}$, $0 < \beta' \leq \frac{\alpha' b}{2}$ iff $\alpha = \alpha'$ and $\beta = \beta'$, see [Zieschang 1977]. The proof of this result seems disproportionately long and a nice short proof is desirable.)

(e) For generators S, T^β with $(\beta,b) = 1$, or S^α, T with $(\alpha,a) = 1$, one defining relation suffices. (In fact, each system of generators with this property is "Nielsen equivalent" to one of the above type [Collins 1978].)

E 2.22 Prove that the group $C = \langle a,b \mid a^4, b^{2k}, (ab)^2, (a^{-1}b)^2\rangle$ $(k \geq 1)$ has order 4k and contains an abelian subgroup A of order 4k. If $k \geq 3$, A is uniqely determined. Calculate A. Discuss the situation for $k = 1,2$ and for $k = 0$.

E 2.23 Let C be as above and $\mathcal{D} = \langle s_1,s_2 \mid s_1^4, s_2^{2k}, (s_1s_2)^2\rangle$, $k \geq 2$.

(a) Show that the kernel of $\rho: \mathcal{D} \to C$, $s_1 \mapsto a$, $s_2 \mapsto b$, has rank $2(k-1)$.

(b) Prove that Kern ρ has a presentation

$$\langle t_1,u_1,\ldots,t_{k-1},u_{k-1} \mid \prod_{i=1}^{k-1} [t_i,u_i]\rangle.$$

(c) Discuss the sitution for $k = 0, 1$.

(The exercises E 2.22-23 are from [Maclachlan 1969]; we will come back to them in 4.14.26.)

3. SURFACES

The first interesting examples of manifolds are the 2-dimensional ones, also called surfaces. In the compact case there is a complete list of the different types and their classification is well-known. We consider this in 3.1 and 3.2 in the spirit of combinatorial topology. Since the Hauptvermutung is true for dimension 2, and each topological surface is triangulable [Rado 1924], this also gives a topological classification. In 3.3 we prove that for closed orientable surfaces each automorphism of the fundamental group can be realized by a homeomorphism. This is the Dehn-Nielsen theorem (see [Nielsen 1927]) and we follow a proof due to [Seifert 1937]. In Chapter 5 we give another proof of this theorem and its generalizations.

3.1 DEFINITIONS

Let C be a 2-complex with the properties:

(a) C is connected.

(b) Each directed edge σ appears in at least one boundary path.

(c) Altogether, each directed edge appears at most twice in boundary paths. (However, it can appear twice in the same boundary path.)

An edge σ is called a *boundary edge* when it appears only once in boundary paths. The directed edge σ' is called the *neighbour* of the directed edge σ if the path $\sigma\sigma'$ appears in a boundary path. A boundary edge has only one neighbour. Conversely, an edge which has only one neighbour is a boundary edge and an edge with two neighbours does not lie on the boundary.

The only exception can occur at a vertex where exactly two edges meet. Then each of the directed edges σ may have one neighbour τ, but $\sigma^{-1}\tau$ occurs twice in the boundaries of faces. An informative example is given by the "projective plane". Let the complex consist of one vertex P, one geometric edge $\{\sigma,\sigma^{-1}\}$ and one geometric face $\{\varphi,\varphi^{-1}\}$ such that $\sigma\sigma$ is the boundary of φ, see figure

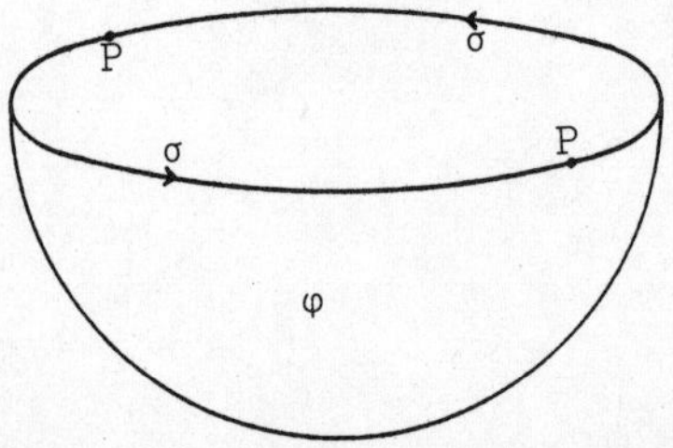

A *star* is a sequence $\sigma_{\alpha_1},\dots,\sigma_{\alpha_n}$ of directed edges with a common initial point, in which the $\sigma_{\alpha_i}^{-1} (1 < i < n)$ have the edges $\sigma_{\alpha_{i-1}}$ and $\sigma_{\alpha_{i+1}}$ as neighbours and $\sigma_{\alpha_{i-1}} \neq \sigma_{\alpha_{i+1}}$. Thus at most σ_{α_1} and σ_{α_n} can be boundary edges.

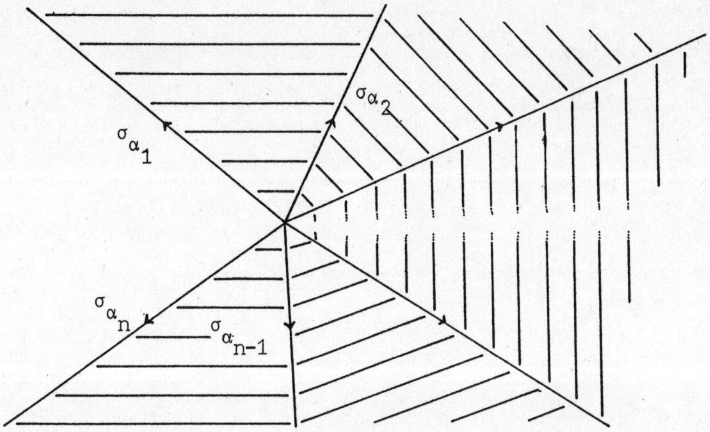

A star is called *closed* when $\sigma_{\alpha_1}^{-1}$ has neighbour σ_{α_n}, it is called *simple* when no directed edge appears twice.

<u>3.1.2 Construction</u>. Given a directed edge σ we now construct a star which contains it. We first add the neighbours σ_{-1},σ_1 of σ^{-1} and obtain $\sigma_{-1},\sigma,\sigma_1$. Then we add, on the left the second neighbour of σ_{-1}^{-1}, and on the right the second neighbour of σ_1^{-1}, and continue this process as long as we can find neighbouring edges. If only finitely many edges emanate from the initial point of σ then we obtain a uniquely determined maximal simple star. Each directed edge belongs to a unique simple closed or maximal open star. Thus the star is uniquely determined up to orientation and cyclic interchange.

<u>3.1.3 Definition</u>. A *surface* is a 2-complex which satisfies 3.1.1 (a) - (c) and the following condition (d):

(d) For any two edges σ_1 and σ_2 with initial point P there is a star $\sigma_1 = \sigma_{\sigma_1},\dots,\sigma_{\alpha_n} = \sigma_2$ around P.

Condition (d) has the consequence that at a vertex P of the surface there is either no closed star or else a uniquely determined (up to orientation and cyclic interchange) closed star. Vertices with closed stars are called *inner vertices*, the others, *boundary vertices*.

<u>3.1.4 Remark</u>. If $\omega\sigma\sigma^{-1}\omega'$ is the boundary path of a face and P is the endpoint of σ, then σ^{-1} constitutes the whole star around P. It then follows from 3.1.3 (d) that P cannot also be the initial point of σ. For the same reason, no boundary path can contain a subpath $\sigma k_1 \ldots k_n\sigma^{-1}$ in which $k_i = \sigma_{i1}\sigma_{i2}\sigma_{i1}^{-1}$ and the segments σ and σ_{ij} have the same initial and final point.

<u>3.1.5 Lemma</u>. *The processes 2.4.5 (a) and 2.4.5 (b) carry surfaces into surfaces; likewise, so do the inverse processes 2.4.5 (c) when they are possible.* We call two surfaces *related* if they can be carried into each other by processes 2.4.5 (a) - (c).

□

In what follows we shall frequently use two further processes which may be obtained from the processes 2.4.5 (a) - (c). We assume that only finitely many edges emanate from each vertex.

3.1.6 (a) Let P be a vertex of the surface F with star $\sigma_1,\ldots,\sigma_n$. Let F' be the surface which contains in addition to the old pieces a new vertex Q and a new edge σ, where the edges $\sigma_j,\ldots,\sigma_k$ ($j \le k$) begin at Q, the remaining σ_i begin at P, and σ leads from P to Q. The stars at P and Q then are $\sigma_i,\ldots,\sigma_{j-1},\sigma,\sigma_{k+1},\ldots,\sigma_n$ and $\sigma^{-1},\sigma_j,\ldots,\sigma_k$ respectively. The only alteration in boundary paths is replacement of terms $(\sigma_{j-1}^{-1}\sigma_j)^{\varepsilon}$ or $(\sigma_k^{-1}\sigma_{k+1})^{\varepsilon}$ by $(\sigma_{j-1}^{-1}\sigma\sigma_j)^{\varepsilon}$ or $(\sigma_k^{-1}\sigma^{-1}\sigma_{k+1})^{\varepsilon}$ respectively ($\varepsilon = \pm 1$).

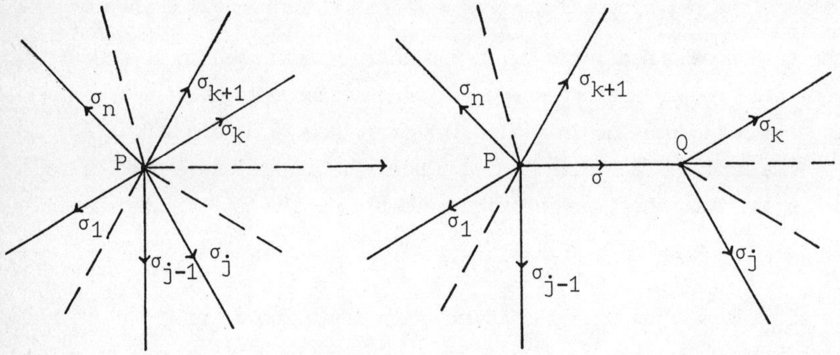

(b) The process inverse to (b), provided it results in a surface. We call (b) *contraction of vertices.*

The contraction process carries surfaces into surfaces as long as the contracted edge σ does not connect two boundary vertices while itself lying in the interior. If σ connects two boundary vertices, then the complex F' which results from F by contraction of vertices violates the condition 3.1.3 (d) for surfaces:

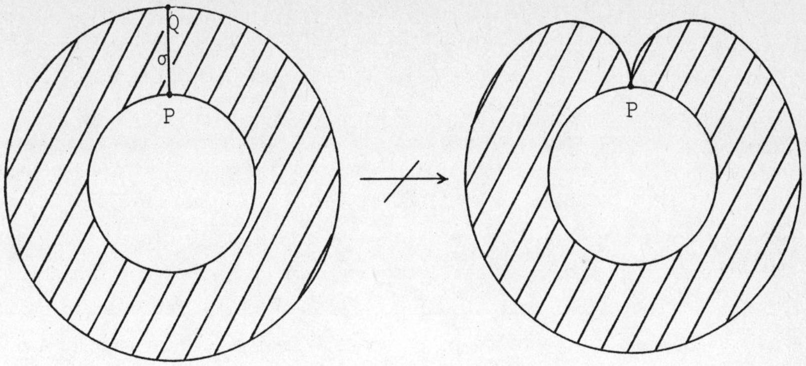

3.1.6 (b) can be obtained from the other processes as is shown in the following figures:

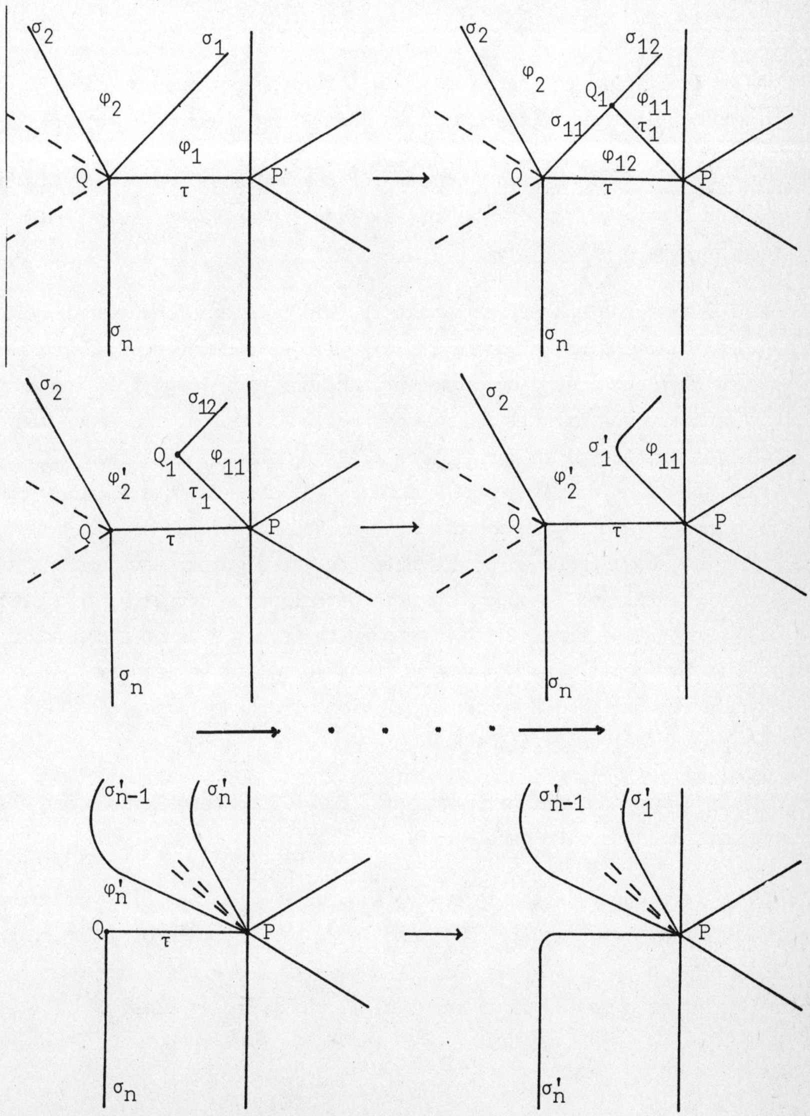

3.2 CLASSIFICATION OF FINITE SURFACES

Let F be a connected surface with only finitely many elements. We first show:

3.2.1 Lemma. *The boundary of* F *consists of finitely many simple closed paths.*

Namely, if τ is a boundary segment and it is closed, then by 3.1.3 it meets no other boundary edges; if its initial and final point, P and Q respectively, are distinct, then the star around Q has first edge τ^{-1} and a last edge τ_1 which is again a boundary edge. If the final point Q_1 of τ_1 equals P we are finished, otherwise we continue. Since at most two boundary edges emanate from a boundary point and F is finite, the process terminates and we obtain a simple closed curve which has no point in common with the rest of the boundary of F.

□

If we amalgamate faces (process 2.4.5 (c)) as often as we can, we finally obtain a surface in which different faces meet along their whole boundary paths.

3.2.2 Proposition. *Since* F *is connected,* F *possesses either two faces which meet along their boundary (this "spherical surface" is uniquely determined up to relatedness) or else a single face.*

□

If F is not the sphere, then with the aid of elementary transformations we can obtain the following situation: F has exactly one inner vertex. There is a vertex and one edge on each boundary component, and one edge leads from the inner vertex to each boundary vertex. All other edges begin and end at the inner vertex. We then say that the surface is in *normal form*. Thus F consists of a face with a boundary path which contains subpaths of the form $\sigma\rho\sigma^{-1}$, where ρ is a boundary curve and σ is the approach path to the boundary curve. Because of the conditions 3.1.1 (b) and 3.1.1 (c), each inner edge, disregarding its sign, appears exactly twice in the boundary path, and the boundary path determines F uniquely. This means that F can be regarded as a face ϕ with its boundary segments denoted consecutively by the edges of F. By identifying edges with the same symbol we recover F. Intuitively speaking, ϕ *results from* F *by cutting* F *along the inner edges, and* F *from* ϕ *by pasting appropriate edges together.*

The following process carries normal forms into normal forms and does not alter F. We describe it for the face ϕ.

Let σ be a curve in the interior, i.e. as well as the first occurrence, σ or σ^{-1} appears again in the boundary of ϕ, and let ω be a subpath of $\partial\phi$ which includes σ but not the other σ or σ^{-1}. Then we cut off a part from ϕ along a new edge σ' from the initial point of σ to the final point of ω, and paste it along σ(or σ^{-1}), again. For F this represents a subdivision of the surface piece by σ^{-1}, then amalgamation

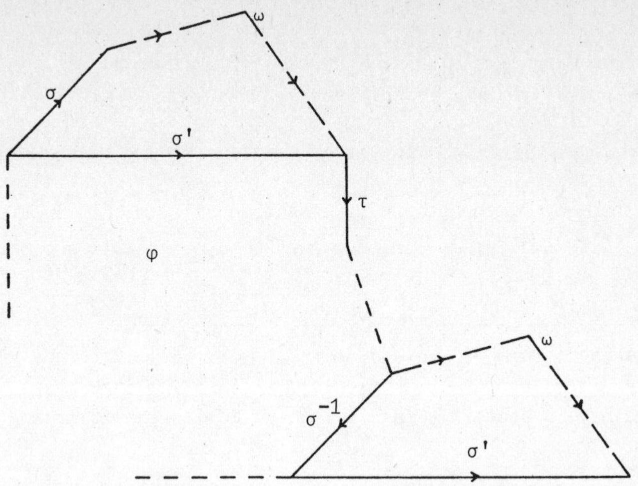

of the new faces by removal of σ. As a result, the boundary path $\ldots \sigma\omega\tau \ldots \sigma^{-1} \ldots$ is converted into $\ldots \sigma'\tau \ldots \omega\delta'^{-1} \ldots$ when the second occurrence is σ^{-1}, or into $\ldots \sigma'\tau \ldots \sigma'\omega^{-1} \ldots$ when the second occurrence is again σ. Instead of replacing $\sigma\omega$ by σ', we can analogously replace $\omega'\sigma$ by σ'. If for example σ is an approach path to a boundary curve ρ, then $\ldots \omega'\sigma\rho\sigma^{-1} \ldots$ is converted into $\ldots \sigma'\rho\sigma'^{-1}\omega' \ldots$.

3.2.3 We call this process *bifurcation* and write $(\omega\underline{\sigma})$ to denote the σ' which replaces $\omega\sigma$ (so σ is replaced by $\omega^{-1}(\omega\underline{\sigma})$). Thus $(\omega\underline{\sigma})$ denoted the boundary of the face which is pasted differently.

Let $\omega = \partial\phi$. Since $\ldots \omega_1\sigma\rho\sigma^{-1} \ldots$ is converted into $\ldots(\omega_1\underline{\sigma})\rho(\omega_1\underline{\sigma})^{-1}\omega_1 \ldots$ we can bring all parts of ω corresponding to boundary curves to the front. In other words, ω takes the form $\omega = \sigma_1\rho_1\sigma_1^{-1} \ldots \sigma_r\rho_r\sigma_r^{-1}\omega'$ where no boundary curves appear in ω'.

If a segment σ appears twice in ω' with the same exponent, then we make the following bifurcations:

$$\omega_1' = \ldots \omega_1\sigma\omega_2\sigma \ldots \to \ldots (\omega_1\sigma)\omega_2\omega_1^{-1}(\omega_1\underline{\sigma}) \ldots \to \ldots ((\omega_1\sigma)\omega_2\omega_1^{-1})((\omega_1\sigma)\omega_2\omega_1^{-1})\omega_1\omega_2^{-1} \ldots$$

Thus we can assume that all edges which appear twice appear as squares, and these squares occur at the beginning of the path which results from ω' by the above bifurcations. Thus we can replace ω' by a boundary $\nu_1 \ldots \nu_k\, \omega''$, where each segment in ω'' occurs with exponents + 1 and - 1.

If there is a section $\ldots \mu_1\sigma\mu_2\tau\mu_3\sigma^{-1}\mu_4\tau^{-1} \ldots$, then by constructing $(\mu_1\underline{\sigma}\mu_2)$ we obtain a section of the form $\ldots \sigma\tau\mu'\sigma^{-1}\mu''\tau^{-1} \ldots$. Taking $(\mu''\underline{\tau}^{-1})$, we then obtain $\ldots \sigma\tau\mu\sigma^{-1}\tau^{-1}$. Using $(\mu\underline{\sigma}^{-1})$ we obtain $\ldots \sigma\mu\tau\sigma^{-1}\tau^{-1}$, and finally by construction of

$(\mu\underline{\tau})$ we obtain the form ... $\sigma\tau\sigma^{-1}\tau^{-1}\mu$ ... . Thus ω finally takes the form $\prod_{i=1}^{i} \sigma\rho\sigma_i^{-1} \prod_{i=1}^{k} \nu_i^2 \prod_{i=1}^{\ell} [\tau_i,\mu_i]\omega'''$ ... where ω''' no longer has sections of the form $\sigma\mu_1\tau\mu_2\sigma^{-1}\mu_3\tau^{-1}$.

If ω''' is non-empty, then there is a section $\sigma\sigma^{-1}$ in ω''', in violation of the star condition 3.1.3 (d), since σ alone would then constitute a closed star (see the remark following 3.1.4).

Thus ω has the form $\omega = \prod_{i=1}^{j} \sigma_i\rho_i\sigma_i^{-1} \prod_{i=1}^{k} \nu_i^2 \prod_{i=1}^{\ell} [\tau_i,\mu_i]$. If ℓ and k are both greater than 0, then we have a section $\nu^2\tau\mu\tau^{-1}\mu^{-1}$ which admits the following bifurcations:

$$\nu^2\tau\mu\tau^{-1}\mu^{-1} \rightarrow \nu(\nu\underline{\tau})\mu(\underline{\tau}^{-1}\nu^{-1})\nu\mu^{-1}$$

$$\rightarrow (\underline{\nu\mu}^{-1})\mu(\nu\underline{\tau})\mu(\underline{\tau}^{-1}\nu^{-1})(\underline{\nu\mu}^{-1})$$

$$\rightarrow (\underline{\nu\mu}^{-1})(\underline{\mu}(\nu\underline{\tau}))(\underline{\mu}(\nu\underline{\tau}))(\nu\underline{\tau})^{-1}(\nu\underline{\tau})^{-1}(\nu\mu^{-1})$$

which therefore has the form $\nu_1'\nu_2'^2\nu_3'^2\nu_1'$, and this converts into $\nu_1''^2\nu_3'^{-2}\nu_2'^{-2}$ as above.

Consequently we have:

3.2.4 Proposition. *If there is a segment σ which appears in ω' twice, then ω can be converted into*

3.2.5 $$\omega = \prod_{i=1}^{r} \sigma_i\rho_i\sigma_i^{-1} \prod_{i=1}^{k} \nu_i^2,\ k > 0$$

by bifurcations. If each segment in ω' appears with the signs + 1 and -1 then ω may be brought into the form

3.2.6 $$\omega = \prod_{i=1}^{r} \sigma_i\rho_i\sigma_i^{-1} \prod_{i=1}^{g} [\tau_i,\mu_i],\ g \geq 0.$$

The decompositions 3.2.5 and 3.2.6 of surfaces are called *canonical normal forms*.

We shall now show that surfaces of type 3.2.5 and 3.2.6 are not related and j,k and g respectively are invariants. We already know that the fundamental group is an invariant. We take the inner point and the approach paths to the boundary as the tree for F. In this way we obtain generators $s_1,\ldots,s_r$ corresponding to the boundaries, generators $v_1,\ldots,v_k$ in the case 3.2.5, $t_1,u_1,\ldots,t_g,u_g$ in the case 3.2.6.

3.2.5 gives the defining relation $\Pi s_i \Pi v_i^2$, and

3.2.6 the defining relations $\Pi s_i \Pi [t_i, u_i]$, so that we have the fundamental groups

3.2.5' $\quad = \langle s_1,\dots,s_r,v_1,\dots,v_k \mid \prod_{i=1}^{r} s_i \prod_{i=1}^{k} v_i^2 \rangle$, $k > 0$, and

3.2.6' $\quad = \langle s_1,\dots,s_r,t_1,u_1,\dots,t_g,u_g \mid \prod_{i=1}^{r} s_i \prod_{i=1}^{g} [t_i,u_i] \rangle$

respectively. If $r = 0$ one can see by abelianization that groups of type 3.2.5' and 3.2.6' are not isomorphic to each other and that g and k are invariants. If there are boundaries then 3.2.5' and 3.2.6' are both free groups, so that matters are not so simple in this case. 3.2.5' and 3.2.6' are called *canonical presentations* of the fundamental groups of surfaces. The generators and relation are also called canonical.

In order to distinguish surfaces of types 3.2.5 and 3.2.6 with boundary we introduce the concept of *orientability*.

<u>3.2.7 Definition</u>. A surface is called *orientable* when one member from each geometric face $\{\phi, \phi^{-1}\}$ can be chosen to be "positive" in such a way that each directed edge of the interior appears exactly once in a "positive" boundary path of a positive face. One easily sees that orientability is invariant with respect to 2.4.5. Since a surface in normal form has only one face, and in 3.2.5 at least one segment appears twice in the boundary path with the same sign, surfaces of type 3.2.5 are not orientable; in case 3.2.6 each segment from the interior of F appears in the boundary path with different signs, so surfaces of this type are therefore orientable.

Since the number of boundaries is trivially an invariant, we have only to show that k and g are invariants. In case 3.2.5 the Euler characteristic (see 2.4.6) $\alpha_0 - \alpha_1 + \alpha_2$ equals $(1+r) - (r+r+k) + 1$, since we have one point in the interior, a point on each boundary, an approach path to each boundary, a curve on each boundary and k curves in the interior. Thus in case 3.2.5 the characteristic equals $2 - k - r$. In case 3.2.6 the characteristic is $(r+1) - (r+r+2g) + 1 = 2 - 2g - r$. So we see that g and k respectively are invariants, they are called the *genus* of the surface.

<u>3.2.8 Theorem</u>. *Orientability, number of boundary curves and genus constitute a complete set of invariante for (finite) surfaces.*

The fundamental group of an orientable surface of genus g *with* r *boundary components has the form*

3.2.5' $\langle s_1,\dots,s_r,t_1,u_1,\dots,t_g,u_g \mid s_1 s_2 \dots s_r \prod_{i=1}^{g} [t_i,u_i]\rangle$,

that of a non-orientable surface of genus k *has the form*

3.2.6' $\langle s_1,\dots,s_r,v_1,\dots,v_k \mid s_1 s_2 \dots s_r v_1^2 \dots v_k^2\rangle$.

□

One may also decide the first part by means of the group. Since the boundary curves of surfaces are given and we can choose at most their order, orientation and approach paths, the generators $s_1,\dots,s_r$ in 3.2.5' and 3.2.6' are determined up to permutation, inversion and conjugation. Therefore the smallest normal subgroup which contains all boundary elements is uniquely determined, and after abelianization becomes a well-defined direct product of infinite cyclic subgroups. In case 3.2.5' the sum of the generators of this subgroup is never zero, whereas with suitable choice of signs the generators in case 3.2.6' have sum zero.

3.2.9 Definition. A finite surface without boundary is called *closed*.

3.2.10 Corollary. *Let* F *be a connected surface.*
Then $H_o(F) = \mathbb{Z}$ ($H_o(F) = \mathbb{Z}^n$ is the case of n components)

$$H_1(F) = \begin{cases} \mathbb{Z}^{2g} & \text{when F is closed orientable of genus g,} \\ \mathbb{Z}^{2g+r-1} & \text{when F is orientable of genus g with } r > 0 \text{ boundaries,} \\ \mathbb{Z}^{k-1+r} & \text{when F is non-orientable of genus k with } r > 0 \text{ boundaries,} \\ \mathbb{Z}_2 \times \mathbb{Z}^{k-1} & \text{when F is non orientable of genus k and closed,} \end{cases}$$

$$H_2(F) = \begin{cases} \mathbb{Z} & \text{when F is closed an orientable,} \\ 0 & \text{otherwise.} \end{cases}$$

□

3.2.11 Notation. The following surfaces are the simplest ones and have special names.

N a m e	(o)	(g)	(b)	(f)
(2)-sphere	+	0	0	1
torus	+	1	0	2
projective plane	-	1	0	3
Klein bottle	-	2	0	4
disk	+	0	1	5
annulus	+	0	2	6
Moebius strip	-	1	1	7

where (o) gives the orientability character (+ for orientable), (g) is the genus, (b) the number of boundary curves and (f) the number in the following figure

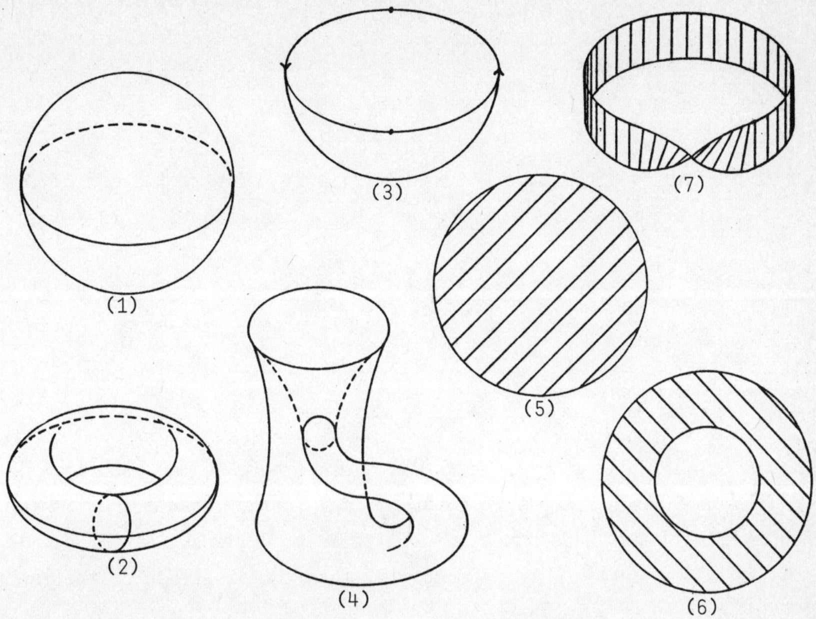

Exercises: E 3.1- 12.

3.3 KNESER'S FORMULA

We introduce the notion of the degree of a mapping and prove that a mapping of degree c between surfaces of (positive) genera g and g' satisfies the inequality

$$g' - 1 \geq |c| (g - 1)$$

Following [Seifert 1937] we then deduce a short proof ot the Dehn-Nielsen theorem.

In this section we admit surfaces which are not connected. This facilitates somewhat the proof of Theorem 3.3.3.

Given a surface mapping $f: F' \to F$ (see 2.7) we can use subdivision of faces of F' to reach the situation where, in case 2.7.4 (b), $f(\partial\phi')$ traverses the boundary of $\partial\phi$ exactly once, possibly with spurs, so in the future we shall assume this is the case. The coverings introduced in 2.4 are dimension-preserving surface mappings which satisfy from the outset the condition we have just achieved. In addition, the images of stars of F' traverse stars of F exactly once.

3.3.1 Definition. A *branched covering* is a dimension-preserving surface mapping with the property that the boundary paths of two faces of F' mapped onto the same boundary path of F have no edge in common. However, stars in F' can multiply cover stars of F. The multiplicity of the covering is called the *branching number of the vertex.*

3.3.2 Definition. Let F and F' be closed orientable surfaces with a fixed choice of orientation, and let F be connected. The difference between the number of positive faces of F' and negative faces of F' mapped by f to positive faces of F does not depend on the choice of face of F (the proof is Exercise at the end of the chapter) and is called the *degree of the mapping* f. In [Kneser 1930] it is proved

3.3.3 Theorem. *If* F *and* F' *are closed orientable surfaces of genera* g *and* g' *respectively* ($g' \geq 1$) *and if* f *is a surface mapping of* F' *onto* F *of degree* c, *then*

$$(g' - 1) \geq |c|(g - 1).$$

Proof. Since $g' \geq 1$ we can also assume $g \geq 1$, since otherwise the assertion is trivial. Thus neither F nor F' is a sphere, and we may further assume that F' contains no spheres (remember that we do not assume F' to be connected). Let n' and n be the Euler characteristics of F' and F respectively, then the assertion of the theorem is that $n' \leq |c| \cdot n$.

The following lemma, which we shall not prove with the means at our disposal, justifies the assumption that F' contains no spheres.

3.3.4 Lemma. *A mapping of the sphere on to a surface of higher genus has degree* 0.[1)]

3.3.5 If $f: F' \to F$ is a covering (unbranched) then over each geometric face geometric edge and vertex of F there are exactly c geometric faces, geometric edges and points respectively, so that $n' = |c| \cdot n$. If f is branched, then again there are $|c|$ geometric faces and geometric edges over geometric faces and geometric edges of F, but at most $|c|$ vertices over a vertex, so that $n' \leq |c| \cdot n$ again holds. In order to prove the theorem we now distinguish several cases.

3.3.6 f never lowers dimension.

By means of subdivision we can first reach a situation where all faces of F are

---

[1)] *Proof of the lemma: The universal covering of the surface F is the plane and therefore contractible. Consequently $\pi_2(F) = 0$ and each image of S^2 is contractible. The mapping degree is preserved under deformations. The degree of the constant mapping is 0 and the lemma follows.*

triangles, no edge appears twice in the boundary of a triangle and no edge is closed, and we lift this subdivision to F'. (f is then a simplicial mapping.) A segment σ of F' is called a *fold* when the two triangles having σ as a boundary edge are mapped onto the same triangle of F.

When folds are present we alter F' in such a way that the mapping degree is preserved, but the Euler characteristic increases or the number of faces decreases. After finitely many applications of this process all folds will have disappeared and a (branched) covering will remain. Thus the theorem will be established in this case.

There are three types of fold.

(a) The two triangles which are mapped onto the same one have only an edge PQ in common

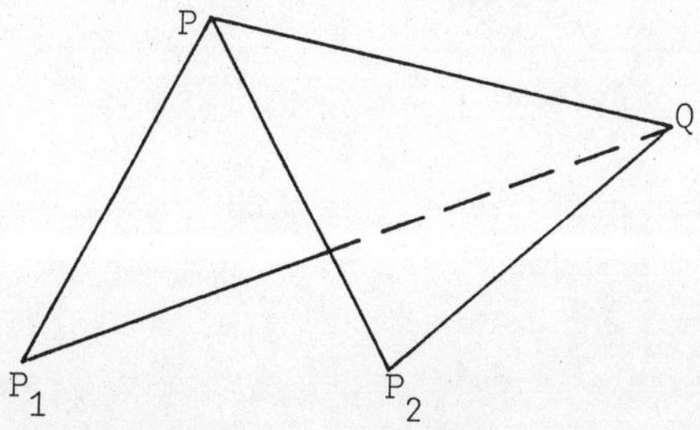

We cut the two triangles from F' along the closed path $P_1QP_2PP_1$ and stitch PP_1 to PP_2 and QP_1 to QP_2 in the surface which remains. Since the two triangles are mapped with different orientations, the mapping degree remains the same. The Euler characteristic is unaltered and the new complex is also a closed surface.

(b) The two triangles have two edges in common.

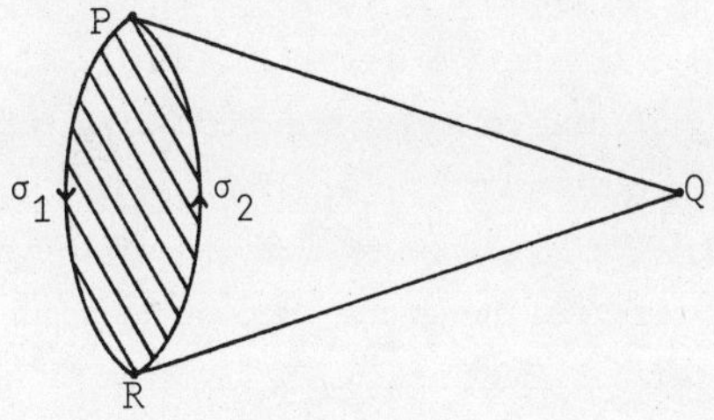

We cut the cone from F' along $\sigma_1\sigma_2$ and stitch together σ_1 and σ_2^{-1} in the remaining complex. Again neither the Euler characteristic nor mapping degree changes, and the complex remains a closed surface.

(c) The two triangles have an edge and the opposite vertex in common. (There are no further cases, for if the two triangles have all edges in common they constitute a sphere)

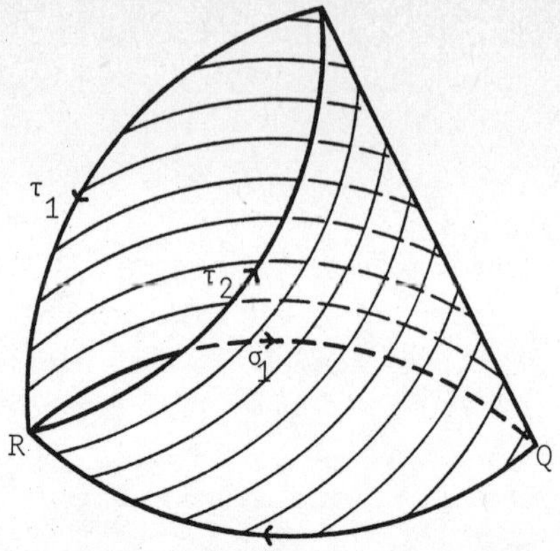

We then cut F' along $\sigma_1\sigma_2$ and stitch σ_1 to σ_2^{-1}.

As a result, the Euler characteristic is raised by 2 and the mapping degree is unaltered. If $\sigma_1\sigma_2$ separates the surface F' then the two new components cannot both be spheres, otherwise the component previously containing the triangles would have been a sphere already. Thus we obtain from F' a new system of surfaces, possibly requiring removal of one sphere, which lowers the characteristic by 2.

Altogether, the Euler characteristic of F' is raised if it is altered at all. After a finite number of such steps we obtain a branched covering, and the theorem follows from 3.3.5.

3.3.7 Let σ' be a segment of F' which is mapped on to a point. If σ' has two distinct endpoints then we contract them into one (Process 3.1.6 (b)). If the two endpoints are identical then we cut along σ' and contract the resulting boundary curves to a point. This raises the Euler characteristic by 2. If σ' separates the surface F' then two components again cannot both be spheres. If one of the new components is a sphere we leave it out. This process therefore does not diminish the Euler characteristic but it preserves the mapping degree.

3.3.8 If a face is mapped onto a vertex, then its boundary edges are mapped onto a point and 3.3.7 may be applied. Finally, it remains only to treat the case where a face ϕ' is mapped on to an edge σ, but no edge is mapped onto a point. Let $\partial\phi' = \sigma_1' \dots \sigma_n'$. Then the σ_i' are mapped on to $\sigma^{\pm 1}$ and there is a subpath $\sigma_i'\sigma_{i+1}'$ of $\partial\phi'$ with image $\sigma^{\varepsilon}\sigma^{-\varepsilon}$ ($\varepsilon = \pm 1$). We subdivide ϕ' with respect to $\sigma_i'\sigma_{i+1}'$

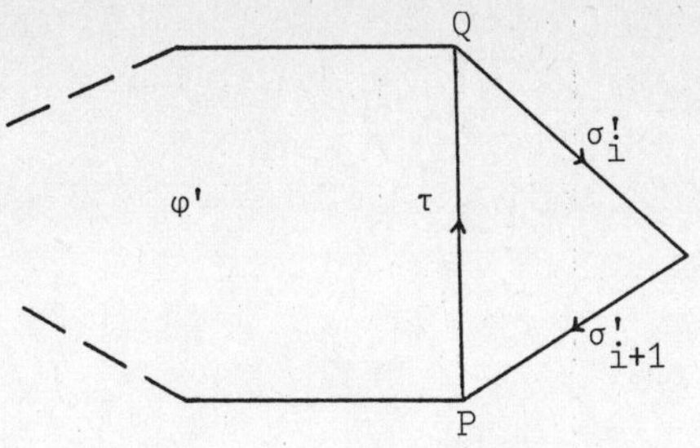

and contract the initial point Q of σ_i' and the final point P of σ_{i+1}' together. After finitely many such steps the face which is mapped onto an edge has exactly two edges in the boundary, which are mapped onto the edge σ and its inverse. We then cut F' along the two boundary edges, leave the face out, and stitch up again. Since this process again does not alter the Euler characteristic, and no face relevant to the computation of the mapping degree is involved, we are able to assume that f does not lower the dimension and the theorem follows from 3.3.6.

□

The rest of this section is not needed for the later text.

We now extract some useful results from the proof of Kneser's formula. Unfortunately, it would be rather complicated and artificial to formulate them in terms of the combinatorial theory we have developed so far (it will be done somewhat in chapter 5), but this can easily be done using basic notions and results of algebraic topology.

3.3.9 Corollary. *Let* F' *and* F *be closed orientable surfaces of genus* g' *and* g *resp., and let* $f\colon F' \to F$ *be a mapping of degree* $c \neq 0$. *If*

$$g' - 1 = |c|\,(g - 1)$$

then f *can be continuously deformed into a covering mapping* $p\colon F' \to F$.

Proof (a bit sketchy). For $g = 0$ it follows that $|c| = 1$ and $g' = 0$. It is known that a mapping $S^2 \to S^2$ of degree ± 1 is homotopic to a homeomorphism, see E 3.29. Now let us assume $g', g \geq 1$. In proving Kneser's formula we altered the mapping f (and possibly F') without changing the degree. If we replaced F' by another surface the Euler characteristic went up. As we now assume that equality holds, F' can never be changed substantially. More precisely: the processes 3.3.6 (a) and (b) can be replaced by a homotopic deformation of F such that the new mapping maps the two triangles to an edge. For process 3.3.6 (c) the curve $\sigma_1\sigma_2$ bounds a disk and we can deform f so that this disk is mapped onto $f(\sigma_1)$. For 3.3.7 the only possibility is the appearance of a sphere and again we can deform f continuously so that the disk is mapped to an edge. 3.3.8 describes a procedure for handling mappings which lower the dimension (translate the process into the language of homotopic deformations). Hence we may assume that f never lowers dimension and has no folds, i.e. f is a branched covering. Since proper branching implies proper inequality in Kneser's formula, it follows that f is a covering.

□

By similar considerations one can prove the following result:

3.3.10 Proposition. *Let* F' *and* F *be as before and let* $f: F' \to F$ *be a mapping of degree* 0. *Fix some complex* C^2 *in* F. *Then* f *can be deformed continuously into a mapping which maps* F *into the* 1*-skeleton* C^1 *of* C^2.

□

From 3.3.9 we can deduce the following important theorem 3.3.11 of Nielsen. In Chapter 5 we give another proof of it which depends only on the material we have developed in these notes and which can be generalized to cases which cannot be covered by the present method.

3.3.11 Theorem ([Nielsen 1927]). *Let* F *be a closed orientable surface,* $\mathfrak{F}$ *the fundamental group of* F *and* $\alpha: \mathfrak{F} \to \mathfrak{F}$ *automorphism. Then there exists a homeomorphism* $\phi: F \to F$ *which induces* α, i.e. $\phi_{\#} = \alpha$.

Proof. Let $\tau_1, \mu_1, \ldots, \tau_g, \mu_g$ be a canonical system of curves on F, see 3.2.6 with r = 0 and genus g. Then their homotopy classes $t_1, u_1, \ldots, t_g, u_g$ generate $\mathfrak{F}$. The curve $\prod_{i=1}^{g} [\tau_i, \mu_i]$ bounds a disk D on F. We select curves $\tau_1', \mu_1', \ldots, \tau_g', \mu_g'$ for $\alpha(t_1), \alpha(u_1), \ldots, \alpha(t_g), \alpha(u_g)$ and define a map $\tau_i \to \tau_i', \mu_i \to \mu_i'$ which maps the basepoint to itself and handles the orientation correctly. Then the curve $\prod_{i=1}^{g} [\tau_i', \mu_i']$ is null-homotopic, so the mapping defined on the curves $\tau_1, \ldots, \mu_g$ can be extended to a mapping $\phi': D \to F$. Since ϕ' maps equivalent points of τ_i and τ_i^{-1} (or μ_i and μ_i^{-1}) to the same point, ϕ factors to a mapping $\phi'': F \to F$ which induces α. Let c be the degree of ϕ''. Kneser's formula 3.3.3 now becomes

$$g - 1 \geq |c| \, (g - 1)$$

Hence either g = 1 or c equals 0 or ± 1. If the degree is 0 the image of F can be deformed onto the 1-skeleton, hence its fundamental group is free. Since α is an automorphism, $\mathfrak{F}$ must be free, which is not true (Proof, see E 2.4). If c = ± 1 the statement 3.3.11 is a consequence of 3.3.9. The case g = 1 needs other considerations.

One way is to use the automorphisms of $\mathbb{Z} \oplus \mathbb{Z}$, which are well known. We give it as exercise E 3.19.

Exercises: E 3.13-19.

3.4 COVERINGS OF SURFACES

As we have seen in section 3.2 surfaces are characterized by their Euler characteristic, the number of boundary components and the orientability behavior. We will now give necessary and sufficient conditions in terms of these invariants for the existence of coverings between two surfaces with a given order.

3.4.1 Notation. In the following, F' and F denote connected surfaces which can be presented by finite complexes. By χ, g, r we denote the Euler characteristic, the genus and the number of boundary components of F, similarly χ', g', r' for F'.

Our aim is the following theorem:

3.4.2 Theorem. *Let* F' *and* F *be as in* 3.4.1 *and* $c \geq 1$, $c \in \mathbb{Z}$. *Then* F' *is a c-fold covering of* F *if and only if* $\chi' = c\chi$ *and one of the following conditions is fulfilled:*

(a) F' *and* F *are both orientable and* $r \leq r' \leq cr$.
(b) F' *and* F *are both non-orientable,* $r \leq r' \leq cr$ *and* $r' \equiv cr \bmod 2$.
(c) F' *is orientable,* F *is non-orientable and* $2r \leq r' \leq cr$ *and* $2|c$.

The proof will be finished in 3.4.12. We follow [Heimes-Stöcker 1978] and look for embeddings of the fundamental group of F' in that of F with index c. This approach is motivated by the facts we have proved in 2.5 and our treatment will give a good exercise in that theory. We have to consider several cases. The first case 3.4.3 is given as exercise E 3.

3.4.3 Case. F and F' are closed surfaces. Then the statement of 3.4.2 is true.

3.4.4 If p: F' → F is an c-fold covering and x_o a vertex of F, then the fundamental group $\mathcal{W}$ of F acts on the set $p^{-1}(x_o)$ of vertices of F' over x_o. This is a consequence of the unique path-lifting property for coverings, see the proof of 1.5.2. Choosing an enumeration of the vertices of $p^{-1}(x_o)$ by 1,2,...,c, we adjoin to each element of $\mathcal{W}$ a permutation from S_c and obtain a homomorphism f: $\mathcal{W} \to S_c$, a so-called representation of $\mathcal{W}$ in the symmetric group S_c. Here we use the following notation: if $x, y \in S_c$ then $xy(i) = x(y(i))$ for $1 \leq i \leq c$. As F' is connected the image $f(\mathcal{W})$ acts transitively on $\{1,\dots,c\}$. If two c-fold coverings over F are isomorphic then their representations in S_c differ only by an isomorphism of S_c, in this case they are called *equivalent*.

On the other hand, to each transitive representation of $\mathcal{W}$ in S_c there corresponds the subgroup of index c of elements which are mapped to permutations that fix 1. To this subgroup corresponds a covering of F, as we have seen in 2.5. If two representations are equivalent then the coverings are isomorphic. Thus we have:

There is a 1-1-correspondence between c-fold (connected) coverings of F *and equivalent transitive representations of* $\mathcal{W}$ *in* S_c.

(This assertion is true for all finite coverings between sufficiently nice topological spaces, see [Heimes-Stöcker 1978] or books on algebraic topology.)

Let γ be a simple closed curve in F. Then γ defines a set of elements in W in the following way: connect the basepoint x_o with the initial vertex of γ by a path ν. Then $\nu\gamma\nu^{-1}$ defines an element $g \in W$. This element is well defined up to conjugation. Homotopic curves define the same conjugacy classes. If $\pi \in S_c$ then we denote by $|\pi|$ the number of elements in the orbit space $\{1,\ldots,c\}/\langle\pi\rangle$ where $\langle\pi\rangle \subset S_c$ is the subgroup generated by π. From the definitions we obtain directly:

<u>3.4.5 Lemma</u>. *With the notation from above,* $p^{-1}(\gamma)$ *consists of* $|f(g)|$ *simple closed curves.*

□

Let us notice that g depends on γ and ν, but f(g) only on γ. By easy calculations we obtain the following from the equations $\chi' = 2 - r' - 2g'$ and $\chi = 2 - k - 2g$:

<u>3.4.6 Lemma</u>. *If* F' *and* F *are orientable surfaces with boundaries such that* $r \le r' \le cr$ *and* $\chi' = c\chi$ *then*

3.4.7 $\quad r' \equiv cr \bmod 2$ *and* $r \le r' \le \begin{cases} c(r-2)+2 & \text{if } g = 0 \\ cr & \text{if } g > 0. \end{cases}$

□

For orientable surfaces theorem 3.4.1 is a consequence of the following proposition.

<u>3.4.8 Proposition</u>. *Let* F *be an orientable surface with boundary and let* r' *and* c *be integers such that 3.4.7 holds. Then there exists a* c*-fold covering over* F *where the cover space has* r' *boundary components.*

We use 3.4.8 to prove 3.4.1 for the case where both surfaces are orientable and have boundaries. As noted above, from the assumptions $r \le r' \le cr$ and $\chi' = c\chi$ we get condition 3.4.7. By 3.4.6, F is c-fold covered by a surface F" with r' boundary components. The Euler characteristic of F" equals $c\chi$, hence, F" and F' have the same Euler characteristic and the same number of boundary components. As both are orientable they are homeomorphic by 3.2.8. Thus F' is a c-fold covering of F.

Proof of 3.4.8. The fundamental group of F has the presentation

$$W = W(r,g) = \langle s_1,\dots,s_r,t_1,u_1,\dots,t_g,u_g \mid \prod_{i=1}^{r} s_i \prod_{j=1}^{g} t_j,u_j \rangle .$$

We look for a representation of W in S_c where some generator is mapped to the cycle $(1,2,\dots,c)$. Then the corresponding curve in F is covered by only one curve, hence, the covering surface is connected. It is orientable as F is.

Since $r \geq 1$ the group W is the free group with free generators $s_2,\dots,s_r,t_1,u_1,\dots,t_g,u_g$. Thus, to construct a representation $f\colon W \to S_c$, it suffices to give the images of these generators. We use the following abbreviation:

$$\Sigma_r = |f(s_1)| + \dots + |f(s_r)| \text{ and } r_{max} = \begin{cases} c(r-2)+2 & \text{if } g = 0, \\ cr & \text{if } g > 0. \end{cases}$$

We say that f *realizes a given integer* r', if $f\colon W \to S_c$ is a transitive representation such that $\Sigma_r = r'$. We prove the statement by induction and start with the induction step:

Let $g \geq 2$ and suppose that $f_o\colon W(k,g-1) \to S_c$ realizes r'. Then we 'extend' f_o to $W(r,g)$ by $f(t_g) = f(u_g) = 1$. Now $f\colon W(r,g) \to S_c$ realizes r', too. Therefore we may assume $g \leq 1$.

Let $r \geq 4$ or $r = 3$, $g = 1$, and suppose that $f_o\colon W(r-2,g) \to S_c$ realizes r'_o. Extend f_o to $W(r,g)$ by defining $f(s_{r-1}) = (1,\dots,m)$ and $f(s_r) = f(s_{r-1})^{-1}$, where $1 \leq m \leq c$. Since this does not change the $f(s_i)$, $1 \leq i \leq r-2$, we get

$$\Sigma_r = \Sigma_{r-2} + |f(s_{r-1})| + |f(s_r)| = r'_o + 2(c-m+1).$$

If 3.4.8 is true for $W(r-2,g)$, then r'_o may be any integer satisfying $r'_o \equiv c(r-2) \bmod 2$ and $r-2 \leq r'_o \leq (r-2)_{max}$. Therefore $r' = r'_o + 2(c-m+1)$ may be any integer such that $r' \equiv cr \bmod 2$ and $r \leq r' \leq r_{max}$. Hence 3.4.8 is true for $W(r,g)$, and besides $g \leq 1$ we may assume that $r \leq 3$, even $r \leq 2$ if $g = 1$. We now consider the remaining cases:

$g = 0$, $r = 1$: 3.4.7 implies that $c = r' = 1$; now 3.4.8 is trivial.
$g = 0$, $r = 2$: 3.4.7 implies that $r' = 2$. Define $f(s_2) = (1,\dots,c)$,

then f realizes r'.

$g = 0$, $r = 3$: 3.4.7 implies that $r' \equiv c \bmod 2$ and $3 \le r' \le c+2$. Define $f(s_2) = (1,\dots,c)$ and $f(s_3) = (1,\dots,c-r'+3)$. Then $|f(s_2)| = 1$ and $|f(s_3)| = r'-2$. Now $f(s_2s_3)$ maps

$$c-r'+3 \mapsto c-r'+4 \mapsto \dots c \mapsto 2 \mapsto 4 \mapsto \dots \mapsto c-r'+2 \mapsto 1 \mapsto 3 \mapsto \dots \mapsto c-r'+3$$

since $c-r' \equiv 0 \bmod 2$. Hence $|f(s_1)| = 1$, thus $\Sigma_3 = r'$.

$g = 1$, $r = 1$: 3.4.7 implies that $r' \equiv c \bmod 2$ and $1 \le r' \le c$. Define $f(t_1) = (1,\dots,c)$ and $f(u_1) = (1,2)(3,4)\dots(c-r'-1,c-r')$. For $s_1 = (t_1u_1t_1^{-1}u_1^{-1})^{-1}$ we have the following cycles

$(1,3,\dots,c-r'-1,c-r',c-r'-2,\dots,2)(c-r'+1)\dots(c-1)$. Hence $\Sigma_1 = |f(s_1)| = r'$.

$g = 1$, $r = 2$: 3.4.7 implies that $r' \equiv 0 \bmod 2$ and $2 \le r' \le 2c$.

Define $f(t_1) = (1,2,\dots,c)$, $f(u_1) = 1$ and $f(s_2) = (1,\dots,m)$ where $m = c+1-\frac{r'}{2}$. From $s_1^{-1} = s_2t_1u_1t_1^{-1}u_1^{-1}$ one obtains $|f(s_1)| = |f(s_2)| = c-m+1$. Therefore $\Sigma_2 = r'$, hence f realizes r'.

This finishes the proof of 3.4.8.

□

For the non-orientable surfaces we take the canonical presentations 3.2.6 of the fundamental group.

<u>3.4.9 Proposition</u>. *Let* F' *and* F *be non-orientable surfaces with boundaries, where* F' *is a* c*-fold covering of* F. *Then*

3.4.10 $$r' \equiv cr \bmod 2 \text{ and } r \le r' \le \begin{cases} c(r-1)+1 & \text{if } g = 1, \\ cr & \text{if } g > 1. \end{cases}$$

Proof. The inequality follows easily from $1 \le r' \le cr$ and the equations $\chi' = c\chi$, $\chi' = 2-r'-g'$ and $\chi = 2-r-g$. But in contrast to the orientable case, see 3.4.6, these equations do not imply the congruence (see remark 3.4.13).

Let $p\colon F' \to F$ be a c-fold covering of F by F'. If $f\colon W \to S_c$ is the representation corresponding to p we get $r' = |f(s_1)| + \dots + |f(s_r)|$ by 3.4.5. For $\pi \in S_c$ let $\operatorname{sgn}\pi$ be 0 if π is an even, and 1 if π is an odd, permutation. Then $c + \operatorname{sgn}\pi \equiv 0 \bmod 2$. Therefore

$$r' \equiv cr + \operatorname{sgn} f(s_1) + \dots + \operatorname{sgn} f(s_r) \equiv cr + \operatorname{sgn} f(s_1\dots s_r) \bmod 2.$$

Since $s_1 \dots s_r = (v_1^2 \dots v_g^2)^{-1}$ the permutation $f(s_1 \dots s_r)$ is even.

□

3.4.11 Proposition. *Let* F *be a non-orientable surface with boundary and let* r', c *be integers such that 3.4.10 is fulfilled. Then there exists a non-orientable surface with* r' *boundaries which is a* c*-fold covering of* F.

□

Proof as exercise.

3.4.12 *Proof of 3.4.1.* We have already proved the theorem for closed surfaces and for orientable surfaces. Now we assume that F' and F are both non-orientable and have boundaries. That the condition 3.4.1 (b) is necessary has been proved in 3.4.9. Now we assume that they are fulfilled. From $\chi' = c\chi$ and $r \leq r' \leq cr$ we obtain the inequality from 3.4.10. The congruence is fulfilled by assumption, hence, by 3.4.11, there exists a non-orientable surface with r' boundary components that covers F with order c. By the same argument as for the orientable case we conclude that this surface is homeomorphic to F'. Thus F' is an c-fold covering of F.

Now let us assume that F' is an orientable surface, but F is non-orientable. Let F_1 be the orientable covering of F. Its Euler characteristic equals $\chi_1 = 2\chi$ and its number of boundary components is $r_1 = 2r$, since each boundary curve preserves orientation. Now, if F' is a c-fold covering of F then $\chi' = c\chi$ and $r \leq r' \leq cr$. Since F' is orientable it covers F_1 in such a way that the composition of the covering maps $F' \to F_1 \to F$ is the given covering map $F' \to F$ (proof!). Therefore $r_1 \leq r'$ and c must be even. Conversely, let us assume $2r \leq r' \leq cr$, $\chi' = c\chi$ and $2|c$. Then $r_1 \leq r' \leq c_1 r_1$ and $\chi' = c_1\chi_1$, where $c_1 = c/2$. As the theorem is already proved for orientable surfaces we know that there exists a c_1-fold covering $F' \to F_1$. Then the composition $F' \to F_1 \to F$ is a c-fold covering as desired.

□

3.4.13 *Remark.* There is a difference between the orientable and the non-orientable cases of 3.4.8 and 3.4.11. If F is an orientable surface and r', c are integers with $r \leq r' \leq cr$, then there is in general no orientable surface with Euler characteristic $r' = cr$ and with r' boundary components; if such a surface exists it is automatically a c-fold covering of F as has been shown after 3.4.8. If F is a non-orientable surface of genus $g > 1$ there exists for every $c > 1$ and every r' with $r \leq r' \leq cr$ a non-orientable surface F' with Euler characteristic $\chi' = c\chi$ and with r' boundary components; but F' is a c-fold covering of F only if in addition the congruence $r' \equiv cr \bmod 2$ is satisfied. (In the orientable case this congruence is a consequence of $\chi' = c\chi$.)

The problem of this section was first considered in [Massey 1974] and was there solved by a geometric construction in the case where both surfaces are orientable. The general theorem is from [Heimes-Stöcker 1978]; our proof follows theirs. Parts of the new results of the paper [Heimes-Stöcker 1978] were independendly obtained by O. Fajuyigbe (Benin City, Nigeria).

3.5 TYPES OF SIMPLE CLOSED CURVES ON SURFACES

We shall find an enumeration of the types of simple closed paths on surfaces. First of all we give a process for deriving the homotopy class (path class) of a closed curve relative to a canonical system of generators for the fundamental group.

3.5.1 Definition. We already know what a canonical curve system $\Sigma = \{\sigma_1 \ldots, \sigma_m, \tau_1, \mu_1, \ldots, \tau_g, \mu_g\}$ or $\{\sigma_1, \ldots, \sigma_m, \nu_1, \ldots, \nu_g\}$ of an orientable or non-orientable F of genus g with m boundaries is. Dual to Σ, there is a system of simple curves $\Sigma^* = \{\sigma_1^*, \ldots, \sigma_m^*, \tau_1^*, \mu_1^*, \ldots, \tau_g^*, \mu_g^*\}$ or $\{\sigma_1^*, \ldots, \sigma_m^*, \nu_1^*, \ldots, \nu_g^*\}$ with the following properties:

(a) All curves have a common initial point P^* and meet nowhere else.

(b) σ_i^* is not closed, and it connects P^* to the ith boundary curve ρ_i of F.

(c) F is decomposed by Σ^* into a disc with the boundary $\prod_{i=1}^{m} \sigma_i^* \rho_i \sigma_i^{*-1} \prod_{j=1}^{g} [\tau_j^*, (\mu_j^*)^{-1}]$ or $\prod_{i=1}^{m} \sigma_i^* \rho_i \sigma_i^{*-1} \nu_1^{*2} \ldots \nu_g^{*2}$ respectively.

(d) The star about P^* has the form $\sigma_1^*, \ldots, \sigma_m^*, \tau_1^*, \mu_1^*, \tau_1^{*-1}, \mu_1^{*-1}, \ldots, \tau_g^*, \mu_g^*, \tau_g^{*-1}, \mu_g$ or $\sigma_1^*, \ldots, \sigma_m^*, \nu_1^*, \nu_1^{*-1}, \ldots, \nu_g^*, \nu_g^{*-1}$ respectively when one considers only the curves of Σ^*.

(e) A curve η_i^* of Σ^* meets the corresponding curve η_i of Σ exactly once, and no other curve of Σ.

A system Σ^* which satisfies 3.5.1 (a) - (e) is called a *canonical dissection* of F.

3.5.2 Definition. We shall now define what it means for a curve ω to *positively intersect* a curve η_i^* of Σ^* in a point P_i. For this purpose we surround η_i by a narrow strip. For curves σ_i^* this strip is a "rectangle", for curves τ_i^*, μ_i^* a cylinder, and for curves ν_i^* a Möbius strip. If the component of ω in the strip, which passes through P_i, is a segment which runs from one side to the other, then in the first two cases we can say whether the segment crosses the strip in the same direction as η_i. If the strip is a Möbius strip then we cut it along an arc through P^* which does not meet η_i, so that it becomes a rectangle and we can again decide whether the segment crosses the rectangle in the same direction as η_i. This direction is called *positive,* and the opposite direction is called *negative.* (In the latter case we are proceeding arbitrarily. The opposite convention will indeed change the description defined below, however only defining relations come into play.)

3.5.3 Definition. We represent the curves $\eta_1^*,\dots,\eta_n^*$ of the canonical dissection by symbols $H_1,\dots,H_n$. Let ω be a closed curve which does not pass through P^*. We associate ω with the word $H_{\alpha_1}^{\varepsilon_1}\dots H_{\alpha_\ell}^{\varepsilon_\ell}$ in the H_i when ω crosses $\eta_{\alpha_1}^*,\dots,\eta_{\alpha_\ell}^*$ successively, and ε_i is + 1 or - 1 according as ω intersects $\eta_{\alpha_i}^*$ positively or negatively. Here we must assume that ω meets each η_i^* properly at a point of intersection, and does not merely touch it. We call the word $H_{\alpha_1}^{\varepsilon_1}\dots H_{\alpha_\ell}^{\varepsilon_\ell}$ the *description* of ω. The curves η_i have the descriptions H_i, and a simple closed curve which runs once around P^* and contains no part $H_i^{\varepsilon_i}H_i^{-\varepsilon_i}$ has the description $\Pi_*(H)$. Conversely it is easy to convince oneself that a simple closed curve with the description $\Pi_*(H)$ bounds a disk around the point P^*.

Naturally a curve has different descriptions, relative to different canonical dissections. We shall give a distinguished description to each simple curve. For this purpose we attach no significance to the initial point of the curve - thus we regard descriptions as cyclic words. Instead of changing the description when we go to a new canonical dissection, we can retain the old dissection and carry the curve into another by a homeomorphism H. This curve has the same description relative to Σ^* as the old curve has relative to the dissection $H^{-1}(\Sigma^*)$, and for each canonical dissection $\Sigma^{*\prime}$ there is a homeomorphism which maps Σ^* onto $\Sigma^{*\prime}$.

3.5.4 Theorem (Enumeration of types of simple curves). *Let ω be a simple closed curve in the interior of* F *which does not bound a disk. Then there is a canonical dissection of* F, *relative to which ω has one of the following descriptions.*

A. F *is orientable.*

If ω does not separate, then the description is

3.5.5 T_1;

if ω does separate, then

3.5.6 $S_{q+1}\dots S_m \prod_{i=1}^{k} [T_i,U_i]$, *where* $(m-q) + 4k \le 1/2(m+4g)$.

The single ambiguous case occurs for

$m = 4g$: $S_1 \ldots S_m$ *or* $\prod_{i=1}^{g} [T_i, U_i]$ *respectively. Then we distinguish the one (say) which has* T_1 *in its description.*

B. F *is not orientable.*

If ω *does not separate and is one-sided, and if by cutting along* ω *we obtain a non-orientable surface, then the word is*

3.5.7 V_1; *in this case we must have* $g > 1$.

If an orientable surface results, then

3.5.8 $V_1 \ldots V_g$ (g odd).

If ω *is two-sided and the surface remains non-orientable after cutting, then*

3.5.9 V_1V_2, $g > 2$.

If a orientable surface results, then

3.5.10 $V_1 \ldots V_g$ (g *even).*

Now if ω *separates, and the resulting components are both non-orientable, then the word is*

3.5.11 $S_{q+1} \ldots S_m \prod_{i=1}^{k} v_i^2$, $k > 0$ *with* $(m-q) + 2k \leq 1/2(m+2g)$.

If one component is orientable, but the other is not, then there are three types

3.5.12 (a) $S_{q+1} \ldots S_m V_1^2 \ldots V_{i-1}^2 V_i \ldots V_g V_i \ldots V_g$

where $(g-i)$ *is even and* > 0, $m - q + 2(i-1) + 1 \leq 1/2(m+2g)$.

If the inequality does not hold for 3.5.12 (a) one replaces $S_{q+1} \ldots S_m V_1^2 \ldots V_{i-1}^2$ *by the inverse of the complement in* Π_* *and reduces cyclically. The word becomes*

(b) $S_q^{-1} \ldots S_1^{-1} V_g^{-2} \ldots V_{i+1}^{-2} V_i^{-1} V_{i+1} \ldots V_g V_i \ldots V_g$ $(q > 0)$

(c) $V_g^{-1} V_{g-1}^{-2} \ldots V_{i+1}^{-2} V_i^{-1} V_{i+1} \ldots V_g V_i \ldots V_{g-1}$ $(q = 0)$.

If two curves have different descriptions under the classification (3.5.5-12) then they cannot be mapped on to each other by a surface homeomorphism.

Proof. We merely remark that curves which have the same description relative to possibly different canonical dissections can be carried into each other by a homeomorphism, and indeed by the one which carries the canonical dissections into each other. (Curves which have the same description relative to the same dissection may be carried into each other by an isotopy, cf.

One can see immediately that curves with different descriptions 3.5.5-12 cannot be carried into each other by a homeomorphism, because of the characterization of the circumstances 3.5.8, 3.5.9 and 3.5.10 relative to each other and in comparison with 3.5.5, 3.5.6 and 3.5.7. On the same basis, 3.5.5, 3.5.6 and 3.5.7 are different from one another. The distinctness of the descriptions in the cases 3.5.6, 3.5.11 and 3.5.12 lies in the fact that the description uniquely determines genus and the number of boundary components.

□

The descriptions 3.5.5-12 have the following properties:

3.5.13 *They are cyclically reduced.*

3.5.14 *They contain no subwords which comprise more than half of the defining relation* $(\Pi_*(H))^{\pm 1}$.

3.5.15 *If a subword occurs which comprises half of a defining relation* $\Pi_*^{\varepsilon}(H)$, $\varepsilon = \pm 1$, *then for* $g = 0$ *this contains the* S_m^{ε}, *and for* $g > 0$ *the* T_1^{ε} *or* V_1^{ε} *respectively.*

3.6 INTERSECTION NUMBERS OF CURVES

In this and the next section we consider some homological properties of orientable surfaces: the intersection numbers of curves and homological properties of mappings between surfaces, in particular, the mapping degree is revisited. Mostly we will restrict ourselves to closed surfaces.

3.6.1 Definition of intersection numbers between curves. Let F be an orientable surface. We choose an orientation and define: The star $\sigma_1,\dots,\sigma_n$ at the vertex Q is called positive if $\sigma_i^{-1}\sigma_{i+1}$ $(1 \le i \le n-1)$ is in the positive boundary of a positive face. Now let ω_1 and ω_2 be paths and assume that there is a situation $\omega_i = \omega_{i1}\sigma_{i1}^{-1}\sigma_{i2}\omega_{i2}$ $(i = 1,2)$ where the edges σ_{ij}, $i, j \in \{1,2\}$, start at one vertex Q. The position gets the number

0 if the pairs $(\sigma_{11},\sigma_{12})$ and $(\sigma_{21},\sigma_{22})$ do not separate themselves in the star at Q,

1 the edges follow in the order $\sigma_{11},\sigma_{21},\sigma_{12},\sigma_{22}$ in the positive star at Q,

$-1 \quad \sigma_{11},\sigma_{22},\sigma_{12},\sigma_{21}$ is in the positive star.

A similar definition is used if the initial or end points of ω_1 or ω_2 are involved.

Now assume that ω_1 and ω_2 are closed curves which have no edge in common. Then the *intersection number* $\nu(\omega_1,\omega_2)$ is by definition the sum of the numbers at all possible positions as considered above.

In a similar way we handle the case where the curves ω_1 and ω_2 have common edges. Assume we have a situation

$$\omega_i^{\varepsilon_i} = \omega'_{i1}(\sigma'^{-1}_{i1}\omega_o\sigma'_{i2})\omega'_{i2} \qquad \varepsilon_1 = 1,\ \varepsilon_2 \in \{1,-1\}$$

where the edges σ'_{1j} and σ'_{2j} are different ($j = 1,2$), and $\omega_o = \tau_1\omega'_o\tau_2^{-1}$ (τ_i edges) leads from Q_1 to Q_2. (If ω_o consists of one edge only, then $\tau_1 = \omega_o = \tau_2^{-1}$.) Now let $\varepsilon_2 = 1$ and consider $\sigma'_{11},\sigma'_{21},\tau_1$ and $\tau_2,\sigma'_{12},\sigma'_{22}$. If both are in the positive stars at Q_1 and Q_2, then we adjoin +1 to the situation; if both belong to the negative stars then -1, and 0 if they belong to stars with different orientations. For $\varepsilon_2 = -1$ we reverse the signs. Now the intersection number is defined as above as the sum of the numbers for all such situations. (See the figure.)

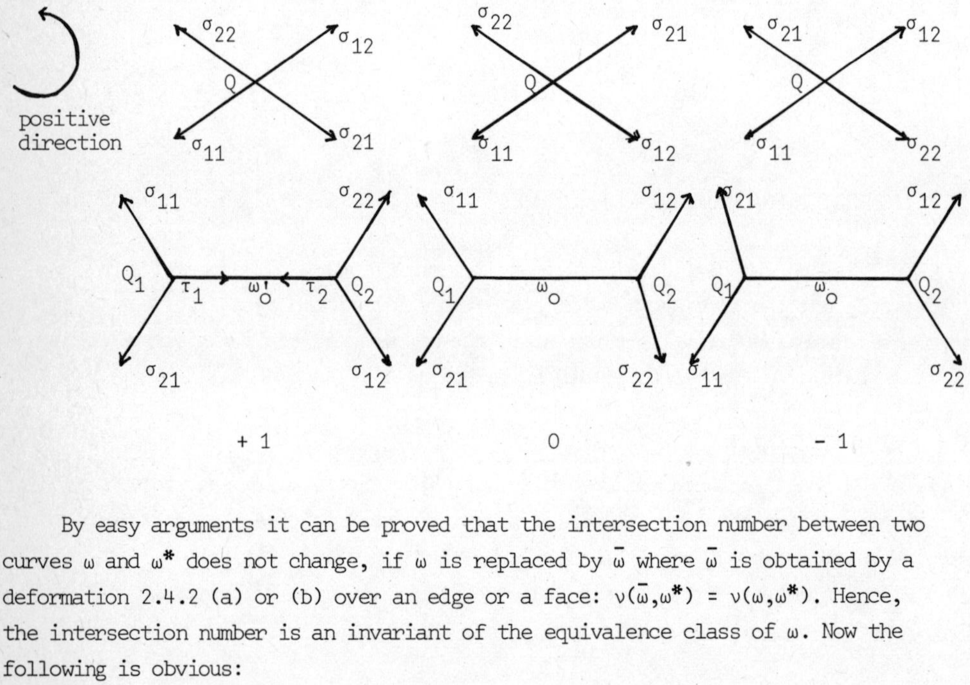

By easy arguments it can be proved that the intersection number between two curves ω and ω^* does not change, if ω is replaced by $\bar{\omega}$ where $\bar{\omega}$ is obtained by a deformation 2.4.2 (a) or (b) over an edge or a face: $\nu(\bar{\omega},\omega^*) = \nu(\omega,\omega^*)$. Hence, the intersection number is an invariant of the equivalence class of ω. Now the following is obvious:

3.6.2 Lemma. *(a)* $\nu(\omega_1,\omega_2) = -\nu(\omega_2,\omega_1)$.
(b) $\nu(\omega_1^{-1},\omega_2) = -\nu(\omega_1,\omega_2) = \nu(\omega_1,\omega_2^{-1})$.
(c) $\nu(\omega_1\omega'_1,\omega_2) = \nu(\omega_1,\omega_2) + \nu(\omega'_1,\omega_2)$.

(d) If ω_i *and* ω_i' *are homotopic, then* $\nu(\omega_1,\omega_2) = \nu(\omega_1',\omega_2')$.

(e) Let F *be the fundamental group of* F. *Then* ν *induces a skew-symmetric bilinear mapping – again denoted by* ν –

$$\nu\colon \mathsf{F} \times \mathsf{F} \to \mathbb{Z},\ (w_1,w_2) \mapsto \nu(\omega_1,\omega_2) \text{ if } \omega_i \in w_i.$$

□

For a fixed $y \in \mathsf{F}$ we obtain a homomorphism $\nu'\colon \mathsf{F} \to \mathbb{Z}$, $x \mapsto \nu(x,y)$. Since $\mathbb{Z}$ is abelian, ν' vanishes on the commutator subgroup of F, hence ν' induces a homomorphism of the abelianized fundamental group, that is of $H_1(F)$, to $\mathbb{Z}$, see 2.7.8. The same argument can be applied to the second variable and we obtain:

3.6.3 Proposition. *(a) The intersection number between curves of the oriented surface* F *induces a bilinear skew-symmetric form*

$$\nu\colon H_1(F) \oplus H_1(F) \to \mathbb{Z}.$$

(b) If $s_1,\dots,s_m,t_1,u_1,\dots,t_g,u_g$ *is a canonical system of generators of* F *then*

$$\nu(s_i,s_k) = \nu(s_i,t_j) = \nu(s_i,u_j) = \nu(t_i,t_\ell) = \nu(u_i,u_\ell) = 0$$
$$(1 \le i,\ k \le m,\ 1 \le j,\ell \le g),$$
$$\nu(t_j,u_\ell) = \delta_{j\ell} = \begin{cases} 0 & \neq j \\ 1 & = j \end{cases} \qquad (1 \le j,\ \ell \le g).$$

Relative to this basis the form is described by the matrix

$$\text{3.6.4} \qquad \tilde{K} = \left(\begin{array}{ccccccc} 0 &&&&&& \\ & \ddots &&&&& \\ && 0 &&&& \\ &&& 0 & 1 && \\ &&& -1 & 0 && \\ &&&&& \ddots & \\ &&&&&& \begin{matrix} 0 & 1 \\ -1 & 0 \end{matrix} \end{array}\right) \begin{matrix} \}\, m \\ \\ \}\, 2g \end{matrix} = \left(\begin{array}{ccc} 0 && \\ & \ddots & \\ && 0 \\ &&& K \end{array}\right) \}\, m$$

i.e. if $\omega_i \sim \sum_{k=1}^{m} c_{ik}\sigma_k + \sum_{j=1}^{g} (a_{ij}\tau_j + b_{ij}\mu_j)$ *and* $\mathbf{x}_i$ *denotes the vector* $(c_{i1},\dots,c_{im},a_{i1},b_{i1},\dots,a_{ig},b_{ig})^t$, *then*

$$\text{3.6.5} \qquad \nu(\omega_1,\omega_2) = \mathbf{x}_1^t\, K\, \mathbf{x}_2 = \sum_{j=1}^{g} (a_{1j}b_{2j} - b_{1j}a_{2j}).$$

□

3.6.6 Remark. For $m = 0$ the quadratic form defines a geometry on $\mathbb{R}^{2g}$: the symplectic geometry. Usually the metric is defined by the form corresponding to the matrix $J = \begin{pmatrix} 0 & E \\ -E & 0 \end{pmatrix}$, where E denotes the g-by-g unit matrix, see [Siegel 1943]. In the following we prove some results of this theory, but for the matrix K instead of J.

Using the group ring of $\pi_1(F)$ a more detailed study of fixed points can be done, see [Baer 1927], which allows a classification of simple closed curves.

3.6.7 Theorem. *Let* F *be an oriented compact surface of genus* g *with* m *boundary components and let* $s_1,\ldots,u_g$ *be a canonical system of generators of* $\mathfrak{F} = \pi_1(F)$. *We use the same symbols for the induced basis of the homology* $H_1(F)$.

(a) Let $\varphi\colon F \to F$ *be a homeomorphism and let the induced mapping* $\varphi_*\colon H_1(F) \to H_1(F)$ *be described relative to the above basis by the matrix* $\tilde{A}$. *Let* $\varepsilon = 1$ *if* φ *preserves the orientation, otherwise* $\varepsilon = -1$. *Then:*

3.6.8 $$\tilde{A} = \begin{pmatrix} A' & 0 \\ 0 & A \end{pmatrix} \}m$$

where A' *is an m-by-m matrix with a coefficient* ε *in each row and column and the integer matrix* A *satisfies*

3.6.9 $$A^t\, K\, A = \varepsilon\, K.$$

(b) Each matrix $\tilde{A}$ *with the above properties is induced by a homeomorphism* $\varphi\colon F \to F$.

Proof. (a): A given orientation of F determines a positive direction on each boundary component. A homeomorphism permutes the boundary curves and either preserves the positive directions for all components or reverses all, hence, the first m columns of the matrix are as in 3.6.8.

Since φ is a homeomorphism we obtain

$$\nu(\varphi_*(\tilde{x}),\varphi_*(\tilde{y})) = \varepsilon\nu(\tilde{x},\tilde{y}) \text{ for } \tilde{x},\ \tilde{y} \in H_1(F),$$

i.e. writing $\tilde{x} = (x',x)^t$ and $\tilde{y} = (y',y)^t$:

$$(Ax)^t\, K\, (Ay) = (\tilde{A}\tilde{x})^t\, \tilde{K}\, (\tilde{A}\tilde{y}) = \varepsilon\tilde{x}^t\, \tilde{K}\, \tilde{y} = \varepsilon x^t\, K\, y,$$

hence $A^t\, K\, A = \varepsilon\, K$.

(b) We will restrict us to the case m = 0. It is convenient to use E 3.19 (a). The following matrices solve the equation $X^t K X = K$ and they are induced by homeomorphisms of F:

(A) $$\begin{pmatrix} B_1 & & & \\ & B_2 & & \\ & & \ddots & \\ & & & B_g \end{pmatrix}$$ where each $B_i \in SL(2,\mathbb{Z})$.

A permutation of the handles is described by a matrix

(B) $$\begin{pmatrix} & & E_1 & \\ & E_2 & & \\ & & & \\ & & & E_g \end{pmatrix}$$ where each E_i equals $\begin{pmatrix} 1 & 0 \\ 0 & 1 \end{pmatrix}$ and different E_i do not have common row or columns.

(C) $$\begin{pmatrix} 1 & 0 & & & & & \\ 0 & 1 & & -1 & & & \\ 1 & & 1 & & & & \\ & & & 1 & & & \\ & & & & 1 & & \\ & & & & & \ddots & \\ & & & & & & 1 \end{pmatrix}$$

This matrix is defined by the automorphism $\alpha: F \to F$

$$t_1 \mapsto t_1 t_2,\ u_1 \mapsto t_2^{-1} u_1 t_2,\ t_2 \mapsto t_2^{-1} u_1 t_2 u_1^{-1} t_2,\ u_2 \mapsto u_2 t_2^{-1} u_1^{-1} t_2$$
$$t_i \mapsto t_i,\ u_i \mapsto u_i \text{ for } i \geq 3.$$

(That α is induced by a homeomorphism can be shown by a direct geometrical construction; it is also a consequence of 3.3.11. Another proof: in the free group with the free generators $t_1, \ldots, u_g$ we have $\alpha(\prod_i [t_i, u_i]) = \prod_i [t_i, u_i]$, hence α is an automorphism, by 5.2.13, and is induced by a homeomorphism, see 5.6.2).

The following matrix solves the equation $X^t K X = -K$:

$$\text{(D)} \qquad \begin{pmatrix} & & & 1 \\ & & 1 & \\ & \cdot^{\cdot^{\cdot}} & & \\ 1 & & & \end{pmatrix}$$

and is induced by the automorphism $t_i \mapsto u_{g+1-i}$, $u_i \mapsto t_{g+1-i}$, $1 \le i \le g$. (In the free group in the generators this automorphism maps $\prod_i [t_i, u_i]$ to its inverse.)

3.6.10 Now let $\mathbf{x}$ be a 2g-column. Then there is a matrix B of type (A) such that $B\mathbf{x}$ has the form $(x_1', 0, x_2', 0, \ldots, x_g', 0)^t$. Next we apply a product C of matrices of type (C) and (B) such that $CB\mathbf{x} = (x_1, 0, \ldots, 0)^t$ where x_1 is the greatest common divisor of the coefficients of $\mathbf{x}$ (the details of these steps: exercise E 3.27).

Next we consider a matrix A such that $A^t\, K\, A = K$. By a multiplication with matrices of type (A-C) we obtain a matrix A_1 where the first column has the form $(x_1, 0, \ldots, 0)^t$ with $x_1 > 0$. Since the determinant of A equals ±1 it follows that $x_1 = 1$. We may add a multiple of the first column to the second, which is given by a multiplication with a matrix of type (A), hence we may assume that the second column starts with the coefficient 0. From $A^t\, K\, A = K$ it follows that the second coefficient of the second column is 1 and that the second coefficients of the other columns vanish, hence

$$A_1 = \begin{pmatrix} 1 & 0 & * & \ldots & * \\ 0 & 1 & 0 & \ldots & 0 \\ \cdot & * & * & \ldots & * \\ \cdot & \cdot & \cdot & & \cdot \\ \cdot & \cdot & \cdot & & \cdot \\ \cdot & \cdot & \cdot & & \cdot \\ 0 & * & * & \ldots & * \end{pmatrix}$$

Next we treat the second column the same way and bring it into $(0, 1, 0, y_2, \ldots, 0, y_g)^t$ and then into $(0, 1, 0, \ldots, 0)^t$. We can apply processes that do not alter the first column. Let the matrix obtained be A_2. From $A^t\, K\, A = K$ it now follows that in the first row of A_2 all coefficients, except the first one, vanish, hence,

$$A_2 = \begin{pmatrix} 1 & 0 & 0 & \dots & 0 \\ 0 & 1 & 0 & \dots & 0 \\ 0 & 0 & & & \\ \vdots & \vdots & & A^* & \\ 0 & 0 & & & \end{pmatrix}$$

and now the proof can be finished by induction.

If we consider a matrix A with $A^t K A = -K$, we multiply A first with the matrix (D) and procede then as before. Now the proof of 3.6.7 (b) is finished for the case $m = 0$.

□

3.6.11 Corollary ([Schafer 1976]). *Let* F *be a closed orientable surface of genus* g *and* $t_1, u_1, \dots, t_g, u_g$ *a canonical basis for* $H_1(F)$, *i.e. obtained from a canonical system of curves. An element*

$$0 \neq x = \sum_{i=1}^{g} (a_i t_i + b_i u_i) \in H_1(F)$$

can be represented by a simple closed curve if and only if $(a_1, \dots, a_g, b_1, \dots, b_g) = 1$.

Proof. If the condition is fulfilled then x can be represented by a simple closed curve, as we have proved in 3.6.10. If x can be represented by a simple closed curve ξ, then ξ is member of some canonical system of curves, hence, x belongs to some basis of $H_1(F) = \mathbf{Z}^{2g}$ and cannot be a proper power of an element of $H_1(F)$.

□

3.6.12 Corollary. *The group* M *of integral matrices* X *that solve the equation*

$$X^t K X = \pm K$$

is generated by the matrices 3.6.9 (A–D). Each element of M *has determinant 1.*

□

It is obvious that the number of generators can be diminished, since the group of integer 2-by-2 matrices, as well as the symmetric group, is generated by 2 elements. A system of generators for M, closely related to the above system, has been determined in [Clebsch-Gordan 1866, pg. 304-308]. That these generators are induced by mappings between canonical system of curves (canonisches Querschnittssystem), has been proved in [Burkhardt 1890, pg. 208-212]. If we replace K by the conjugate matrix J the corresponding group is called *Siegel's modular group* or *group of integral symplectic matrices*. In this form the group has often been investigated in recent times. Generators have been given in [Witt 1941], [Hua-Reiner 1949] and [Klingen 1956]. In [Klingen 1961] a method is described for finding a finite system of defining relations. [Birman 1971] describes a system of defining relations explicitly.

It is obvious that, if genus $g \geq 2$, there are autormorphisms of $\pi_1(F_g)$ that induce the identity on $H_1(F)$. The group of these automorphisms has been studied in [Powell 1978] and [Johnson 1979]; they determine infinite sets of generators but which are from few conjugacy classes: from 3 in [Powell 1978], from 1 in [Johnson 1979].

Exercises: E 3.23-28.

3.7 ON CONTINUOUS MAPPINGS BETWEEN SURFACES

In this section we follow [Hopf 1931]. For convenience and simplification of the notation etc. we will use some elementary concepts and results from algebraic topology and we will not translate them into the language of combinatorial topology.

3.7.1 Assumptions. Let F_g denote the closed orientable surface of genus g and $\mathcal{F}_g = \pi_1(F_g)$ the fundamental group of F_g. Let $\tau_1,\mu_1,\ldots,\tau_g,\mu_g$ be a canonical system of curves of F_g; then $t_1,u_1,\ldots,t_g,u_g$ denote at the same time the induced generators from $\mathcal{F}_g$ (here the product is written multiplicatively) as well as those from $H_1(F_g)$ (here the operation is written additively). Now we will consider continuous mappings $\varphi\colon F_g \to F_h$ and we will denote the curves of the image F_h and their homotopy classes by $\tau_1',\ldots,\mu_h'$, $t_1',\ldots,u_h'$.

First we repeat some definitions from algebraic topology.

3.7.2 Definition. (a) Two continuous mappings $\varphi, \psi\colon F_g \to F_h$ belong to the same *homotopy class* if φ is homotopic to ψ, i.e. if there is a continuous mapping $\Phi\colon F_g \times [0,1] \to F_h$ such that

$$\Phi(p,0) = \varphi(p),\ \Phi(p,1) = \psi(p).$$

The homotopy class is denoted by $[\varphi]$, the set of all homotopy classes by $[F_g,F_h]$.

(b) Two homomorphisms $\alpha, \beta\colon \mathcal{F}_g \to \mathcal{F}_h$ belong to the same *homomorphism class* if there is an inner automorphism ι of F_h such that $\beta = \iota \circ \alpha$. The class of α is denoted by $[\alpha]$ and the set of all homomorphism classes by $[\mathcal{F}_g, \mathcal{F}_h]$.

If $\varphi, \psi\colon F_g \to F_h$ map the basepoint of F_g to that of F_h, they induce homomorphisms φ , ψ : $\mathcal{F}_g \to \mathcal{F}_h$. If the homotopy is constant on the basepoint then $\varphi_{\#} = \psi_{\#}$. If the image of the basepoint moves during the homotopy then $\varphi_{\#}$ and $\psi_{\#}$ differ by the inner automorphism of $\mathcal{F}_h$ which corresponds to the path of the image of the basepoint. Anyway, the homotopy class of φ induces the homomorphism class of $\varphi_{\#}$. Now we can drop the assumption that φ maps basepoint to basepoint.

<u>3.7.3 Proposition.</u> $T\colon [F_g, F_h] \to [\mathcal{F}_g, \mathcal{F}_h]$, $[\varphi] \mapsto [\varphi_{\#}]$, *is bijective if* $h \geq 1$.

Proof. First we show that T is surjective. Let $\alpha\colon \mathcal{F}_g \to \mathcal{F}_h$ be a homomorphism. The curves τ_i, μ_i can be considered as continuous mappings of the interval $[0,1]$ to F_g; we will use the same letter for these mappings. We choose curves $\xi_i \in \alpha(t_i)$, $\eta_i \in \alpha(u_i)$ and consider the mappings $\xi_i \circ \bar{\tau}_i$, $\eta_i \circ \bar{\mu}_i$ $(1 \leq i \leq g)$, where $\bar{\tau}_i$, $\bar{\mu}_i$ denote the "inverse mappings" to τ_i, μ_i. They define a continuous mapping

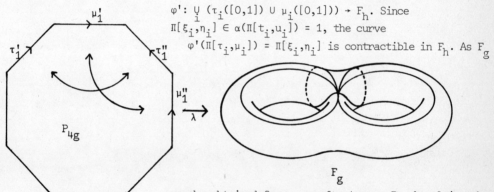

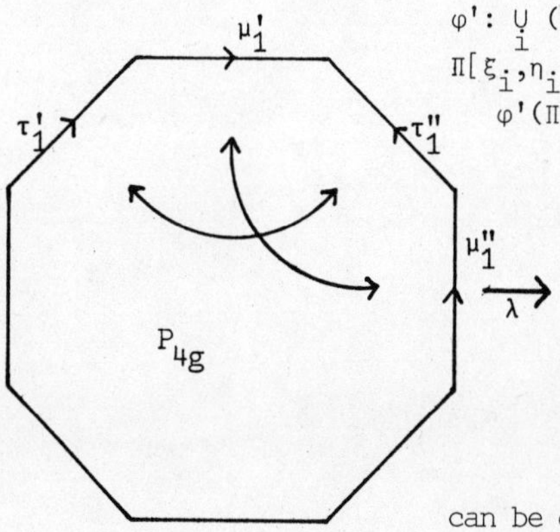

$\varphi'\colon \bigcup_i (\tau_i([0,1]) \cup \mu_i([0,1])) \to F_h$. Since $\Pi[\xi_i, \eta_i] \in \alpha(\Pi[t_i, u_i]) = 1$, the curve $\varphi'(\Pi[\tau_i, \mu_i]) = \Pi[\xi_i, \eta_i]$ is contractible in F_h. As F_g

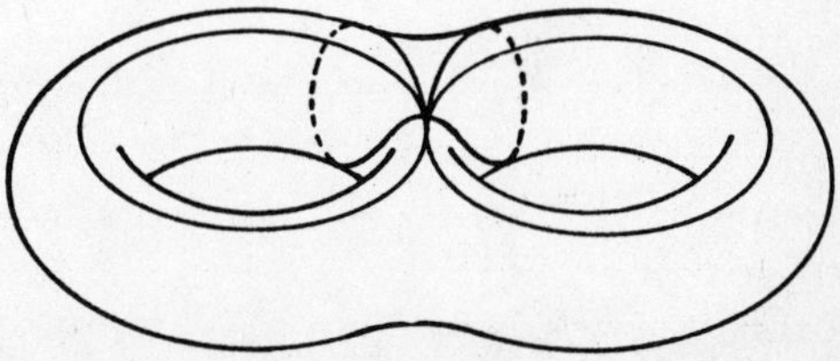

can be obtained from a regular $4g$-gon P_{4g} by gluing the sides τ_j' and τ_j'' as well as μ_j' and μ_j'' together - $F_g = \lambda(P_{4g})$, see the figure - it follows that φ' can be extended to a continuous mapping $\varphi\colon F_g \to F_h$. By construction: $\varphi_{\#} = \alpha$.

The proof that T is injective is more complicated. It can be proved with the help of some fundamentals of homotopy theory (which we will pose as exercise E 3.30). The approach of [Hopf 1931] is of a more geometrical nature; unfortunately, we do not describe the geometrical tools till chapter 6, in particular see 6.3, but we will use them already here.

The surface F_h can carry a complex analytic structure and becomes a Riemann surface. If we lift this structure to the universal cover $\tilde{F}_h$ of F_h we obtain a simply connected Riemann surface, hence, it is (conformally equivalent to) the euclidean or non-euclidean plane as follows from the Riemann mapping theorem - remember: $h \geq 1$. The covering transformations are now motions in the corresponding geometry.

Now let φ, ψ: $F_g \to F_h$ be continuous mappings which induce the same homomorphism class. By a deformation of ψ the situation can be improved such that φ and ψ map the basepoint of F_g to the same point of F_h and induce the same homomorphism $F_g \to F_h$. Now we lift the mappings $\varphi\lambda$, $\psi\lambda$: $P_{4g} \to F_h$ to the universal cover (this is possible since P_{4g} is simply connected) and obtain mappings $\Phi, \Psi : P_{4g} \to \tilde{F}_h$. (We may assume, though it is not necessary here, that one corner of P_{4g} is mapped by Φ and Ψ to the same point.)

A homotopy Φ_t: $P_{4g} \to \tilde{F}_h$ can be defined as follows: For $p \in P_{4g}$, we parametrize the segment from $\Phi(p)$ to $\Psi(p)$ linearly by $[0,1]$; here segment and length are taken in the appropriate geometry. We denote the point with parameter value t by $\Phi_t(p)$. Now Φ_t is well defined and continuous, depends continuously on t and $\Phi_0 = \Phi$, $\Phi_1 = \Psi$.

Each pair (τ_j', τ_j'') (or $(\mu_j', \mu_j''))$ of edges of P_{4g} determines two covering transformations x_j^Φ, x_j^Ψ (or y_j^Φ, y_j^Ψ, resp.) such that

$$x_j^\Phi(\Phi\tau_j') = \Phi(\tau_j''), \quad x_j^\Psi(\Psi\tau_j') = \Psi(\tau_j'')$$
$$(\text{or } y_j^\Phi(\Phi\mu_j') = \Phi(\mu_j''), \quad y_j^\Psi(\Psi\mu_j') = \Psi(\mu_j'')).$$

By assumption, φ and ψ differ by an inner automorphism, hence, there is a covering transformation z such that

$$z^{-1}x_j^\Phi z = x_j^\Psi, \quad z^{-1}y_j^\Phi z = y_j^\Psi \quad (1 \le j \le g).$$

We replace Ψ by $\Psi' := z\Psi$: $P_{4g} \to F_h$ and obtain

$$x_j^\Psi(z^{-1}\Psi'\tau_j') = z^{-1}\Psi'(\tau_j'') \text{ or } \Psi'(\tau_j'') = zx_j^\Psi z^{-1}(\Psi'\tau_j') = x_j^\Phi(\Psi'\tau_j')$$

etc. Hence, we may assume that $x_j^\Psi = x_j^\Phi =: x_j$ and $y_j^\Psi = y_j^\Phi =: y_j$.

Now let $p' \in \tau_j'$ and p" the corresponding point in τ_j''. Then

$$x_j(\Phi(p')) = \Phi(p'') \text{ and } x_j(\Psi(p')) = \Psi(p'').$$

Since the covering transformation x_j is a motion in the corresponding geometry it follows that

$$x_j(\Phi_t(p')) = \Phi_t(p'') \text{ for } 0 \le t \le 1.$$

Hence, the homotopy Φ_t: $P_{4g} \to \tilde{F}_h$ is compatible with the identification λ: $P_{4g} \to F_g$ if we project the image to F_h. Thus we obtain a homotopy from φ to ψ.

□

<u>3.7.4 Proposition</u>. *Let* $\varphi\colon F_g \to F_1$ *be a continuous mapping and* $\varphi(t_i) = a_i t + b_i u$, $\varphi(u_i) = c_i t + d_i u$; *here* $(t_1,\ldots,u_g)$ *is a canonical basis for* $H_1(F_g)$ *and* (t,u) *is a basis for* $H_1(F_1) = \pi_1(F_1)$. *Then*

$$\varphi \mapsto M_\varphi := \begin{pmatrix} a_1 & c_1 & \ldots & a_g & c_g \\ b_1 & d_1 & \ldots & b_g & d_g \end{pmatrix}$$

defines a bijective mapping between $[F_g, F_1]$ *and the system of integral 2-by-2g matrices.*

Proof. Since $\pi_1(F_1)$ is commutative the matrix M_φ is a homotopy invariant of φ. It remains to prove that each matrix is induced by a homomorphism: A mapping φ' from $\{t_1,u_1,\ldots,t_g,u_g\}$ into a group G can be extended to a homomorphism of $\mathcal{F}_g$ if and only if $\Pi[\varphi'(t_i), \varphi'(u_i)] = 1$ in G. If G is abelian this is no restriction. Now the assertion is a consequence of 3.7.3.

□

Since the homotopy class of the mapping is determined by the matrix M_φ it must be possible to calculate the mapping degree from M_φ. Here we have a simple result:

<u>3.7.5 Proposition</u>. *Let the assumptions be as in 3.7.4, and let* c *be the mapping degree of* φ. *Then*

$$c = \sum_{i=1}^{g} \begin{vmatrix} a_i & c_i \\ b_i & d_i \end{vmatrix}$$

Proof. We consider F_g as a "sphere with g handles": we take g simple closed curves $\kappa_1,\ldots,\kappa_g$ that divide F_g into g tori $T_1,\ldots,T_g$, each with a hole, and a sphere S with g holes, see figure (a). The curve κ_i shall be homotopic to the commutator $[\tau_i,\mu_i]$, hence $[\varphi(\kappa_i)] = 0$ in the abelian $\mathcal{F}_1 = \pi_1(F_1)$. Therefore φ can

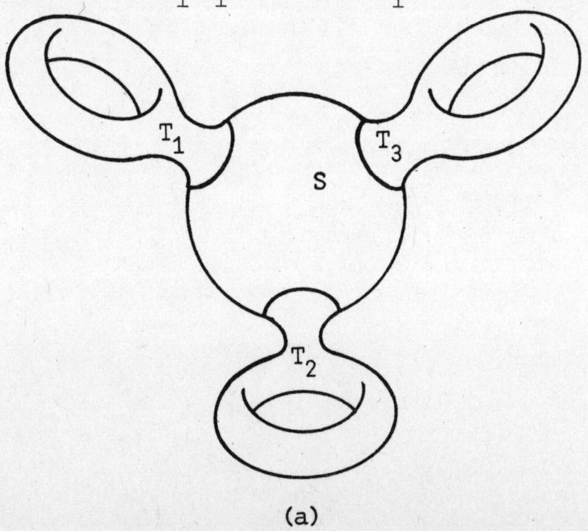

(a)

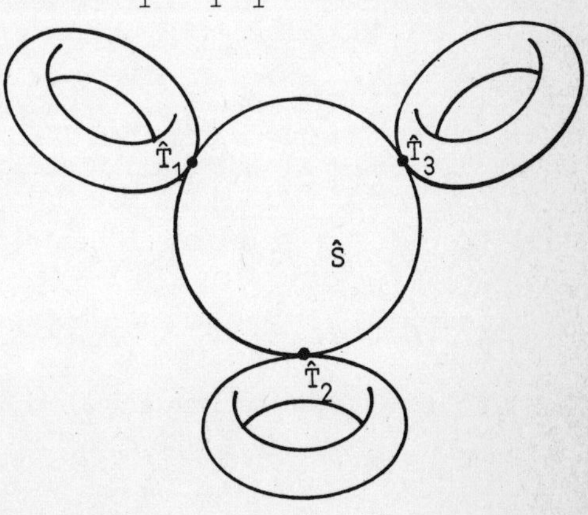

(b)

be deformed into $\bar{\varphi}$ such that $\bar{\varphi}$ maps each κ_i to a point. Let $\hat{F}_g$ be a sphere $\hat{S}$ with g tori $\hat{T}_1,\ldots,\hat{T}_g$, each attached with a point to $\hat{S}$, see figure (b). On $\hat{F}_g$ we take the induced orientation. Then $\bar{\varphi}$ factors through a mapping $\hat{\varphi}\colon \hat{F}_g \to F_1$,

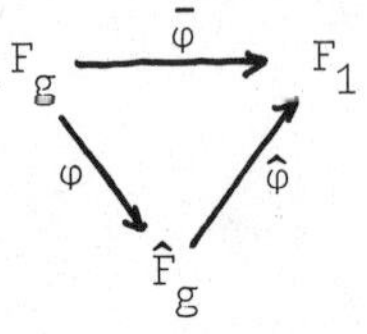

and φ and $\hat{\varphi}$ have the same degree.

The mapping of the torus $\hat{T}_j$ to F_1 has degree $\begin{vmatrix} a_j & c_j \\ b_j & d_j \end{vmatrix}$, see E 4.18. Now the assertion is a consequence of the fact that the mapping degree of $\hat{\varphi}$ is the sum of the degrees of the $\hat{\varphi}|\hat{T}_j$ $(1 \le j \le g)$ and $\hat{\varphi}|\hat{S}$ where the last degree vanishes, see 3.3.4.

□

Next we apply 3.7.5 to calculate the degree of a mapping $\varphi\colon F_g \to F_h$ for $h \ge 1$.

<u>3.7.6 Theorem ([Hopf 1931])</u>. *Let* $\varphi\colon F_g \to F_h$ *induce on homology*

$$\varphi_*(t_i) = \sum_{j=1}^{h} (a_{ij}t'_j + b_{ij}u'_j),\quad \varphi_*(u_i) = \sum_{j=1}^{h} (c_{ij}t'_j + d_{ij}u'_j) \text{ for } 1 \le i \le g.$$

Let c *be the degree of* φ. *Then*

$$c = \sum_{i=1}^{g} \begin{vmatrix} a_{ij} & c_{ij} \\ b_{ij} & d_{ij} \end{vmatrix}$$

where j *is anyone of the numbers* 1,2,...,h.

Proof. In addition to F_g and F_h we consider a torus F_1; the basis for $H_1(F_1)$ is again {t,u}. Let $\varphi_1\colon F_h \to F_1$ be a continuous mapping such that $\varphi_{1*}(t'_j) = t$, $\varphi_{1*}(u'_j) = u$ and $\varphi_{1*}(t'_k) = \varphi_{1*}(u'_k) = 0$ for $k \ne j$. We have seen in 3.7.4 that such a mapping exists and it follows from 3.7.5 that its degree is 1. The mapping $\varphi_1\varphi\colon F_g \to F_1$ induces the matrix $\begin{pmatrix} a_{1j} & c_{1j} & \cdots & a_{gj} & c_{gj} \\ b_{1j} & d_{1j} & \cdots & b_{gj} & d_{gj} \end{pmatrix}$, hence, by 3.7.5, it has the degree $\sum_{i=1}^{g} \begin{vmatrix} a_{ij} & c_{ij} \\ b_{ij} & d_{ij} \end{vmatrix}$. Since the degree behaves multiplicatively, see E 3.17, it follows that the above sum is also the degree of φ.

□

The sum in the formula from 3.7.6 does not depend on j, hence, the theorem gives us necessary conditions on the coefficients $a_{ij},\dots,d_{ij}$. By similar arguments the following can be proved:

3.7.7 Corollary. *Let the conditions be as in 3.7.6. Then:*

(a) $$\sum_{i=1}^{g} \begin{vmatrix} a_{i1} & c_{i1} \\ b_{i1} & d_{i1} \end{vmatrix} = \sum_{i=1}^{g} \begin{vmatrix} a_{i2} & c_{i2} \\ b_{i2} & d_{i2} \end{vmatrix} = \dots = \sum_{i=1}^{g} \begin{vmatrix} a_{ih} & c_{ih} \\ b_{ih} & d_{ih} \end{vmatrix};$$

(b) $$\sum_{i=1}^{g} \begin{vmatrix} a_{ij} & c_{ik} \\ b_{ij} & d_{ik} \end{vmatrix} = 0 \quad \text{for } 1 \le j, k \le h,\ j \ne k;$$

(c) $$\sum_{i=1}^{g} \begin{vmatrix} a_{ij} & c_{ij} \\ a_{ik} & c_{ik} \end{vmatrix} = 0 \quad \text{for } 1 \le j, k \le h;$$

(d) $$\sum_{i=1}^{g} \begin{vmatrix} b_{ij} & d_{ij} \\ b_{ik} & d_{ik} \end{vmatrix} = 0 \quad \text{for } 1 \le j, k \le h.$$

Proof as exercise E 3.31.

□

The conditions (a-d) can be expressed in another way. We consider the homomorphism $\varphi'\colon H_1(F_h) \to H_1(F_g)$ defined by

3.7.8 $$\varphi'(t_j') = \sum_{i=1}^{g} (d_{ij}t_i - b_{ij}u_i),\quad \varphi'(u_j') = \sum_{i=1}^{g} (-c_{ij}t_i + a_{ij}u_i) \quad (1 \le j \le h).$$

By an easy calculation we obtain

$$\varphi_*\varphi'(t_j') = ct_j,\quad \varphi_*\varphi'(u_j') = cu_j \quad (1 \le j \le h),$$

hence:

3.7.9 Proposition. *The conditions are as in 3.7.6 and the homomorphism* $\varphi'\colon H_1(F_h) \to H_1(F_g)$ *is defined by 3.7.8. Then*

$$\varphi_*\varphi' = c\cdot \mathrm{id}_{H_1(F_h)}.$$

□

3.7.9 connects the results of 3.7.7 and 8. From linear algebra we conclude:

3.7.10 Corollary. *(a) If* $c \neq 0$, *the linear transformation* φ_* *has rank* $2h$.
(b) If $c = 0$ *the rank of* φ_* *is at most* g.

□

Using intersection numbers we may characterize the puzzling mapping φ' by:

3.7.11 Lemma. $\nu(\varphi'(x'),y) = \nu(x',\varphi_*(y))$ for $x' \in H_1(F_h)$, $y \in H_1(F_g)$.
This condition characterizes φ'.

Proof. It suffices to check the formula for the elements of basis of $H_1(F_g)$ and $H_1(F_h)$. We obtain from 3.6.3 and the definitions of the coefficients in 3.7.6 and of φ' in 3.7.8:

$$\nu(t_j',\varphi_*(t_i)) = b_{ij} = \nu(\varphi'(t_j'),t_i) \text{ etc.}$$

□

3.7.12 Corollary. *(a)* $\nu(\varphi'(x'), \varphi'(y')) = c\cdot\nu(x',y')$ *for* $x', y' \in H_1(F_h)$.
(b) If $x = \varphi'(x')$, $y = \varphi'(y')$, $x',y' \in H_1(F_h)$, *then* $c\cdot\nu(x,y) = \nu(\varphi_*(x),\varphi_*(y))$.
In words: the intersection number of two elements of $\varphi'(H_1(F_h))$ *is multiplied by* c *if the mapping* φ_* *is applied.*
(c) If $c \neq 0$ *then* φ_* *maps* $\varphi'(H_1(F_h))$ *injectively into* $H_1(F_h)$; *by 3.7.9 the image consists of* $c\cdot H_1(F_h)$.
(d) If $c \neq 0$ *then* $\varphi_*^{-1}(cH_1(F_h)) = \text{Kern}\,\varphi_* \oplus \varphi'(H_1(F_h))$.

Proof as exercise E 3.32.

□

Now we can mainly calculate the effect of φ_* to the elements of $H_1(F_g)$ and their intersection numbers, since the p-th multiple of any element of $H_1(F_g)$ is contained in $\varphi_*^{-1}(cH_1(F_h))$.

3.7.13 Remark. What was done in 3.7.4-12 only by the use of linear algebra also has an important topological background: the cohomological theory of the mapping degree and the intersection number and the Poincaré duality. The cohomology theory makes the given constructions and conclusions more transparent.

Sufficient conditions for 2g-by-2h matrices to be induced by continuous mappings do not seem to be given in the literature.

Exercises: E 3.29-35.

3.8 REMARKS

The first subjects which were studied from a topological point of view were graphs and 2-complexes. The discovery of the Euler characteristic by Euler (1752) may be considered as the first major result of topology. (This polyhedron formula is already contained in an unpublished fragment of Descartes.) Surfaces became interesting because of the work of [Riemann 1851, 1857]. In systematic studies of [Möbius 1866, 1886], [Jordan 1866], [Schäfli 1872, 1873], [Dyck 1888] the classification of the closed surfaces became known: they are classified by the orientation property and the Euler characteristic. The classification by fundamental groups is from [Reidemeister 1932].

The definition of a surface and combinatorial equivalence for 2-complexes in 3.1 already shows the difficulty of classification problems and shows that invariants are needed to solve them. In dimension 2 the fundamental group, or even the first homology group, is strong enough to do the work. Unfortunately these problems become much more complicated in higher dimensions, and in fact unsolvable from dimension 4 onwards. The proof of this is due to [Markov 1958] and depends on the unsolvability of the isomorphism problem for groups, cf. 2.12.

The classification in dimension 2 can be extended to non-compact surfaces with finitely generated fundamental group. These surfaces can be obtained from compact surfaces by "deleting the boundary". In a sense, all non-compact surfaces can be classified (see [Richards 1963]) but naturally in a less effective way.

Additional literature to chapter 3:
[Brown-Messer 1979], [Edmonds 1979], [Goldman 1971], [Hansen 1974], [Hempel 1972], [Hirsch 1976], [Johansson 1931], [Jucovič-Trenkler 1973], [Luft 1971], [Mennicke 1961], [Pettey 1972].

EXERCISES

E 3.1 Prove that the only subdivisions of the sphere into n-gons, m of which surround each vertex, are those given by the five regular polyhedra.

E 3.2 Which of the following surfaces are related?

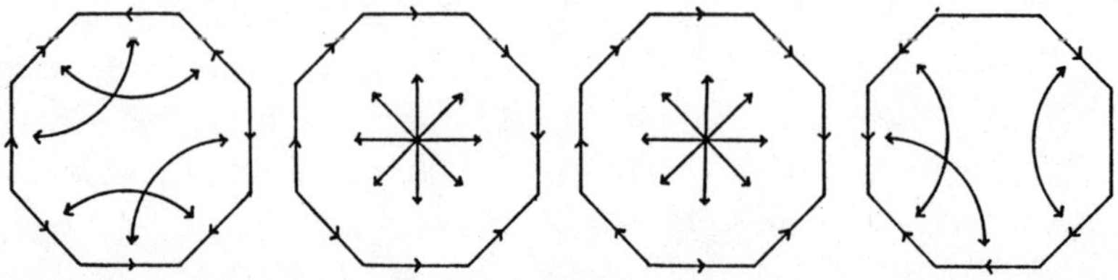

E 3.3 Let $f\colon F' \to F$ be a covering. Prove that if one of F and F' is a surface then the other is too (cf. remark after Definition 2.5.1 (c)).

E 3.4 Prove that a closed surface has a symmetric normal form with boundary path

$$\sigma_1 \ldots \sigma_{-N+2}\, \sigma_1^{-1} \ldots \sigma_{-N+1}^{-1}\, \sigma_{-N+2}^{\varepsilon}$$

$$\text{Here } \varepsilon = \begin{cases} -1 & \text{if the surface is non-orientable,} \\ 1 & \text{otherwise} \end{cases}$$

and N is the Euler characteristic.

E 3.5 Determine which of the following groups are isomorphic to fundamental groups of closed surfaces and determine the type of the surface in such cases:

$$\langle a_1, \ldots, a_n \mid a_1 \ldots a_n a_1^{-1} \ldots a_n^{\varepsilon} \rangle \qquad \varepsilon \in \{-1,1\}$$

E 3.6 Prove that each orientable surface of genus $g \geq 2$ covers the pretzel surface (orientable surface of genus 2), with finite order.

E 3.7 Prove (a) Each non-orientable surface N has a unique orientable double covering F (cf. example 2.2.2).
(b) How are the genera related if N, hence F, is closed?
(c) If N is closed and its genus is greater than 2 then there is only one covering of N by F (up to isomorphism). Formulate the corresponding statement for the fundamental groups of N and F.
(d) What is the situation when N has genus 1 or 2?

E 3.8 Prove that a simple null-homotopic curve on a surface bounds a disc. (Use the Seifert-Van Kampen theorem. Another way to do it will emerge in the next chapter, where the exercise will be set again.)

E 3.9 Prove that surface groups other than that of the projective plane contain no non-trivial elements of finite order.

E 3.10 Let F be an orientable surface. Prove that the homotopy class of a simple closed curve S cannot be a proper power in the fundamental group. (Hint: Consider the cases where S is separating and non-separating differently, finding simple expressions for S in terms of canonical generators.)

E 3.11 Let F be a non-orientable surface and S a simple closed curve on F. Then $[S]$ is a proper power of the homotopy class of another curve β iff β is orientation reversing and S bounds a Möbius strip with center curve β.

E 3.12 Define the rank of a group as in 2.1.11. Find the ranks of the fundamental groups of closed surfaces.

E 3.13 What does the Kneser formula say about rank when the = sign holds? Compare with the Schreier index formula and explain the difference.

E 3.14 Prove that every closed orientable surface is a branched covering of the sphere .

E 3.15 Prove that the mapping degree is well-defined in 3.3.2.

E 3.16 Let F' and F be closed orientable surfaces of genera g' and g and suppose $g' - 1 \geq |c|(g-1)$ for some $c \in \mathbb{Z}$. Prove that there exists a mapping $f\colon F' \to F$ of degree c.

E 3.17 Prove that a surface mapping $f\colon F' \to F$ induces homomorphisms $f_{i*}\colon H_i(F') \to H_i(F)$, and that the image of the generator of $H_2(F') \cong \mathbb{Z}$ is an element of $H_2(F) \cong \mathbb{Z}$ which equals the mapping degree in absolute value.

E 3.18 Let $f\colon T \to T$ be a mapping of the torus T into itself. We fix a basis for the fundamental group, which we then write as $\mathbb{Z} \oplus \mathbb{Z}$, and let f be determined by a 2×2 integer matrix A. Prove that det A is the degree of f.

E 3.19 Write the elements of the group $\mathbb{Z} \oplus \mathbb{Z}$ as columns $\binom{x}{y}$ with $x, y \in \mathbb{Z}$. The integer matrices with determinant ± 1 form a group under multiplication, denoted by $GL(2,\mathbb{Z}) = Aut(\mathbb{Z} \oplus \mathbb{Z})$. It contains the subgroup $SL(2,\mathbb{Z})$ of matrices with determinant 1. Prove the following statements:

(a) $SL(2,\mathbb{Z})$ is generated by the matrices

$$A = \begin{pmatrix} 0 & -1 \\ 1 & 0 \end{pmatrix} \text{ and } B = \begin{pmatrix} 0 & 1 \\ -1 & 1 \end{pmatrix}$$

A system of generators for $GL(2,\mathbb{Z})$ is obtained if we add

$$C = \begin{pmatrix} 0 & 1 \\ 1 & 0 \end{pmatrix}.$$

(b)* $SL(2,\mathbb{Z}) = \langle A,B \mid A^2B^{-3}, A^4 \rangle = \langle A \mid A^4 \rangle *_{\mathbb{Z}_2} \langle B \mid B^6 \rangle$ where $\mathbb{Z}_2 = \langle A^2 \rangle = \langle B^3 \rangle$ is the group generated by A^2,

$$GL(2,\mathbb{Z}) = \langle A,B,C \mid A^2B^{-3}, A^4, C^2(CA)^2, (CB)^2 \rangle$$
$$= \langle A,C \mid A^4, C^2, (CA)^2 \rangle *_{\mathbb{D}_2} \langle B,C \mid B^6, C^2, (CB)^2 \rangle = \mathbb{D}_2 *_{\mathbb{D}_2} \mathbb{D}_6$$

where $\mathbb{D}_2 = \langle C,A^2 \rangle = \langle C,B^3 \rangle = \mathbb{Z}_2 \oplus \mathbb{Z}_2$, $\mathbb{D}_4$ and $\mathbb{D}_6$ are the dihedral groups of orders 4, 8 and 12.

(c) Considering $\mathbb{Z} \oplus \mathbb{Z}$ as the fundamental group of the torus, prove that each automorphism of $\mathbb{Z} \oplus \mathbb{Z}$ is induced by a homeomorphism.

E 3.20 Prove 3.4.3.

E 3.21 Prove Proposition 3.21.

E 3.22 Let T be a torus and s,t a canonical system of generators for the fundamental group. Prove

(a) If $a,b \in \mathbb{Z}$, $(a,b) = 1$ then the class of $s^a t^b$ contains a simple closed curve.

(b) If $a,b,c,d \in \mathbb{Z}$ and $ad-bc = \pm 1$ then there is a canonical pair of curves with the homotopy classes of $s^a t^b$ and $s^c t^d$.

(c) Converse to (b).

E 3.23 (a) Give examples of simple closed curves γ_1, γ_2 on an oriented closed surface of genus $g \geq 2$ that are not homotopic, but for an arbitrary curve δ satisfy

$$\nu(\gamma_1,\delta) = \nu(\gamma_2,\delta).$$

(b) Show that γ_1 and γ_2 are homologous.

E 3.24 Fill in the details to 3.6.1.

E 3.25 Do the calculations explicitly to obtain the formula 3.6.5.

E 3.26 Describe homeomorphisms of the surface from 3.6.7 that induce the matrices (A - D). In case (A) you may restrict yourself to the case where $B_2,\ldots,B_g$ are unit matrices and B_1 is $\begin{matrix} 0 & 1 \\ -1 & 0 \end{matrix}$ or $\begin{matrix} 0 & 1 \\ -1 & 1 \end{matrix}$, see E 3.19 (a).

E 3.27 Give the details for 3.6.10.

E 3.28 Prove 3.6.7 (b) also for $m > 0$.

E 3.29 Prove: A continuous mapping $f: S^2 \to S^2$ of degree 1 or -1 is homotopic to a homeomorphism, in fact, either to the identity - if the degree is 1 - or to the reflection at the equator: $(x,y,z) \to (x,y,-z)$, if the degree is -1. (Hence S^2 has been identified with $\{(x,y,z) \in \mathbb{R}^3 \mid x^2 + y^2 + z^2 = 1\}$.)

E 3.30 Prove 3.7.3 with tools of algebraic topology.

E 3.31 Proof of 3.7.7.

E 3.32 Proof of 3.7.8.

E 3.33 Give the cohomological interpretation of section 3.7.

E 3.34 Prove that there is no mapping $F_g \to F_h$, $g < h$, of non-zero degree.

E 3.35 Discuss $T: [F_g,F_o] \to [F_g,F_o]$, see 3.7.3.

4. PLANAR DISCONTINUOUS GROUPS

In this chapter we consider finitely generated groups that act on planar complexes, construct their canonical fundamental domains and presentations, and classify them with respect to geometric equivalence and isomorphism. Then we show that each group with a presentation in one of the canonical forms can be realized as a group acting on a planar complex. The interest in those groups arises from complex analysis where they occur as discontinuous groups of motions of the non-euclidean plane, often called Fuchsian groups; we will deal with this side in chapter 6. Our treatment here is purely combinatorial and we prove the group theoretic properties of these groups using only combinatorial arguments.

4.1 PLANAR NETS

In this section we classify the plane among surfaces and prove combinatorial versions of the Jordan curve theorem and the Schönflies theorem. For the famous topological versions of these theorems and some remarks on their history, see [Newman 1951] and chapter 7.

4.1.1 Definition. Let F be a surface. We say that two faces ϕ, ϕ' are *connected* by faces when there is a sequence $\phi_1 = \phi, \phi_2, \ldots, \phi_n = \phi'$ of faces in which ϕ_i and ϕ_{i+1} have at least one boundary edge in common. F is called *2-connected* if any two faces of F can be connected by faces. A curve ω is called *separating* when there exist faces ϕ, ϕ' of F which cannot be connected by a sequence of faces where consecutive have a common edge not from ω in the boundary.

4.1.2 Lemma. *Let* F *be a surface and* ω *a null-homologous (see 2.7.1) simple closed curve in the interior of* F. *Then* F *is separated by* ω.

Proof. Let $\omega = \sigma_1 \ldots \sigma_n$, and let $w = s_1 + \ldots + s_n$ be the associated chain of C_1 (see 2.7). Since ω is simple-closed all the s_i are distinct and since ω is null-homologous there is a chain $f \in C_2$ with $\partial_2 f = w$. Let $f = \Sigma a_i f_i$ where the f_i correspond to the faces ϕ_i of F. Naturally only finitely many of the a_i are non-zero. If two faces ϕ_i, ϕ_j have a common edge which does not appear in ω, then since $\partial f = \partial(\Sigma a_i f_i) = w$ we have the equation $a_i = -\varepsilon a_j$, where the common edge appears in the boundary of ϕ_i with exponent + 1 and in the boundary of ϕ_j with exponent ε. If ω does not separate, one can connect any two faces without "crossing" ω. But that means that all the coefficients a_i have the same absolute value, so that the coefficients of the s_i in w are divisible by 2. But this contradicts the fact that $w = s_1 + \ldots + s_n$ and the s_i are distinct. □

4.1.3 Definition. By a *line* we mean a connected infinite graph in which two edges emanate from each vertex. A surface equivalent to a complex with one geometric face $\phi^{\pm 1}$, one boundary geometric edge $\sigma^{\pm 1}$ and one vertex is called a *disk*. A *planar net* is a connected surface $\mathbb{E}$ with the following properties:

(a) $\mathbb{E}$ is open, i.e. $\mathbb{E}$ has infinitely many faces and no boundary.
(b) Each vertex has a finite star.
(c) Each simple closed path bounds a disk.

It follows immediately from (c) that $\pi_1(\mathbb{E}) = 1$ and $H_1(\mathbb{E}) = 0$. The converse also holds:

4.1.4 Lemma. *Let* $\mathbb{E}$ *be a connected open surface with finite stars. If* $H_1(\mathbb{E}) = 0$ *then* $\mathbb{E}$ *is a planar net.*

Proof. Let ω be a simple closed curve, w the corresponding element of C_1. Then there are finitely many faces ϕ_i such that $\partial(\Sigma a_i f_i) = w$, with $a_i \neq 0$, for the associated f_i. Since $\mathbb{E}$ is infinite and ω is simple, the $a_i = \pm 1$. The ϕ_i constitute a connected surface with the one boundary curve ω. One easily reaches the conclusion that the homology of this complex in dimension 1 is trivial. Thus by corollary 3.2.10 it constitutes a surface of genus 0 with boundary curve ω. It is therefore a disk.

□

4.1.5 Lemma. *A line in a planar net separates it.*

Proof. If not, consider two faces which meet along a segment σ of the line, and take two vertices P_1, P_2 (possibly after subdivision) which lie on their boundaries but not on the line. We connect them by a simple curve ω which does not meet the line. Further, after subdividing the faces by two edges σ_1, σ_2 where σ_1 goes from P_1 to a point Q on σ, and σ_2 goes from Q to P_2, we add σ_1 and σ_2 to ω. Then $\sigma_1\sigma_2\omega$ bounds a (finite!) disk containing one (infinite!) half of the line.

□

4.1.6 Lemma. *If* $\mathbb{E}$ *is a planar net and* $\mathbb{E}'$ *is related to* $\mathbb{E}$, *then* $\mathbb{E}'$ *is a planar net.*

Since $\mathbb{E}'$ is related to $\mathbb{E}$, $\mathbb{E}'$ is open and the stars of $\mathbb{E}'$ are finite. In addition, $H_1(\mathbb{E}') = 0$, so the lemma follows from lemma 4.1.4.

□

We recall that two nets $\mathbb{E}$ and $\mathbb{E}'$ are called isomorphic when there is a one-to-one mapping $f\colon \mathbb{E} \to \mathbb{E}'$ which maps faces to faces, edges to edges and vertices to vertices, in such a way that the boundary relations are preserved, and each part of $\mathbb{E}'$

has a pre-image.

4.1.7 Theorem. *Up to isomorphism, any two planar nets are related.*

Proof. Let $\mathbb{E}$ be a planar net. By subdividing segments, $\mathbb{E}$ can be converted into a related net in which no segment is closed. This net will also be denoted by $\mathbb{E}$. Let ϕ_o be a face of $\mathbb{E}$ with boundary ω_o. The 2-complex K containing all faces, and their boundaries, which have an edge or a vertex in common with ϕ_o is a finite complex, since at most finitely many faces meet at each vertex. Consider all the simple closed curves on its boundary. Each bounds a disk. If we fill in all disks which contain no face of the complex K in their interior, then only a single simple closed boundary curve ω_1 remains, since ϕ_o can be connected with each face of the new complex without crossing a boundary curve of this complex. Let the disk bounded by ω_1 be ϕ_1. By repeating this process with ϕ_1, and continuing, we obtain a system of disks $\phi_o \subset \phi_1 \subset \ldots$ of which the ith contains all the preceding in its interior as well as all edges emanating from ϕ_{i-1}. Since $\mathbb{E}$ is connected it is clear that $\bigcup_{i=1}^{\infty} \phi_i$ contains all edges and hence equals $\mathbb{E}$. Furthermore, we have a system of simple closed paths $\omega_o, \omega_1, \omega_2, \ldots$ which do not meet, and the ith of which contains all the preceding ones in the interior of its disk. ϕ_i is related to a disk; thus if one removes the disk ϕ_i from ϕ_{i+1}, $\phi_{i+1} - \phi_i$ is related to a surface of genus 0 with two boundaries. Thus we can convert $\phi_{i+1} - \phi_i$ into the complex shown in Figure A by elementary transformations, and by induction $\mathbb{E}$ has the form given in Figure B.

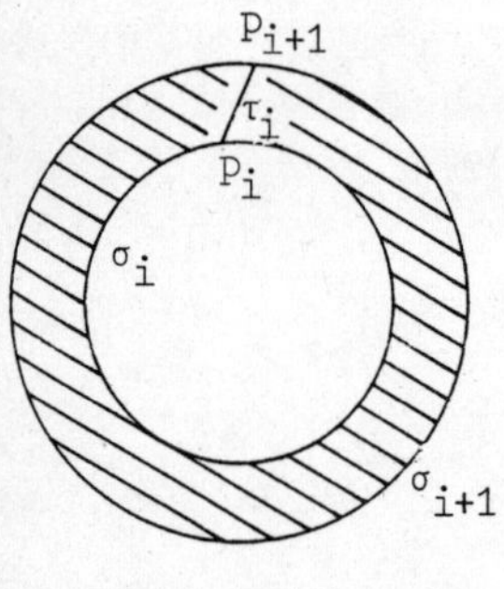

Figure A

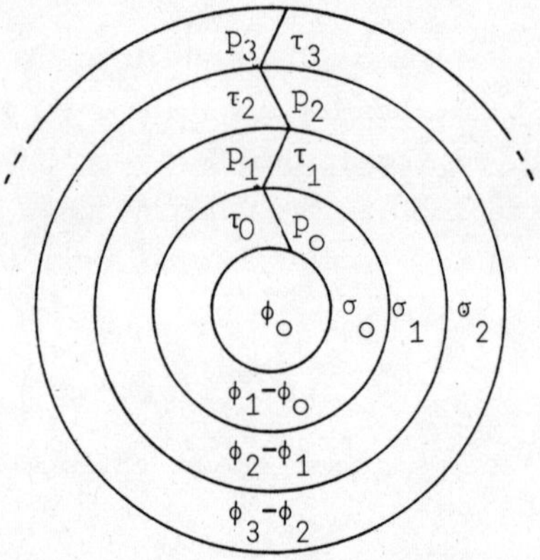

Figure B

We make the analogous construction with $\mathbb{E}'$ and obtain disks $\phi'_o, \phi'_1, \ldots$ with boundaries $\sigma'_o, \sigma'_1, \ldots$, connecting edges $\tau'_o, \tau'_1, \ldots$ and vertices $P'_o, P'_1, \ldots$. The mapping from unprimed elements to primed elements is then an isomorphism between planar nets.

□

4.1.8 Definition. By a *subcomplex* of a planar net we mean a set of geometric faces together with their boundary curves and vertices. In what follows it will be called *connected* when any two faces may be connected by faces with common edges. (This definition differs from the old definition, which said that a 2-complex is connected when the underlying graph is connected.) The subcomplex is called *simply-connected* when it is connected and each simple closed curve bounds a disk.

4.1.9 Lemma.

(a) *A simply connected non-empty subcomplex* K *of a planar net* $\mathbb{E}$ *is either* $\mathbb{E}$ *itself, a disk, or it has a number of simple open "infinitely long" boundary curves which do not intersect each other.*

(b) *A subcomplex* K *of* $\mathbb{E}$ *which has no edges that are in the boundary of only one face of* K *is either empty or equals* $\mathbb{E}$.

The proof is Exercise E 4.1

□

In particular, a closed boundary curve gives a disk, an infinite boundary curve a "half-plane", two boundary curves a "strip" etc..

4.2 AUTOMORPHISMS OF A PLANAR NET

We study the elementary properties of automorphisms of planar nets.

4.2.1 Definition. An *automorphism* of a planar net $\mathbb{E}$ is an isomorphism of $\mathbb{E}$ onto itself.

As we have seen in the proof of Theorem 4.1.7, a planar net is related to an ascending sequence of disks. One sees immediately from this construction that each planar net is orientable. We choose a fixed orientation, so that a positive member and a positive boundary path for each disk is established once and for all.

4.2.2 Definition. If an automorphism maps one positive boundary path onto a positive boundary path, then all positive boundary paths are mapped to positive boundary paths, and we call the automorphism *orientation-preserving*. Otherwise we call it *orientation-reversing*.

4.2.3 Lemma. *An orientation-preserving automorphism which leaves an edge fixed is the identity.*

Proof. Let x be an orientation-preserving automorphism with fixed edge σ. Let ϕ_1 and ϕ_2 be the two faces which have σ in their boundary paths. Since boundary relations are preserved, ϕ_1 can only be mapped onto ϕ_1 or ϕ_2. The second possibility is excluded since x is orientation preserving. Thus ϕ_1 and ϕ_2 are fixed faces. At most, x could cyclically move the boundary paths of ϕ_1 and ϕ_2; but that does not happen since σ is fixed. Thus the boundary paths of ϕ_1 and ϕ_2 are fixed, and the lemma follows by extending the argument over the whole net.

□

4.2.4 Lemma. *An automorphism of infinite order has neither fixed vertices nor fixed faces.*

Proof. Suppose that x leaves the vertex P fixed and let σ be an edge of the star of P. Since each star is finite there is a power n of x which maps σ on to itself. Then x^{2n} is an orientation preserving automorphism with fixed edge σ, hence the identity. If x leaves a face ϕ fixed, then some power of x leaves a boundary edge of ϕ fixed, and hence is the identity.

□

4.2.5 Theorem. *An orientation preserving automorphism of finite order (unequal to the identity) has exactly one fixed vertex or fixed face. (We can regard such automorphisms as rotations.)*

Proof. a) Let the automorphism x fix vertices P and Q, connected by a path ω of minimal length. Then each intersection of ω and ωx is a fixed vertex of x. Thus we can assume that there are two fixed vertices P and Q of x connected by a path ω such that $\omega(\omega x)^{-1}$ is a simple closed path. This bounds a disk, and the images of the disk under all the powers of x constitute a finite (!) subcomplex without boundary, in contradiction to lemma 4.1.9.

When we subdivide the fixed faces by connecting midpoints to vertices and extend the automorphism x in the natural way to the new net, we see that each fixed face corresponds to a fixed vertex in the new net. Thus the uniqueness of the fixed element is established. It follows also that a fixed vertex or face for a non-trivial power of x is also fixed by x. For this reason we need only carry out the existence proof for powers of x, without mentioning how it extends to x.

b) Let n be the order of x, P a vertex and ω a path of minimal length from x to Px. The path $\omega(\omega x)(\omega x^2) \ldots (\omega x^{n-1})$ is closed. If the path $\omega(\omega x)(\omega x^2)\ldots(\omega x^{n-1})$ is not simple and Q is an intersection of ω and ωx^j $(0 < j < n)$ (excepting the vertices Px and P for $j = 1$ and $n - 1$) then we consider the element x^j instead of x, the vertex Q instead of the vertex P, and the part of ωx^j which runs from Q to Qx^j

instead of ω. This path is shorter than ω or else we would already have a fixed vertex for a power of x, so after a finite number of such steps we find either a fixed vertex or a simple closed path $\omega'(\omega'x^{i})(\omega'x^{2i})\ldots$, which bounds a disk S.

In the latter case, let P_1 be a vertex in the interior of S and ω_1 a path of minimal length from P_1 to P_1x in the interior of S. As above we find a path ω_1' and a power x^{ij} ($ij \not\equiv 0 \bmod n$) such that $\omega_1'(\omega_1'x^{ij})(\omega_1'x^{2ij})\ldots$ is the boundary of a disk S_1. This lies wholly in the interior of S and is mapped on to itself by x^{ij}. This process terminates with a disk having no interior vertex, which is therefore a fixed face, or else we arrive at a fixed vertex.

□

4.2.6 Lemma. *An orientation-reversing automorphism of finite order has order 2.*

Proof. If x is an orientation-reversing automorphism of finite order, then x^2 is orientation-preserving, of finite order, and thus has a fixed vertex P (possibly after subdivision of a fixed face). If x^2 were not equal to 1 as claimed, then because of $(Px)x^2 = (Px^2)x = Px$ and Theorem 4.2.5, P would be a fixed vertex of x also. Now x maps the star of P into itself with a reversal of orientation. Thus if x maps the edge σ of the star of P to σx, the right neighbour of σ is mapped to the left neighbour of σx. If we proceed to the next neighbour on the right, and continue, we eventually find either a pair of successive edges in the star of P exchanged by x, or else a fixed edge for x. In either case x^2 has a fixed edge, so by lemma 4.2.3 it is the identity.

□

We shall now show that an orientation-reversing automorphism x of order 2 is a *"reflection"*. Let P be a vertex of the net and ω a path from P to Px. Since $Px^2 = P$, ωx connects Px with P. By replacing P by intersection points of ω and ωx we can reach the conclusion that either $\omega \cdot \omega x$ is a simple closed path, or P is a fixed vertex, or ω is an edge with $\omega x = \omega^{-1}$. The first case cannot occur.

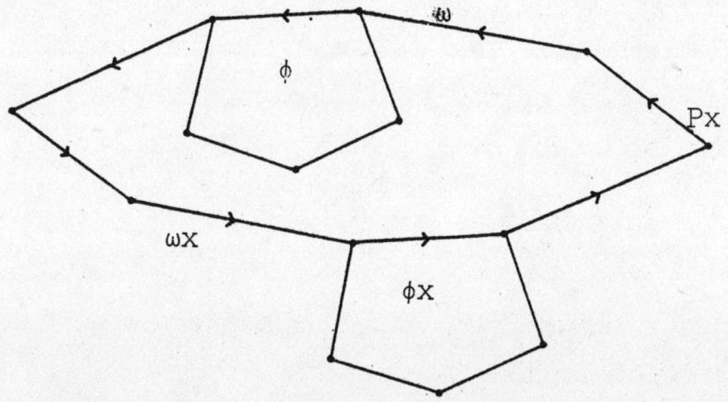

Namely, let ϕ be a face of the disk bounded by $\omega \cdot \omega x$ which meets this boundary. Then ϕx lies outside the disk, since x reverses orientation. Thus the disk together with its image constitutes a finite closed subcomplex of the plane net, which is impossible by 4.1.9.

Therefore, each orientation-reversing automorphism of order 2 either leaves a vertex fixed or carries an edge into its inverse. By subdividing the edge we can conclude that x has a fixed vertex Q. In the star of Q there are either two fixed edges, or a fixed edge and a pair of successive edges exchanged by x, or two such pairs. Let σ_1, σ_2 be a pair exchanged by x. Then $\sigma_1^{-1}\sigma_2$ is in the boundary of a face which is mapped on to itself. In its boundary there is either a second pair which is exchanged, or an edge which is mapped into its inverse. In either case we find, subdividing an edge if necessary, another fixed vertex Q_1 in the boundary. We subdivide the face by a fixed edge from Q to Q_1. We can therefore assume that there are at least two fixed edges in the star of Q, and no pair of successive edges which are exchanged. Let σ_1 and σ_2 be two fixed edges. Since one face meeting a σ_i is mapped on to the other, each edge of the star of Q which lies between σ_1 and σ_2 is mapped onto an edge between σ_2 and σ_1. Thus only σ_1, Q, σ_2 are fixed in the star of Q.

The endpoints of σ_1, σ_2 likewise are fixed, and, by making further subdivisions if necessary, we obtain two further fixed edges. We continue this process and obtain a curve. It has no double points, since at most two fixed edges emanate from each point, and it bounds no disk, otherwise the "finite interior" would be mapped on to the "infinite exterior". It is therefore a *fixed line*, which therefore separates the planar net, and the two halves are exchanged by x. Thus there are no other fixed elements. This proves:

4.2.7 Theorem. *Each orientation-reversing automorphism* x *of finite order has exactly one fixed line, after suitable subdivision. One may regard* x *as a reflection in this line.*

□

Exercise: E 4.2.

4.3 AUTOMORPHISM GROUPS OF PLANAR NETS

For groups of automorphisms of planar nets we show that the "quotient space" is a surface and we construct a special complex on it. In the next section we lift this complex to the plane and introduce the notion of a fundamental domain.

Let $\mathbb{E}$ be a planar net, G a group of automorphisms of $\mathbb{E}$.

4.3.1 Definition. Two vertices, edges or faces of $\mathbb{E}$ are called *equivalent* when there is an automorphism in G which carries one object into the other.

4.3.2 Assumption. We shall assume that, *under an automorphism different from 1, no face is mapped to itself or its inverse, and no edge is mapped to its inverse.* Thus a rotation in G takes place about a vertex in $\mathbb{E}$, and a reflection in G has a full axis of reflection. We can always reach this state of affairs by suitable subdivision. An extension (subdivision of an edge by a vertex or of a face by an edge) is reproduced at all places equivalent under G , and G is extended to the new net $\mathbb{E}'$. Reduction is defined as the converse process, assuming that the new pair $(\mathbb{E}',G')$ has no fixed face etc.

4.3.3 Definition. Two pairs $(\mathbb{E},G)$, $(\mathbb{E}',G')$ are called *equivalent* when one can be converted into the other by a chain of extensions and reductions. Rotation centers and lines of reflection are always preserved by the above processes. A class of equivalent pairs is called a *planar discontinuous group*. However, we shall also use this expression for a single representative.

If we now identify equivalent vertices, edges and faces respectively and retain the boundary relations, we obtain a system $\mathbb{E}/G$ of vertices, edges and faces, the so-called *quotient space*.

4.3.4 Theorem. $\mathbb{E}/G$ *is a surface.*

Proof. We denote classes of equivalent vertices, edges or faces by PG, σG etc. Now if two segments are equivalent, so are their initial and final points. Further, σG and $\sigma^{-1}G$ are different, and $\sigma^{-1}G$ can be viewed as $(\sigma G)^{-1}$. Thus we obtain a graph C/G from the graph C of $\mathbb{E}$. Since no faces are fixed under G, ϕG and $\phi^{-1}G$ are distinct for each face ϕ, and no σG appears in the boundaries of faces ϕG more than twice.

The star condition is also satisfied. Let P be a vertex of $\mathbb{E}$ and let $\sigma_1,\ldots,\sigma_n,\sigma_1$ be the star around P. If P lies on lines of reflection, then there is a substar σ_j, $\sigma_{j+1},\ldots,\sigma_k$ of non-equivalent edges, of which σ_j and σ_k lie on lines of reflection (naturally the remainder do not). Then $\sigma_jG,\ldots,\sigma_kG$ is the star around PG. If P does not lie on a line of reflection, then there is a substar $\sigma_1,\ldots,\sigma_j$ where σ_1 and σ_j are equivalent but the others are not, and $\sigma_1G,\ldots,\sigma_jG$ is the star around PG.

□

There is an obvious mapping $\mathbb{E} \to \mathbb{E}/G$; we say that an object of E lies "over" its image. The boundary of $\mathbb{E}/G$ consists of the images of edges on the lines of reflection. An elementary transformation of $\mathbb{E}/G$ can be lifted to $\mathbb{E}$ when the images of rotation centers remain fixed. In $\mathbb{E}/G$ there are images of two different kinds of rotation centers, namely those which lie on lines of reflection and those which do not. The images of the first kind occur on the boundary of $\mathbb{E}/G$, while those of the second kind do not. (If a vertex P of $\mathbb{E}$ lies on two different lines of reflection, and $c_1, c_2 \in G$ are the corresponding reflections, then c_1c_2 is a rotation about P.) Images of rotation centers are likewise called rotation centers.

4.3.5 Assumption. We now assume that there are only finitely many faces of $\mathbb{E}$ inequivalent with respect to G, and say that the group has *compact fundamental domain*.

Then $\mathbb{E}/G = F$ is a finite complex (a compact surface) and we can apply our classification to it. However, we must now note the appearance of distinguished vertices, the rotation centers. As vertices of $\mathbb{E}/G$ we take the images of rotation centers, plus an extra vertex on each boundary curve and an interior vertex $\bar{Q}$ distinct from them all. Edges $\bar{\sigma}_1,\dots,\bar{\sigma}_m$ lead from $\bar{Q}$ to the rotation centers which are not on the boundary.
In addition there is an edge $\bar{\eta}_k$ to each boundary curve ending in the vertex which is not a rotation center. The kth boundary curve is equal to a path $\bar{\gamma}_{k,1}\cdots\bar{\gamma}_{k,m_k+1}$ $(m_k \geq 0)$ where $\bar{\gamma}_{k,1}$ begins at the final point of $\bar{\eta}_k$ and $\bar{\gamma}_{k,m_k+1}$ ends there. For $i = 1,\dots,m_k$ the final point of $\bar{\gamma}_{k,i}$ coincides with the initial point of $\bar{\gamma}_{k,i+1}$ and is a rotation center (if $m_k \geq 1$). By extending the curves already obtained by cuts $\bar{\tau}_1, \bar{\mu}_1,\dots,\bar{\tau}_g,\bar{\mu}_g$ or $\bar{\nu}_1,\dots,\bar{\nu}_g$ respectively, beginning and ending at Q, $\mathbb{E}/G$ becomes simply connected. We then have a single face for $\mathbb{E}/G$ which we can give the boundary

$$\prod_{i=1}^{m} \bar{\sigma}_i\bar{\sigma}_i^{-1} \prod_{j=1}^{g} \bar{\tau}_j\bar{\mu}_j\bar{\tau}_j^{-1}\bar{\mu}_j^{-1} \prod_{j=1}^{q} \bar{\eta}_k\bar{\gamma}_{k,1}\cdots\bar{\gamma}_{k,m_k+1}\bar{\eta}^{-1} \quad \text{or}$$

$$\prod_{i=1}^{g} \bar{\sigma}_i\bar{\sigma}_i^{-1} \prod_{j=1}^{g} \bar{\nu}_j\bar{\nu}_j \prod_{k=1}^{q} \bar{\eta}_k\bar{\gamma}_{k,1}\cdots\bar{\gamma}_{k,m_k+1}\bar{\eta}_k^{-1} \quad \text{respectively, the same way as in 3.2.}$$

All the processes necessary to obtain this *canonical normal form* may be lifted to $\mathbb{E}$ and we obtain

4.3.6 Theorem. *A planar discontinuous group with compact fundamental domain may be realized by a pair* $(\mathbb{E},G)$ *in which any two faces are equivalent, only the identity mapping from G leaves a face fixed, and the boundary path of a face has the form:*

4.3.7 $$\prod_{i=1}^{m} \sigma_i'\sigma_i^{-1} \prod_{j=1}^{g} \tau_j'\mu_j^{-1}\tau_j^{-1}\mu_j' \prod_{k=1}^{q} \eta_k'\gamma_{k,1}\cdots\gamma_{k,m_k+1}\eta_k^{-1}$$

or

4.3.8 $$\prod_{i=1}^{m} \sigma_i'\sigma_i^{-1} \prod_{j=1}^{g} \nu_j\nu_j' \prod_{k=1}^{q} \eta_k'\gamma_{k,1}\cdots\gamma_{k,m_k+1}\eta_k^{-1} \quad \textit{respectively.}$$

Here edges denoted by the same Greek letter and index (e.g. σ_i' *and* σ_i*) are equivalent.*

The endpoint of $\sigma_i^!$ *is a rotation center (of order* $h_i \geq 2$*). If* $m_k > 0$ *then the final points of* $\gamma_{k,i}$ $(1 \leq i \leq m_k)$ *are likewise rotation centers (of orders* $h_{k,i} \geq 2$*). If* $m_k = 0$ *the initial vertex of* $\gamma_{k,1}$ *is not a rotation centre. Any two of these rotation centers and any two segments not denoted by the same Greek letter are inequivalent.*

The path 4.3.7 or 4.3.8 is simple-closed.

□

The proof of the latter assertion will occur incidentally in what follows.

Exercises: E 4.3,4.

4.4 FUNDAMENTAL DOMAINS

4.4.1 Definition. A connected subcomplex of $\mathbb{E}$ containing exactly one face from each equivalence class, together with their boundaries, is called a *fundamental domain* of $(\mathbb{E},G)$.

4.4.2 Theorem. *Each pair* $(\mathbb{E},G)$ *has a fundamental domain, and fundamental domains are simply connected.*

Proof. a) Let ϕ be a face. By proceeding from ϕ to other faces in succession, each having a boundary edge in common with the complex K already constructed, but inequivalent to any face of K, we obtain a connected subcomplex F. Then any face ϕ' inequivalent to the faces of F can only meet faces inequivalent to those of F. Since ϕ and ϕ' may be connected by faces, such a ϕ' does not exist.

b) Let ω be a simple closed curve in a fundamental domain F which bounds a disk S not lying wholly in F. Then at least one face ϕ of this disk S does not belong to F. Let ϕ' be the face in F equivalent to ϕ and x the automorphism which maps ϕ' to ϕ. Now ϕ is in Fx, and Fx is connected. Since F and Fx have at most boundary edges in common, and each edge from $\omega = \partial S$ is in F, Fx, and hence ωx, is contained in S and we have $Sx \subsetneqq S$. This contradicts the fact that S and Sx contain the same, finite, number of faces.

□

It follows likewise that two images of a fundamental domain have at most a simple, non-closed (possibly infinite) path in common. If F is finite, then F is a disk. In particular, its boundary is simple and closed. One obtains $\mathbb{E}/G$ from F by identifying equivalent boundary edges. The boundary path of F uniquely determines

$\mathbb{E}/G$ (as a surface). We provide the boundary path with an orientation.

4.4.3 Theorem. *At most two directed edges from the same equivalence class appear in the boundary path of a fundamental domain. If there are two, then they are equivalent under an orientation reversing automorphism, and no element from the class of their inverse edges appears. If only one edge* σ *appears from an equivalence class, then there is an edge equivalent to* σ^{-1} *in the boundary path, or else* σ *is a fixed edge.*

□

The proof is Exercise E 4.5.

By means of elementary transformations, carried out simultaneously in all Fx for $x \in G$, we can reach the situation where F is a face with boundary path 4.3.7 or 4.3.8. Among other things, it follows that this path is simple and closed.

4.5 THE ALGEBRAIC STRUCTURE OF PLANAR DISCONTINUOUS GROUPS

The algebraic structure of a planar discontinuous group with compact fundamental domain may be easily derived from the "dual net".

4.5.1 Definition. The *surface* F^* *dual to a surface* F without boundary is defined in the following way:

(a) Each geometric face $\phi^{\pm 1}$ of F corresponds to a vertex ϕ^* of F^*.

(b) Each geometric edge $\sigma^{\pm 1}$ of F corresponds to a geometric edge $\sigma^{*\pm 1}$ of F^*.

(c) Each vertex P of F corresponds to a geometric face $P^{*\pm 1}$ of F^*.

(d) The boundary relations are carried over. This means in particular: if P is a boundary vertex of σ, then σ^* is a boundary edge of the face P^* (if $\sigma_1,\ldots,\sigma_n,\sigma_1$ is the star of P in F, then $\sigma_1^*\ldots\sigma_n^*$ is a positive boundary path of the pair $P^{*\pm 1}$, the oppositely oriented star of P corresponds to a positive boundary path of the oppositely oriented face). If σ is in the positive boundary path of ϕ, then ϕ^* is the initial point of σ^* and the final point of $(\sigma^{-1})^* = \sigma^{*-1}$. If P is in the boundary of ϕ, then ϕ^* is in the boundary of P^*.

4.5.2 Lemma. *The dual complex* F^* *of a surface* F *without boundary can be identified with a complex of the same surface.*

Proof (see [Reidemeister 1932, pp. 133-136]). We first construct a subdivision of F; this is described intuitively rather than formally. Each edge σ is divided by a new vertex σ° into two edges σ', σ''. Then in each face ϕ we take a vertex ϕ^* and connect ϕ^* with all the new vertices σ° which belong to edges in the boundary of ϕ. The complex F' obtained is equivalent to F.

Let P be a vertex of F. Then there are as many faces containing P in their boundary as the star of P in F contains edges. We join all the faces of F' which have P in its boundary into one face P^*. Here we loose all edges of F' with P in its boundary, as well as P itself. A vertex of the form σ° meets exactly two edges, say from ϕ_i^* and ϕ_j^* in the faces ϕ_i, ϕ_j which have σ° in their boundary. We drop σ° and combine these two edges into an edge σ^*. The construction is clear from the figure.

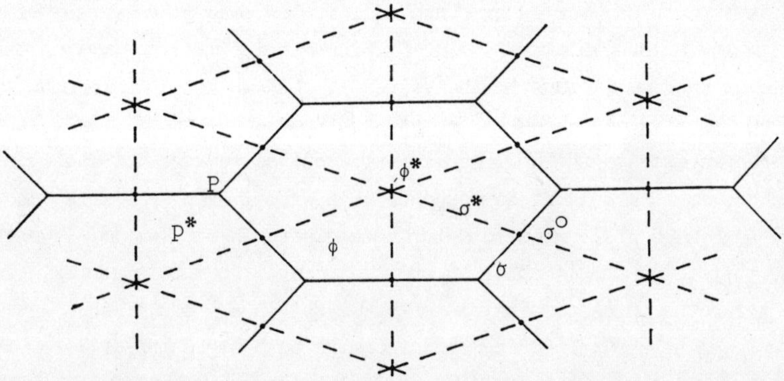

□

We now assume that the group G acts on $\mathbb{E}$ simply transitively on the faces, in particular, that G has a compact fundamental domain. The group G acts on $\mathbb{E}^*$, but now simply transitively on vertices and perhaps rotationally on faces. At each vertex we label the edges originating there with symbols $S_i^{\pm 1}, T_j^{\pm 1}, U_i^{\pm 1}, V_j^{\pm 1}, E_k^{\pm 1}, C_{k,j}$ according to the following rule, where the upper line contains the symbol from 4.3.7 or 4.3.8 and the lower line contains the symbol for the corresponding dual edge:

$$\sigma_i' \;\; \sigma_i^{-1} \;\; \tau_j . \mu_j^{-1} \;\; \tau_j^{-1} \;\; \mu_j' \;\; \nu_j \;\; \nu_j' \;\; \eta_k' \;\; \gamma_{k,j} \;\; \eta_k^{-1}$$

$$S_i \;\; S_i^{-1} \;\; T_j \;\; U_j^{-1} \;\; T_j^{-1} \;\; U_j \;\; V_j \;\; V_j^{-1} \;\; E_k \;\; C_{k,j} \;\; E_k^{-1} .$$

The star of edges emanating from a vertex then reads:

4.5.4 $$S_1, S_1^{-1}, \ldots, S_m, S_m^{-1}, T_1, U_1^{-1}, T_1^{-1}, U_1, \ldots, T_g, U_g^{-1}, T_g^{-1}, U_g,$$

$$E_1, C_{1,1}, \ldots, C_{1,m_1+1}, E_1^{-1}, \ldots, E_q, C_{q,1}, \ldots, C_{q,m_q+1}, E_q^{-1}$$

or

4.5.5 $S_1, S_1^{-1}, \ldots, S_m, S_m^{-1}, V_1, V_1^{-1}, \ldots, V_g, V_g^{-1}$,

$E_1, C_{1,1}, \ldots, C_{1,m_1+1}, E_1^{-1}, \ldots, E_q, C_{q,1}, \ldots, C_{q,m_q+1}, E_q^{-1}$

respectively.

In what follows we use X to denote a "general" symbol.

A pair of oppositely directed edges from $\mathbb{E}^*$ receives either inverse symbols or else the single symbol $C_{k,j}$, because G acts simply transitively on the vertices of $\mathbb{E}^*$ and has no fixed edges other than the $\gamma_{k,i}$. Therefore, only edges denoted by the same symbol (including exponent) are equivalent.

We now distinguish one vertex of $\mathbb{E}^*$ and call it 1. Every other vertex of $\mathbb{E}^*$ is associated with the unique automorphism of G which carries it into 1. When the edge labelled X emanating from 1 leads to the vertex x, then we let X correspond to the element x of G. The edge labelled X^{-1} emanating from 1 then leads from 1 to x^{-1}, and thus X^{-1} corresponds to the inverse element. Equivalent edges of $\mathbb{E}^*$ obtain the same symbol. Thus each path in $\mathbb{E}^*$ corresponds to a word in the symbols X, and conversely, each such word W(X) uniquely determines a path when the initial vertex is given. We say that this path is obtained by tracing W(X) from this vertex. If we now always write X for the corresponding element x we obtain a product W(x) of elements of G, hence an element of G. The final point of the path beginning at 1 for the word W(X) is the vertex W(x); namely, if $W(X) = X_1 \ldots X_n$, then $x_1 \in G$ maps the final vertex of the edge denoted by X_1, which begins at 1, to 1. As a result, the path denoted by $X_2 \ldots X_n$ beginning at x_1 is mapped to the path denoted by $X_2 \ldots X_n$ beginning at 1. It follows that the elements x corresponding to the X are generators of G and the relations correspond to the closed paths emanating from 1.

Since only edges in $\mathbb{E}^*$ denoted by the same symbol X are equivalent under G, when the word W(X) is traced from a vertex we either obtain a closed path on all occasions, or never. This now facilitates the determination of defining relations. Namely, a closed path may be decomposed into simple closed subpaths, possibly with approach and return paths, and spurs. But a simple closed path in a planar net is a product of paths consisting of an approach path and the boundary of a face. Spurs each contain a subpath which runs out and back across a single edge. Therefore we obtain the defining relations for G by traversing a path out and back across an edge or by traversing the boundary paths of the faces of $\mathbb{E}^*$, and replacing the symbols X in the resulting word W(X) by the generators x or a path out and back across a segment. Of course the path out and back across a segment only makes a contribution when the two directions are not denoted by inverse symbols, i.e. if it is on an axis of reflection. Since G acts simply transitively on vertices we need only consider faces and edges which have 1 in their boundary. Even now, some W(X) can appear several times (up to cyclic interchange). We do not attempt to abstractly characterize a minimal system of defining relations, but determine only the words W(X) which appear.

We call the star 4.5.4 or 4.5.5 positive when its center vertex corresponds to an orientation-preserving automorphism. At all other vertices the oppositely oriented star will be reckoned positive. We then obtain the boundary paths of faces by following each successive edge with the neighbour of its inverse, taking the positive sense in the star of its final point. In addition, of course, we must consider the orders of rotations, as a result of which the boundary path of a surface piece has the form $(R^*(S))^k$.

From 4.5.4 and 4.5.5 we obtain the boundary paths

$$S_i^{h_i},\ (C_{k,j+1}C_{k,j})^{h_{k,j}}\ 1 \le j \le m_k,\ C_{k,1}E_kC_{k,m_k+1}E_k^{-1},$$

$$S_1S_2\ldots S_mT_1U_1T_1^{-1}U_1^{-1}\ldots U_g^{-1}E_1E_2\ldots E_q$$

or

$$S_1S_2\ldots S_mV_1^2V_2^2\ldots V_g^2E_1E_2\ldots E_q$$

respectively.

In addition we still have the spurs $C_{k,j}^2$. Thus we have proved the following theorem giving the *canonical presentation:*

<u>4.5.6 Theorem</u>. *A planar discontinuous group has one of the following structures:*

4.5.7 *Generators:*

(a) $s_1,\ldots,s_m \quad m \ge 0$

(b) $t_1,u_1,\ldots,t_g,u_g \quad g \ge 0$

(c) $e_1,\ldots,e_q \quad q \ge 0$

(d) $c_{1,1},\ldots,c_{1,m_1+1},\ldots,c_{q,m_q+1} \quad m_k \ge 0.$

4.5.8 *Defining relations:*

(a) $s_i^{h_i} = 1 \quad i = 1,\ldots,m$

(b) $c_{i,j}^2 = 1 \quad i = 1,\ldots,q,\ j = 1,\ldots,m_i$

(c) $(c_{i,j+1}c_{i,j})^{h_{i,j}} = 1 \quad i = 1,\ldots,q,\ j = 1,\ldots,m_i$

$c_{i,1}e_ic_{i,m_i+1}e_i^{-1} = 1 \quad i = 1,\ldots,q$

(d) $\prod_{i=1}^{m} s_i \prod_{i=1}^{g} [t_i,u_i] \prod_{i=1}^{q} e_i = 1$

B) *As in* A) *but without the generators* t_i,u_i; *in their place*

4.5.7 (b') $v_1,\ldots,v_g^2 \quad g > 0$

and, instead of the last relation,

4.5.8 (d') $\prod_{i=1}^{m} s_i \prod_{j=1}^{g} v_j^2 \prod_{k=1}^{q} e_k = 1.$

Among the generators only the $c_{i,j}$ *and the* v_j *are orientation reversing. No proper subword of a defining relation is a defining relation.*

□

Remark: The last assertion is not claimed here for groups given by generators and relations abstractly in the above form, but only for planar discontinuous groups.

Exercises: E 4.6-9

4.6 CLASSIFICATION OF PLANAR DISCONTINUOUS GROUPS

Two types of equivalences present themselves for planar discontinuous groups:

4.6.1 The groups have the same algebraic type *(algebraic isomorphism).*

4.6.2 There are realisations $(\mathbb{E},G)$, $(\mathbb{E}',G')$ and an isomorphism $h: \mathbb{E}' \to \mathbb{E}$ such that $x \mapsto h^{-1}xh$ defines an isomorphism from G to G' *(geometric isomorphism).*

The geometric isomorphism obviously implies the algebraic. It follows easily from the preceding that:

4.6.3 Theorem. *For the geometric isomorphism of two planar discontinuous groups* $(\mathbb{E}, G)$ *and* $(\mathbb{E}',G')$ *with compact fundamental domains it is necessary and sufficient that:*

(a) The surfaces $\mathbb{E}/G$ *and* $\mathbb{E}'/G'$ *are isomorphic, i.e. both are orientable or both not (thus* $(\mathbb{E}, G)$ *and* $(\mathbb{E}', G')$ *are both of type A or both of type B), have equal genus (i.e.* $g = g'$*) and the same number of boundary components (i.e.* $q = q'$ *).*

(b) The number of inequivalent rotation centers not lying on the boundary is the

same (m = m') *and the orders of the rotations are the same.*

(c) On each boundary curve of $\mathbb{E}/G$, $\mathbb{E}'/G'$ *respectively there is a cycle of rotation centers with corresponding orders* $h_{i,1},\ldots,h_{i,m_i}$. *If we are dealing with groups of type A then either* $(\mathbb{E},G)$ *and* $(\mathbb{E}',G')$ *have the same cycles or all those of* $(\mathbb{E}',G')$ *are inverse to those of* $(\mathbb{E},G)$. *If the groups are of type B then the cycles of* $(\mathbb{E},G)$ *may be put in one-to-one correspondence with those of* $(\mathbb{E}',G')$ *where image and preimage may have the same or opposite orientation.*

The assertion 4.6.3 (c) results from the classification of surfaces. The orientations of all boundary curves of an orientable surface are uniquely determined by the orientation of the surface itself, whereas for non-orientable surfaces there are automorphisms which map one boundary curve into its inverse but leave the others fixed. (The proof of this assertion is Exercise E 4.9.)

□

Further, we have

Theorem 4.6.4. *If there is an algebraic isomorphism between two planar discontinuous groups with compact fundamental domains then there is also a geometrical isomorphism between them.*

This was proved in general by A.M. Macbeath [Macbeath 1967]. There it is shown that each isomorphism is geometrically realisable, see also 5.6-8. If no reflections occur then the statement of Theorem 4.6.4 follows easily: the group in this case has the form

$$\langle s_1,\ldots,s_m,t_1,u_1,\ldots,t_g,u_g \mid s_1^{-h_1},\ldots,s_m^{-h_m}, \prod_{i=1}^{m} s_i \prod_{j=1}^{g} [t_j,u_j]\rangle$$

or

$$\langle s_1,\ldots,s_m,v_1,\ldots,v_g \mid s_1^{-h_1},\ldots,s_m^{-h_m}, \prod_{i=1}^{m} s_i \prod_{j=1}^{g} v_j^2 \rangle$$

respectively. Its elements of finite order are conjugate to the powers of the s_i. Then $h_1,\ldots,h_m$ are the orders of the maximal finite cyclic subgroups of G which are not conjugate to each other. If we cancel out the elements of finite order and abelianise we obtain either $\mathbb{Z}^{2g}$ or $\mathbb{Z}_2 \oplus \mathbb{Z}^{g-1}$. This uniquely determines the type A or B and the genus g.

However, the proof in the general case uses difficult analytic methods and no algebraic proof is known to us, cf. Corollary 6.6.10. Weaker results are in section 4.8.

□

Exercises: E 4.9-12

4.7 EXISTENCE PROOF

In this section we show that each presentation of the form <4.5.7|4.5.8> actually occurs as canonical presentation of a group of automorphisms of a planar complex.

4.7.1 Theorem. *A group defined abstractly by generators and defining relations of the type A or B given in Theorem 4.5.6 and of infinite order, exists as a planar discontinuous group with compact fundamental domain.*

The finite groups with presentation as in theorem 4.5.6 are:

Groups of type A:

(a) $g = 0$, $q = 0$, $m \leq 2$
(b) $g = 0$, $q = 0$, $m = 3$, $1/h_1 + 1/h_2 + 1/h_3 > 1$
(c) $g = 0$, $q = 1$, $m = 1$, $m_1 = 1$, $1/h_1 + 1/h_1 + 1/h_{1,1} > 1$
(d) $g = 0$, $q = 1$, $m = 0$, $m_1 = 3$, $1/h_{1,1} + 1/h_{1,2} + 1/h_{1,3} > 1$
(e) $g = 0$, $q = 1$, $2m + m_1 \leq 2$.

Groups of type B:

(f) $g = 1$, $q = 0$, $m \leq 1$.

The groups (b) with rotation orders (h_1,h_2,h_3) are the wellknown platonic groups:

(2,2,n) dihedral group
(2,2,3) tetrahedral group
(2,3,4) octahedral group
(2,3,5) dodecahedral group.

Proof. We shall assume that no proper subword of a defining relation is itself a defining relation, and prove this fact later (Lemma 4.7.5). First, we let each group element correspond to a vertex and construct an edge emanating from it for each of the generating symbols S_i, S_i^{-1}, T_j, T_j^{-1}, U_j, U_j^{-1}, V_j, V_j^{-1}, E_k, E_k^{-1}, $C_{k,\ell}$. The edge labelled X leads from the vertex y to the vertex yx^{-1}, and the inverse edge, beginning at yx^{-1}, is labelled X^{-1}. (If $X = C_{k,\ell}$ we set $X^{-1} = C_{k,\ell}$). Let C be the graph obtained. We can again trace words in the X from each point and obtain paths as a result. A reflection relation $C_{k,j}^2$ now means running out and back across an edge. We exclude these relations from the considerations which follow.

We now trace each of the defining relations from each vertex. To each of the paths obtained we associate a face, and we say that the path bounds it positively. The reverse path bounds it negatively, and we associate it with a face inverse to the first. Faces associated in this way with paths that differ only by cyclic interchange are identified. The 2-complex obtained in this way is called D.

4.7.3 Proposition. *Each directed edge appears in the boundaries of exactly two faces and is traversed once in each boundary.*

This is clear for generators which only appear twice (including inverses) in the relations. For the generators in the power relations 4.5.8 (a), (c) it follows from the yet unproved assertion that proper subwords of relations are not relations, for then the same edge cannot be multiply traversed by a power relation. Thus a face corresponding to one of these relations has a given edge appearing at most once in its boundary. A boundary path including an E_k has the form $\prod_{i=1}^{m} S_i \prod_{j=1}^{g} [T_j,U_j] \prod_{k=1}^{q} E_k$, $\prod_{i=1}^{m} S_i \prod_{j=1}^{g} V_j^2 \prod_{k=1}^{q} E_k$ or $E_k C_{k,m_k+1} E_k^{-1} C_{k,1}$ or the inverse $C_{k,1} E_k C_{k,m_k+1} E_k^{-1}$, which is the same, up to cyclic interchange.

4.7.4 Proposition. *The segments emanating from each point constitute a star.* The neighbours of a symbol X are defined to be the symbols which follow X^{-1} in the relations, or the inverses of the symbols which precede X. A symbol then has two neighbours. If one writes them in succession one obtains the cycles 4.5.3 and 4.5.4 respectively, in which all symbols X appear.

Thus the canonical complex D is a surface. Since each closed path is a relation and this is a product of conjugates of defining relations, the fundamental group of the surface D is trivial, so D must be planar net or a net on the sphere. The group G acts on this net in the obvious way. Thus Theorem 4.7.1 is proved.

□

4.7.5 Lemma. *In an infinite group defined by the generators and defining relations of Theorem 4.5.6 no proper subword of a defining relation is a relation.*

Proof. We first show this for groups of type A without the generators 4.5.7 (c), (d) (i.e. without reflection generators). Let

$$G = \langle s_1,\ldots,s_m,t_1,u_1,\ldots,t_g,u_g \mid s_1^{-h_1},\ldots,s_m^{-h_m}, \prod_{i=1}^{m} s_i \prod_{i=1}^{g} [t_i,u_i]\rangle$$

and let F be the free group on the generators S_i, T_j, U_j. The kernel of the canonical mapping $F \to G$ consists of the relations. We decompose G into a free product with amalgamation. In the group

$$\langle s_1,s_2 \mid s_1^{-h_1},s_2^{-h_2}\rangle = \langle s_1 \mid s_1^{-h_1}\rangle * \langle s_2 \mid s_2^{-h_2}\rangle$$

the element s_1s_2 has infinite order. Similarly, in

$$\langle s_3,\ldots,s_m,t_1,u_1,\ldots,t_g,u_g \mid s_3^{-h_3},\ldots,s_m^{-h_m}\rangle$$

the element $s_3 \ldots s_m \prod_{i=1}^{g} [t_i, u_i]$ has infinite order for $g > 0$ or $m \geq 4$. Thus G is the free product of $\langle s_1, s_2 | s_1^{-h_1}, s_2^{-h_2} \rangle$ and $\langle s_3, \ldots, s_m, t_1, u_1, \ldots, t_g, u_g | s_3^{-h_3}, \ldots, s_m^{-h_m} \rangle$ with the cyclic subgroups generated by $s_1 s_2$ and $(s_3 \ldots s_m \prod_{i=1}^{g} [t_i, u_i])^{-1}$, respectively, amalgamated. The exceptions are:

4.7.6 $g = 0$ and $m \leq 3$

4.7.7 $m = 1$, $g > 0$

4.7.8 $m = 0$, $g > 0$.

Thus when G is not one of the exceptional cases one sees immediately from this representation that no proper subword of a defining relation is mapped by the canonical homomorphism onto the unit element of a factor or the amalgamated subgroup. If $g > 1$ in case 4.7.7 then one decomposes G into

$$\langle s_1, t_1, u_1 | s_1^{-h_1} \rangle \quad * \quad \langle t_2, u_2, \ldots, t_g, u_g \rangle$$

$$\langle s_1 t_1 u_1 t_1^{-1} u_1^{-1} = (\prod_{i=2}^{g} [t_i, u_i])^{-1} \rangle$$

and the assertion follows as above.

For $g = 1$ we introduce the relations

$$u_1^2 = 1, \; s_1 = t_1^{-2}$$

and then obtain the group $\langle t_1, u_1 | t_1^{2k}, u_1^2, t_1^{-1} u_1 t_1^{-1} u_1^{-1} \rangle$ which is the dihedral group with $4k$ elements. No subword of the defining relations equals 1 here, so the assertion is also proved for this case.

In the case 4.7.8 the assertion again follows from a suitable decomposition or else is trivial (for $g = 1$). Thus we have only to consider 4.7.6. Then G has the form $\langle s_1, s_2, s_3 | s_1^{h_1}, s_2^{h_2}, s_3^{h_3}, s_1 s_2 s_3 \rangle$. It is called a *"triangle group"*. According as $1/h_1 + 1/h_2 + 1/h_3$ is greater than, equal to, or less than 1, one considers a triangle with angles π/h_1, π/h_2, π/h_3 on the sphere, plane, or non-euclidean plane. By repeated reflections in the sides a, b, c we obtain a net on the sphere, plane or non-euclidean plane. The associated group is generated by the reflections, which we denote by A, B, C, in the edges a, b, c respectively. CA, AB, BC correspond to rotations through the angles $2\pi/h_1$, $2\pi/h_2$, $2\pi/h_3$ respectively, and (CA)(AB)(BC) is the identity. The mapping $s_1 \mapsto CA$, $s_2 \mapsto AB$, $s_3 \mapsto BC$ therefore de-

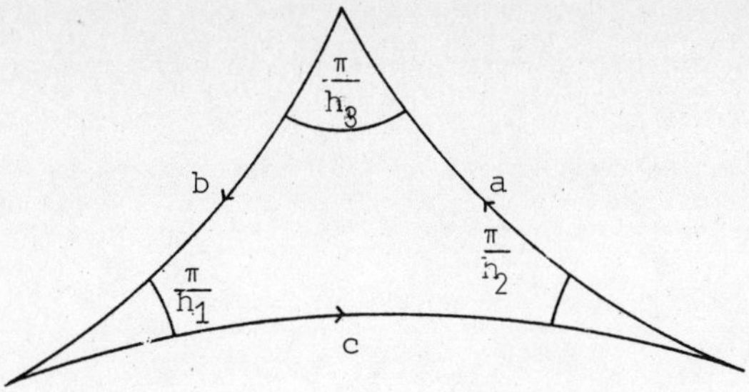

fines a homomorphism of G into the group just described. If a subword of a relation was itself a relation, then its image in the new group would be the identity. However, it follows immediately from geometrical considerations that no such subword exists.

The proof for groups G with "orientation-reversing" elements is set as Exercise E 4.13. One considers the subgroup of "orientation-preserving" elements and proves that it is a group of type (A) without reflections. "Orientation-reversing" generators are the $c_{i,k}$, v_j. □

The complex constructed above is a so-called *planar group diagram:* the group elements are represented by the vertices, the generators by edges (hence, this gives a group diagram or Cayley diagram of the group) and the relations by faces. This approach to the existence theorem for discontinuous groups arose in discussions with K. Reidemeister and was first published in [Zieschang 1966]. For an alternative to the combinatorial treatment see 6.4.

Exercises: 4.13, 14

4.8 ON THE ALGEBRAIC STRUCTURE OF PLANAR GROUPS

Using the presentation as a free product with amalgamated subgroup we determine the elements of finite order in planar groups and show that groups which contain orientation reversing elements are not isomorphic to those which do not.

By Theorems 4.2.5 and 4.2.7 and Lemma 4.2.4 the elements of finite order are characterized as those which have a fixed vertex (rotations) or a fixed line (reflections). If a vertex P or an edge σ is fixed under the transformation $b \in G$ then Px^{-1}, σx^{-1} respectively are fixed under $a = xbx^{-1}$ for any $x \in G$. If P or σ is fixed under b and Px^{-1}, σx^{-1} respectively are fixed under a, then xbx^{-1} and a have the same fixed vertex or fixed edge respectively, and consequently they are powers of the same element. If the fixed point of a rotation does not lie on a line of reflection then xax^{-1} has the same rotation center as a only when x itself is a rotation about this vertex, so x and a are powers of the same element and hence commute. The different powers of a are therefore not conjugate to each other. If the rotation center of a lies on the line of reflection for c, so that a has the form cc_o, where c_o is another, suitably chosen, reflection, then $a^{-1} = c_o c = c(cc_o)c$. Thus a and a^{-1} are conjugate. Further possibilities of conjugacy do not exist, since the possible conjugation factor differs from c by at most a rotation. Thus we have:

4.8.1 Theorem. *a) An element of finite order is conjugate to a power of* $s_i, c_{k,j}$ *or* $(c_{k,j}\, c_{k,j+1})$.

b) Among the different powers of s_i *and* $(c_{k,j}\, c_{k,j+1})$ *only the elements* $(c_{k,j}\, c_{k,j+1})^{\alpha}$ *and* $(c_{k,j}\, c_{k,j+1})^{-\alpha}$ *are conjugate, and none of the rotations is conjugate to a reflection* $c_{k,j}$.

c) If a transformation commutes with an element of finite order and they are not powers of the same element, then one of them is a reflection.

d) A torsionfree planar discontinuous group is the fundamental group of a surface. □

Remark: If $h_{k,j}$ is odd = 2n + 1, then since
$(c_{k,j}\, c_{k,j+1})^n\, c_{k,j} (c_{k,j+1}\, c_{k,j})^n\, c_{k,j+1} = 1$ the $c_{k,j}$ and $c_{k,j+1}$ are conjugate.

The assertion c) may be considerably sharpened. If we assume that no reflections are present, then in general two elements commute only when they are powers of the same element. The exceptions are a small number of groups such as the fundamental groups of the torus and the Klein bottle . Finding the proof and the exceptional cases is Exercise E 4.15.

4.8.2 Theorem. *Planar groups of orientation-preserving transformations are not isomorphic to planar groups with orientation-reversing elements.*

Proof. For our initial considerations we shall exclude the following groups:

4.8.3 Groups, containing only orientation preserving elements, with g = 0.

4.8.4 Groups with reflections, for which g = 0 and q = 1.

If we annihilate the elements of finite order, then by Theorem 4.8.1 we obtain fundamental groups of closed surfaces of genus greater than 0 in the case of planar groups of orientation-preserving transformations, and fundamental groups of non-orientable surfaces or non-trivial free groups in the case of groups containing orientation-reversing transformations. Since this annihilation results in the trivial group in the cases of type 4.8.3 and 4.8.4 it remains only to distinguish the groups in these two exceptional cases. In case 4.8.3 we have

$$4.8.3' \quad \langle s_1,s_2,\ldots,s_m \mid s_1^{-h_1},\ldots,s_m^{-h_m}, s_1 s_2 \ldots s_m \rangle$$

and the groups in case 4.8.4 have the form

$$4.8.4' \quad \langle s_1,\ldots,s_{m'},c_1,\ldots,c_{r'+1},e \mid s_1^{-h'_1},\ldots,s_{m'}^{-h'_{m'}}, c_1^2,\ldots,c_{r'+1}^2,$$

$$(c_1c_2)^{h'_{1,1}},\ldots,(c_{r'}c_{r'+1})^{h'_{1,r'}}$$

$$c_1 e c_{r'+1} e^{-1}, s_1 \ldots s_{m'} e \rangle.$$

If $r' > 0$ for 4.8.4' then there are elements of order 2 which are not powers of the same element and which have a product of finite order. This is true of the groups 4.8.3' only when $m = 3$, as one sees from the decomposition into a free product with amalgamation (see the proof of Lemma 4.7.5).

If $r' = 0$ then c_1 and e commute and hence by Theorem 4.8.1 they must be powers of the same element if we are dealing with a group of the form 4.8.3'. In the case $m' \geq 1$, e is non-trivial and unequal to c_1, as one sees by annihilation of c_1. However c_1 is a power of itself only. If r' and m' equal 0 then the group is finite. Thus it remains only to compare groups 4.8.3' with $m = 3$ and 4.8.4' with $r' > 0$.

The classification of elements of finite order by their fixed point behaviour (Theorem 4.8.1) can also be expressed algebraically. In fact, two elements of finite order belong to the same "class" when conjugates of them are powers of the same element. For groups 4.8.3' the number of such classes equals m, for groups 4.8.4' it is at least $m' + r' + 1$.

The number of these classes and the "maximal" order of elements of a class are algebraic invariants. For us the interesting case is $m = 3$, and a possibly isomorphic group 4.8.4' with numbers m', r' such that $m' + r' + 1 \leq 3$, $r' > 0$. There are three cases

(1) $m' = 0$, $r' = 1$. The group is finite ($= \mathbb{Z}_2$).
(2) $m' = 0$, $r' = 2$. The group is finite (dihedral group).
(3) $m' = 1$, $r' = 1$. The groups to be compared have the forms

$$4.8.3'' \quad \langle s_1,s_2,s_3 \mid s_1^{-h_1}, s_2^{-h_2}, s_3^{-h_3}, s_1 s_2 s_3 \rangle$$

4.8.4" $\langle s,c_1,c_2,e \mid c_1^2, c_2^2, s^{-H_1}, (c_1c_2)^{-H_2}, c_1ec_2e^{-1}, se\rangle$

By means of Tietze transformations we convert them into

4.8.3"' $\langle s_1,s_2 \mid s_1^{-h_1}, s_2^{-h_2}, (s_1s_2)^{-h_3}\rangle$

4.8.4"' $\langle s,c \mid c^2, s^{-H_1}, (cs^{-1}cs)^{-H_2}\rangle$

Since we can distinguish the maximal order of elements in each class, and for 4.8.3" they are h_1,h_2,h_3, while for 4.8.4" they are $2,H_1,H_2$, without loss of generality, we can assume $h_1 = 2$, $h_2 = H_1$ and $h_3 = H_2$. 4.8.4"' contains a normal subgroup of index 2, namely the subgroup of orientation-preserving elements. It has the presentation

4.8.4^{IV} $\langle v,w \mid v^{H_1} = w^{H_1} = (vw)^{H_2}\rangle$.

A subgroup of index 2 for groups of type 4.8.3"' is uniquely determined as the kernel of a homomorphism onto $\mathbb{Z}_2$. Such a homomorphism is determined by the images of s_1 and s_2 in $\mathbb{Z}_2$. Under a mapping of 4.8.4 onto $\mathbb{Z}_2$ at least one element of order 2 must be mapped onto -1. Since h_2 and h_3 in 4.8.3"' are both greater than 2 (otherwise we would have a dihedral group), there are essentially only two homomorphisms in question, namely $s_1 \mapsto -1$, $s_2 \mapsto +1$ and $s_1 \mapsto -1$, $s_2 \to -1$. In the first case $h_3 = H_2$ must be even, and in the second $h_2 = H_1$. If we now compute the subgroups we obtain the presentation

$\langle x,y \mid x^{H_1}, y^{H_1}, (xy)^{\frac{1}{2}H_2}\rangle$ in the first case,

$\langle x,y \mid x^{H_2}, y^{H_2}, (xy)^{\frac{1}{2}H_1}\rangle$ in the second.

But neither of these is isomorphic to 4.8.4^{IV}, thus we have also distinguished 4.8.3' and 4.8.4' in this case.

□

4.8.5 Corollary. *The subgroup of orientation-preserving elements of a planar group is characteristic.*

□

4.9 THE WORD AND CONJUGACY PROBLEMS

Next we describe the method of M. Dehn for the solution of the word and conjugacy problems.

4.9.1 Theorem. *Let $\mathbb{E}$ be a planar net such that $p \geq 6$ edges emanate from each vertex, and let D be a disk of $\mathbb{E}$. Then one of the following cases occurs:*

(a) ∂D is the boundary of a face of $\mathbb{E}$.

(b) ∂D contains disjoint connected subpaths, one of which is the boundary path of a face minus at most one edge, the other of which is the boundary path of a face minus at most two edges.

(c) ∂D contains at least three disjoint connected subpaths which are boundaries of faces minus at most two edges.

Proof. Let $\phi_1,\ldots,\phi_g$ be the geometric faces of D, k_i the number of geometric edges of $\partial\phi_i$ which lie in the interior of D, and ℓ_i the number of those which lie on ∂D. f and e are the numbers of geometric edges and vertices of D respectively.

The Euler characteristic of a disk is 1, so

(1) $$e-f+g = 1.$$

The number of vertices in the interior of D is $e - \Sigma\ell_i$; furthermore

(2) $$f = \frac{1}{2}\sum_{i=1}^{g} k_i + \sum_{i=1}^{g} \ell_i .$$

Since p edges emanate from each vertex in the interior, and at least two from each vertex on the boundary, we have

(3) $$f \geq \frac{p}{2}(e - \sum_{i=1}^{g} \ell_i) + \sum_{i=1}^{g} \ell_i ,$$ hence by (2),

$$\frac{p}{2}(e - \sum_{i=1}^{g} \ell_i) \leq \frac{1}{2}\sum_{i=1}^{g} k_i ,$$ and by (1)

$$\frac{p}{2}(1 + f - g - \sum_{i=1}^{g} \ell_i) \leq \frac{1}{2}\sum_{i=1}^{g} k_i .$$

Since $f - \Sigma\ell_i = \frac{1}{2}\Sigma k_i$ we finally obtain

$$p(g-1) \geq (\frac{p}{2} - 1) \sum_{i=1}^{g} k_i .$$

If $g = 1$ then 4.9.1 (a) holds. Otherwise $g > 1$ and $k_i \geq 1$ for all i. If two $k_i > 1$ and all other $k_j > 2$ then

$$2p(g - 1) \geq (p - 2)(3(g - 2) + 4)$$

so $\qquad 6g \geq pg + 4$, which is impossible since $p \geq 6$.

We obtain the same inequality when at most one k_i is ≤ 2. Therefore there are always two $k_i \leq 2$. If exactly two $k_i \leq 2$ then one of them equals 1. If no k_i equals 1 at least three $k_i = 2$. Thus if D is not a face, D either contains two boundary paths of faces, one of them missing an edge, the other possibly two, or D contains three boundary paths each missing at most two edges. When sub-boundary paths with $k_i = 2$ are not connected in ∂D there is an edge in the interior of D which runs from boundary to boundary. This edge divides D into two disks. We apply the above argument again to each component and find "large" subpaths in the boundaries of faces. But only one of the components contains the dividing edge, so it follows by induction that we can find connected sub-boundary paths of the desired length. They can be chosen to be disjoint, since at most one face of D meets a given boundary edge of D and the sub-boundary paths belong to different faces of D.

□

In the considerations which follow we assume that the planar discontinuous group G is given by at least three canonical generators and all defining relations have length at least 5. (In particular no reflections occur.) The generators of G will be denoted by $h_1,\ldots,h_n$ and $H_1,\ldots,H_n$ will be regarded as free generators. Each element $x \in G$ may be represented by a word $W(H)$ (i.e. $W(h) = x$) and we may also assume - reckoning the defining relations to be cyclic words and also allowing passage to inverses:

4.9.2 (a) $W(H)$ is a reduced word.

(b) $W(H)$ contains no more than half of any defining relation.

Analogously, we can represent conjugation classes by words with the following properties:

4.9.3 (a) $W(H)$ is cyclically reduced.

(b) Regarded as a cyclic word, $W(H)$ contains no more than half of any defining relation.

4.9.4 Theorem. *Let G be a planar discontinuous group with at least three generators, all relations of which have length at least 5. Let u and v be elements of G, written as words U and V. Then u and v are equal in G if and only if we can reduce $U^{-1}V$ to the empty word by replacing subrelations (of the defining relations) by their shorter complements, and by free cancellations.*

Proof. As in the proof of Theorem 4.7.1 we construct a planar net, the vertices of which correspond to the group elements, with the directed edges around each vertex corresponding to the generators and their inverses, and the boundary paths of faces corresponding to defining relations. A word is then equal to 1 if and only if it corresponds to a closed path of the net. After removal of spurs the curve contains a simple closed subpath (or else is trivial). Since G has at least three generators and none of them is a reflection, we can apply Theorem 4.9.1 and find at least one subpath which is a relation with at most two symbols missing. Since the length of relations is ≥ 5, this subrelation is longer than its complement.

□

<u>4.9.5 Theorem</u>. *Let G be a planar discontinuous group with at least three generators, and let all defining relations have length at least 8. Let* u *and* v *be conjugate elements of* G, *written as words* U *and* V *which satisfy 4.9.3 (a) and (b). Then after suitable cyclic interchange of* U *and* V *we have* $U(h) = V(h)$ *in G.*

Proof. Since U and V are conjugate there is a word $W(H)$ with $U^{-1}WVW^{-1}(h) = 1$ which satisfies 4.9.2 (a) and (b). If $U^{-1}WVW^{-1}(H) = 1$ already holds in the free group then U and V result from each other by cyclic interchange. Otherwise we apply Theorem 4.9.1 to find "large" parts of the defining relations in $U^{-1}WVW^{-1}$. If $W = 1$ we have nothing to show.

a) Suppose that among the "large" subwords of defining relations (i.e. at most two symbols missing) there is one which meets at most one of W and W^{-1}, so we have, say, the situation:

$W = W'X$, $V = YV'$ and $(XYA)^{\pm 1}$ is a defining relation, where $\ell(A) \leq 2$.

If we replace the X in W by $A^{-1}Y^{-1}$ then we get

$U^{-1}\cdot W'A^{-1}Y^{-1}\cdot YV'\cdot YAW'^{-1}(H) = U^{-1}\cdot W'A^{-1}\cdot V'Y\cdot AW'^{-1}(H)$. If $\ell(A) < \ell(X)$ then, after a cyclic interchange of V, we have found a shorter conjugation factor. Since $\ell(Y)$ and $\ell(X)$ are at most $\frac{1}{2}\ell(XYA)$, $2 \geq \ell(A) \geq \ell(X)$ only holds when $\ell(XYA) = 8$, $\ell(Y) = 4$, $\ell(X) = \ell(A) = 2$. In this case one replaces the Y in V by $X^{-1}A^{-1}$ and again cyclically interchanges V, which shortens W by 2.

The conjugating factor can be shortened analogously when the "large" subword still meets only one of W, W^{-1}, but meets both U and V.

b) If a situation of the form a) does not occur, then each "large" subword of a relation must meet both W and W^{-1}. It follows immediately that we do not have the case of three large subwords, since at most two disjoint words can meet both W and W^{-1}. Also, $U^{-1}WVW^{-1}$ cannot be a full relation, since the subword of a relation between two inverse symbols (in particular the last of W and the first of W^{-1}) is either a single symbol or more than half the relation. The former implies that U and V are single symbols, hence conjugate only if identical, and the latter contradicts the assumption that U,V satisfy 4.9.2 (b).

There must therefore be two large subwords of relations, each meeting both W, W^{-1}, so that one of them includes U^{-1} and the other includes V (when $U^{-1}WVW^{-1}$ is written cyclically). This case is disposed of by the same argument about the subwords of a relation between inverse symbols.

Because of a) and b), W(H) = 1.

□

The above solution of the word and conjugacy problems was first given in [Dehn 1912, 1912'] for fundamental groups of orientable surfaces, and by [Reidemeister 1932] in the generality above. Our approach is close to that of [Lyndon 1966], [Schupp 1968]. Both word and conjugacy problems are solved by repeated replacement of a "long" part of a defining relation by the shorter complementary part. The question arises: under what conditions on a group is this method of solution possible? This is treated in the so-called *small cancellation theory;* for the description of its results and literature see [Lyndon-Schupp 1977, chap. V].

Another approach to the word and conjugacy problems can be made using the decomposition into a free product with amalgamation from the proof of 4.7.5. We use this in 5.14.

4.10 SURFACE SUBGROUPS OF FINITE INDEX

Subgroups of planar discontinuous groups are themselves planar discontinuous groups. We already know that planar groups without elements of finite order are the fundamental groups of surfaces. In this section we shall prove:

4.10.1 Theorem. *Each planar discontinuous group has a surface group as a subgroup of finite index.*

Theorem 4.10.1 was conjectured by Fenchel and first proved in [Bungaard-Nielsen 1951], and [Fox 1952]. The proof below was initiated by A.M. Macbeath. Another proof, using 3 x 3 matrices, is in [Mennicke 1967, 1968]. For triangle groups see [Feuer 1971]. The result is a special case of the Selberg lemma, see [Selberg 1960].

To prove this theorem we need a few facts from the theory of finite fields (Galois fields) (cf. e.g. [Van der Waerden 1955 ,§ 40] . In a finite field GF(q) of q elements the integral multiples of the field unit constitute a field, the prime field P; for:The integral multiples of the 1 constitute a homomorphic image of the ring $\mathbb{Z}$ of integers. Thus the ring of integral multiples of the 1 is isomorphic to $\mathbb{Z}/(p)$, where p is the smallest integer for which $p\cdot 1 = 0$ in GF(q). Since the field GF(q) has no zero divisors, p is prime, so $\mathbb{Z}/(p)$ is a field. Consequently P is isomorphic to the field of residue classes of integers modulo p. If $\alpha_1,\dots,\alpha_m$ is a basis for GF(q) over P, then $q = p^m$.

Since the order of the multiplicative group is $q - 1$, each element $x \in GF(q)$ satisfies the equation $x^q - x = 0$. Consequently all field elements are roots of the polynomial $X^q - X$, and consideration of degree shows $X^q - X = \prod_{y \in GF(q)} (X-y)$. Thus GF(q) results from P by adjunction of all roots of the polynomial $X^q - X$ and therefore:

4.10.2 Proposition. *All finite fields of order* q *are isomorphic.*

□

The existence of $GF(p^m)$ follows from the fact that the derivative of $X^{p^m} - X$ equals $p^m X^{p^m-1} - 1$ which is therefore always -1, since $p^m \equiv 0 \pmod p$. Thus if we adjoin to the residue class field modulo p all roots of $X^{p^m} - X$ then these are all distinct. On the other hand they already constitute a field, since

$(x-y)^{p^m} = x^{p^m} - y^{p^m}$ and $(\frac{x}{y})^{p^m} = x^{p^m}/y^{p^m} = x/y$ (for all x and y with $x^{p^m} = x$, $y^{p^m} = y$).

4.10.3 Proposition. *For each prime power* p^m *there is exactly one field of order* p^m.

□

For all elements x unequal to 0 we have $x^{q-1} - 1 = 0$. Therefore we can write the multiplicative group G in the form

$$G = \mathbb{Z}_{d_1} \oplus \mathbb{Z}_{d_2} \oplus \dots \oplus \mathbb{Z}_{d_n}$$

where d_i divides the numbers d_{i+1} and $q - 1$ and we assume $1 < d_1$. If $n > 1$ then there are d_1^2 elements in $\mathbb{Z}_{d_1} \times \mathbb{Z}_{d_2}$ which satisfy $x^{d_1+1} = x$. But this means that $d_1^2 \leq d_1 + 1$, so $d_1 = 1$. Thus:

4.10.4 Proposition. G *is a cyclic group, isomorphic to* $\mathbf{Z}_{q-1}$.

□

4.10.5 Proposition. Let p be a prime > 2. *The quadratic equation* $X^2 + rY^2 + s = 0$ *has solutions in* $GF(p^m)$ *for* $m \geq 1$.

For there are exactly $\frac{p^m+1}{2}$ elements in $GF(p^m)$ of the form x^2. If all of these elements are substituted for X^2 we obtain $\frac{p^m+1}{2}$ expressions for Y^2, so one of them must be a square. Thus $X^2 + rY^2 + s = 0$ always has a solution.

□

The group of special linear fractional transformations over GF(q), SLF(2,q), is defined as the set of functions $w = \frac{az+b}{cz+d}$ where $ad - bc = 1$ and $a,b,c,d \in GF(q)$. Each element is represented by a matrix $A = \begin{pmatrix} a & b \\ c & d \end{pmatrix}$. We are interested in the orders of linear fractional transformations. The eigenvalue λ of A satisfies the equation $\lambda^2 - (a+d)\lambda + 1 = 0$. If $a + d \neq \pm 2$ then we obtain two different solutions λ and $1/\lambda$. Thus in general there is a transformation T from $SLF(2,q^2)$ such that $TAT^{-1} = \begin{pmatrix} \lambda & 0 \\ 0 & 1/\lambda \end{pmatrix}$. The transformation corresponding to the latter matrix is $w = \lambda^2 z$, and the order of this transformation equals the order of λ^2. If λ (and hence $\frac{1}{\lambda}$) lies in $GF(p^m)$ then the order of A divides the number $\frac{p^m-1}{2}$. If λ does not lie in $GF(p^m)$ it lies in $GF(p^{2m})$, which is an extension of $GF(p^m)$ of degree 2. The group of automorphisms of $GF(p^m)$ is given by the mappings $x \mapsto x^{p^k}$ for $k = 0,1,\ldots,m-1$. They leave fixed the elements of GF(p). The automorphism $x \mapsto x^{p^m}$ of $GF(p^{2m})$ is the identity on $GF(p^m)$, though it is not on $GF(p^{2m})$. If we apply the automorphism to $\lambda^2 - (a+d)\lambda + 1 = 0$ then we obtain $(\lambda^{p^m})^2 - (a+d)\lambda^{p^m} + 1 = 0$. Thus λ and λ^{p^m} are the different solutions λ and $\frac{1}{\lambda}$. Thus we have shown:

4.10.6 Lemma. *For* $a + d \neq 2$ *the order of* A *divides either* $\frac{p^m-1}{2}$ *or* $\frac{p^m+1}{2}$.

□

4.10.7 Lemma. *If* d *is a number which divides* $\frac{1}{2}(p^m-1)$ *or* $\frac{1}{2}(p^m+1)$, *then there is an element of* $SLF(2,p^m)$ *which has order* d.

Namely, if d divides the number $\frac{1}{2}(p^m-1)$ then there is a $\lambda \in GF(p^m)$ with order 2d. But then $\begin{pmatrix} \lambda+1/\lambda & 1 \\ -1 & 0 \end{pmatrix}$ has the given order d. If d divides $\frac{1}{2}(p^m+1)$ then we take a λ from $GF(p^{2m})$ which has order 2d. But then $\lambda^{p^m} = \frac{1}{\lambda}$, so that $\lambda + \frac{1}{\lambda} \in GF(p^m)$. Consequently $\begin{pmatrix} \lambda+1/\lambda & 1 \\ -1 & 0 \end{pmatrix} \in SLF(2,p^m)$ is a transformation of order d.

□

Proof of Theorem 4.10.1.

1st case: Let G be a planar group of the form

$$\langle t,u,v \mid t^{-\ell}, u^{-m}, v^{-n}, tuv\rangle$$

We now seek a suitable $q = p^k$ so that there is a homomorphism φ: $G \to$ SLF(2,q) which maps t,u,v on to elements of orders ℓ,m,n respectively. Thus we seek elements t', u' $\in$ SLF(2,q) with o(t') = ℓ,o(u') = m and o(t'u') = n, where o(x) denotes the order of x. We choose p so that $(p,2\cdot\ell\cdot m\cdot n) = 1$. Then we look for a k with the following property: ℓ,m and n divide $\frac{1}{2}(p^k-1)$. We need only find a k which satisfies the congruence $(p^k-1) \equiv 0 \bmod 2\ell mn$; remember p is relatively prime to $a = 2\ell mn$. The residue classes modulo a which are relatively prime to a constitute a finite group under multiplication and p represents an element of this group. If the order of this element is k, then $p^k \equiv 1 \bmod a$ and our search is over. Thus we have elements λ,μ,ν in GF(p^k) of orders 2ℓ, 2m, 2n. If we set $\alpha = \lambda + \frac{1}{\lambda}$, $\beta = \mu + \frac{1}{\mu}$, $\gamma = \nu + \frac{1}{\nu}$ then we know that elements of SLF(p^k) with traces α,β,γ have orders ℓ,m,n respectively.

We set $t' = \begin{pmatrix} 0 & 1 \\ -1 & \alpha \end{pmatrix}$ and seek $u' = \begin{pmatrix} x & y \\ z & w \end{pmatrix}$ of order m. Then we must have

$$x + w = \beta, \quad wx - yz = 1.$$

Since t'u' is to have order n we obtain the further equations

$$z - y + \alpha w = \gamma, \quad -y(z+\alpha w) + w(x+\alpha y) = 1.$$

The last equation is a consequence of xw - yz = 1, so our system of conditions has only three equations, and the solubility of the system results from the solubility of quadratic equations (see 4.10.5). The existence of the homomorphism φ described above is therefore established.

None of the powers of t, u, v different from the identity lies in the kernel, and since elements of finite order in G are conjugate to powers of t, u, v, the kernel contains no elements of finite order. It is not trivial, since G is infinite while SLF(2,q) is finite, and it has finite index.

2nd case: Let G be a planar discontinuous group. Since each planar group contains a subgroup of finite index with only orientation-preserving members we can assume that G has the form

$$\langle s_1,\ldots,s_m,t_1,u_1,\ldots,t_g,u_g \mid s_i^{-h_i}, \prod_{i=1}^{m} s_i \cdot \prod_{i=1}^{g} [t_i,u_i]\rangle.$$

We have already proved Theorem 4.10.1 for $g = 0$ and $m = 3$. Suppose $m \geq 3$. As before we take a p and a k such that $(p, 2h_1 \dots h_m) = 1$ and $(p^k - 1) \equiv 0 \bmod 2h_1 \dots h_m$. Then in $SLF(2,p^k)$ we take elements $s'_3, s'_4, \dots, s'_m, u'_1, t'_1, \dots, u'_g, t'_g$ such that s'_i has order h_i and $\pi = s'_3 \dots s'_m \prod_{i=1}^{g} [t'_i, u'_i]$ is unequal to 1. As in the foregoing proof we find s''_1, s''_2 in $SLF(2,p^k)$ where s''_1 and s''_2 have orders h_1, h_2 respectively and $s''_1 s''_2$ has the same trace as π^{-1}. Then π^{-1} is conjugate to $s''_1 s''_2$, so $x s''_1 s''_2 x^{-1} \pi = 1$ or in other words, $x s''_1 x^{-1} \cdot x s''_2 x^{-1} \cdot \pi = 1$. If we set $s'_1 = x s''_1 x^{-1}$, $s'_2 = x s''_2 x^{-1}$, then by mapping unprimed symbols to primed symbols we have a homomorphism of G into $SLF(2,p^{2k})$ with the same properties as in case 1.

If $m = 2$ and $g > 0$ ($g = 0$ means that G is finite and the theorem is trivial), then we map G homomorphically onto

$$H = \langle s_1, s_2, t, u \mid s_1^{k_1}, s_2^{k_2}, s_1 s_2 \, t u t^{-1} u^{-1} \rangle$$

We define a homomorphism $H \to \mathbb{Z}_3$ by $s_1 \mapsto 1$, $s_2 \mapsto 1$, $t \mapsto 1$, $u \mapsto -1$, the kernel of which has the presentation

$$\langle s_1, s_2, s'_1, s'_2, x, y \mid s_1^{k_1}, s_1'^{k_1}, s_2^{k_2}, s_2'^{k_2}, s_1 s_2 s'_1 s'_2 \, x y x^{-1} y^{-1} \rangle$$

For this group, $m = 4$. Thus we have a subgroup U of finite index in H which contains no element of finite order. But then the preimage of U in G has finite index. It contains no element of finite order, since the kernel of $G \to H$ contains no elements of finite order.

If $m = 1$ we apply this construction twice, and when $m = 0$ the Theorem is trivial.

□

Remark. Theorem 4.10.1 is trivial for groups on the sphere. They are finite and therefore have the fundamental group of the sphere as a subgroup of finite index. For the index of the subgroup see 4.14.25-26.

When one recalls that the intersection of two subgroups of finite index is again of finite index, and that there are only finitely many subgroups with a given number as index (because our groups are finitely generated), then one sees that the intersection of all surface subgroups of the same finite index is a characteristic subgroup of finite index. Hence:

4.10.8 Corollary. *Each planar discontinuous group contains the fundamental group of a closed orientable surface as characteristic subgroup of finite index.*

□

4.11 ON PLANAR GROUPS WITH NON-COMPACT FUNDAMENTAL DOMAIN

In this section we present some results from the literature, without proof, but we give hints how they can be proved in a similar way as for the compact case in the preceding part of the chapter.

4.11.1 As before let the group G act on a net of the plane $\mathbb{E}$. Then G has a fundamental domain F, i.e. a connected subcomplex of $\mathbb{E}$ which contains exactly one copy from each class of equivalent faces. $\mathbb{E}$ is covered by the images of F under G and two images Fx and Fy have no face in common if $x \neq y$. Let us call two domains Fx and Fy $(x \neq y)$ neighbours if they have at least one boundary edge in common. (If they have only a vertex from the boundary in common they are not called neighbours.) We restrict ourselves to the case where the domain F has only a finite number of neighbours. Using the Jordan curve theorem 4.1.2, the Seifert-van Kampen theorem 2.8.2 and the Grushko theorem 2.9.1 it is easily proved:

4.11.2 Proposition. *For a group G of automorphisms of a planar net the following two statements are equivalent:*
(a) G is finitely generated.
(b) G has a fundamental domain F that has only a finite number of neighbours.

The proof of this statement is more complicated if the action is not on a complex, see [Marden 1967]. If G has a fundamental domain with a finite number of neighbours one can proceed as in the section 4.3 to get a normal form. Let us describe one way to come back to the compact case: We consider the quotient surface $\mathbb{E}/G$. It can be proved that there are finitely many simple closed curves $\gamma_1,\dots,\gamma_n$ on $\mathbb{E}/G$ which decompose the quotient surface into connected surfaces $S_0, S_1,\dots, S_n$ where each S_i, $1 \le i \le n$, is a cylinder with one boundary curve and one end, and S_0 is a compact surface. Furthermore,

$$S_i \cap S_j = \begin{cases} \emptyset & \text{if } 1 \le i < j \le n \\ \gamma_j & \text{if } i = 0,\ 1 \le j \le n. \end{cases}$$

The images of fixed vertices of some transformation $x \neq 1$ which are not on a line whose vertices are all fixed by x (a reflection line) are contained in the interior of S_0 (these are the "rotation" centers). The general scheme of the quotient surface is described in the figure

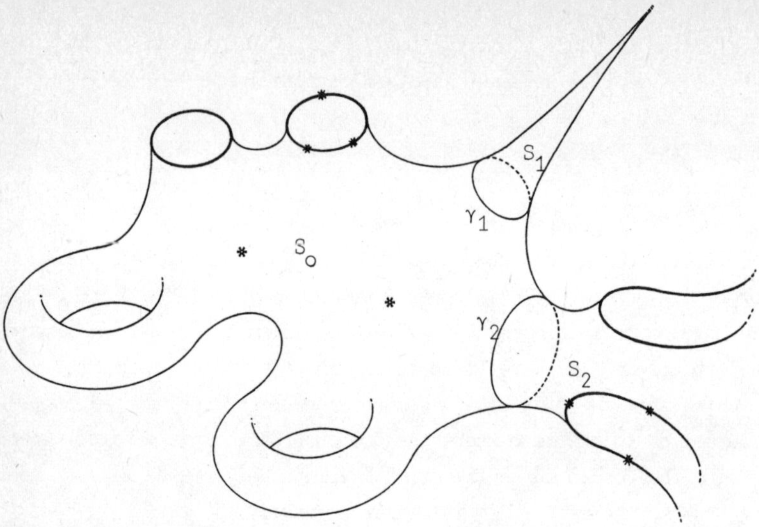

Here the marked points correspond to the centers of rotations and the dark curves to lines of reflection. Marked points on dark lines denote centers of rotations which are the products of two reflections from G. The number of marked points and dark curves is finite.

The surface S_o together with the marked points and dark lines in it represents a group as before, where the curves $\gamma_1,\dots,\gamma_n$ define additional free generators which only occur in the "long" relation. The subsurfaces S_i, $1 \leq i \leq n$, with their markings give further generators (if there are dark curves) and relations. A full system of generators and defining relations is given in 4.11.3,4. Here the generators 4.11.3 (c,f) and defining relations 4.11.4 (d,f) belong to the "holes".

<u>4.11.3 Generators.</u>

(a) s_i $\quad i = 1,\dots,m,$

(b,+) t_k, u_k $\quad k = 1,\dots,g,$

(b,-) v_k $\quad k = 1,\dots,g,$

(c) e_j $\quad j = 1,\dots,r,$ (here $r_1 > 0$ if the fundamental domain is not compact)

(d) c_j $\quad j = r_1+1,\dots,r_1+r_2,$

(e) c_{jp} $\quad j = r_1+r_2+1,\dots,r_1+r_2+r_3;\ p = 1,\dots,m_j+1,$

(f) c_{jpq} $\quad j = r_1+r_2+r_3+1,\dots,r;\ p = 1,\dots,k_j;\ q = 1,\dots,m_{jp}.$

4.11.4 Defining Relations.

(a) $s_i^{h_i} = 1$ $\quad j = 1,\dots,m;\ h_i \geq 2,$

(b) $c_j^2 = 1$ $\quad j = r_1+1,\dots,r_1+r_2,$

(c) $c_{jp}^2 = 1$ $\quad j = r_1+r_2+1,\dots,r_1+r_2+r_3;\ p = 1,\dots,m_j+1,$

(d) $c_{jpq}^2 = 1$ $\quad j = r_1+r_2+r_3+1,\dots,r;\ p = 1,\dots,k_j;\ q = 1,\dots,m_{jp},$

(e) $(c_{jp}c_{jp+1})^{h_{jp}} = 1$ $\quad j = r_1+r_2+1,\dots,r_1+r_2+r_3;\ p = 1,\dots,m_j;$
$h_{jp} \geq 2,\ m_j \geq 1,$

(f) $(c_{jpq}c_{jpq+1})^{h_{jpq}} = 1$ $\quad j = r_1+r_2+r_3+1,\dots,r;\ p = 1,\dots,k_j;$
$q = 1,\dots,m_{jp}-1;\ k_j \geq 1,\ m_{jp} \geq 1,$

(g) $c_je_jc_je_j^{-1} = 1$ $\quad j = r_1+1,\dots,r_1+r_2,$

(h) $c_{j1}e_jc_{j,m_j+1}e_j^{-1} = 1$ $\quad j = r_1+r_2+1,\dots,r_1+r_2+r_3,$

(i,+) $\prod_{i=1}^{m} s_i \prod_{j=1}^{r} e_j \prod_{k=1}^{g} [t_k,u_k] = 1,$

(i,-) $\prod_{i=1}^{m} s_i \prod_{j=1}^{r} e_j \prod_{k=1}^{g} v_k^2 = 1.$

Here (b,+) and (i,+) or (b,-) and (i,-) are valid together. The + denotes the case where the surface E/G is orientable, in the "-" -case the surface is non-orientable.

4.11.5 Theorem. *G has the canonical presentation* $\langle 4.11.3 \mid 4.11.4 \rangle$.

The proof of this theorem is left to the reader. The algebraic structure of the groups has been described in [Macbeath-Hoare 1976] and [Zieschang 1976]. From the geometrical interpretation of the generators and the numbers $m,g,\dots$ involved it is clear that presentations with essentially different numbers describe inequivalent groups of actions. Of course, one may permute the orders h_i, and the systems $(h_{j1},\dots,h_{jm_j})$ or $(h_{j11}, h_{j12},\dots,h_{jk_jm_{jk_j}})$. Another question is the classification with respect to (algebraic) isomorphism. From the presentation 4.11.5 it becomes clear that the group is the free product of cyclic groups and groups which are generated by reflections connected by rotations, (the joining long relations 4.11.4 (i) can be omitted). Hence, the classification can be done by going to a free product decomposition with indecomposable factors, which are uniquely determined up to isomorphisms and permutation, see 2.3.14.

Exercise: E 4.16

4.12 ON DECOMPOSITIONS OF DISCONTINUOUS GROUPS OF THE PLANE

The decomposition of a discontinuous group of the plane into free products mentioned at the end of the preceding section does not correspond to a 'geometric decomposition of the action'. But the decompositions into free products with amalgamated subgroups which we have used in the sections 4.7,8 have geometric interpretations: The subgroup common to both factors arises from a simple closed curve γ on the quotient surface $\mathbb{E}/G$ and consists of the homotopy class of this curve and its powers. The curve γ separates $\mathbb{E}/G$ and the parts define the factors. All proper decompositions into a free product with an infinite cyclic amalgamated subgroup arise in this way. To get a simple formulation we restrict ourselves to the case of discontinuous group of $\mathbb{E}$ with compact fundamental region where all elements preserve orientation.

<u>4.12.1 Theorem</u>. *Let* $G <s_1,\ldots,s_m,t_1,u_1,\ldots,t_g,u_g \mid s_i^{h_i}, \prod_{i=1}^{m} s_i \prod_{j=1}^{g} [t_j,u_j]>$

and assume that $G = G_{-1} *_A G_1$ *where* $\mathbb{Z} \cong A \neq G_i$, $i = -1,1$. *(a) Then there is a system of generators and defining relations for G such that*

$$G_i = <x_i,s_{i1},\ldots,s_{im(i)},t_{i1},u_{i1},\ldots,t_{ig(i)},u_{ig(i)} \mid s_{ij}^{h_{ij}}, x_i \prod_{j=1}^{m(i)} s_{ij} \prod_{j=1}^{g(i)} [t_{ij},u_{ij}]>$$

for $i = -1,1$. *The amalgamating relation is* $x_{-1}x_1^{-1}$, *and* x_1 *is a generator of* A.

(b) On the quotient surface $\mathbb{E}/G$ *the element* x_i *is represented by a simple closed curve* γ *which does not meet the images of rotation centers and which separates* $\mathbb{E}/G$ *into two quotient surfaces.*

The Seifert-van Kampen theorem 2.8.2 gives the decomposition $G = G_{-1} *_A G_1$ from the decomposition of $\mathbb{E}/G$.

Not only is the formulation of the theorem a bit vague, but in addition we cannot give a full proof based on the theory already developed. There are two ways to prove it: one uses the generalized Nielsen method and the theorem from [Zieschang 1970] which was described in 2.10 to get statement (a), and then (b) follows from the Dehn-Nielsen theorem 3.3.11 and the Baer theorem 5.11.1 . This way of proof can be generalized to all finitely generated discontinuous groups of the plane and even to free products with arbitrary amalgamated subgroups [Zieschang 1980]. This proof is an unpleasant combinatorial one. A pleasant topological proof of the theorem for surfaces was given by [Hendriks-Shastri 1978]. We will generalize their proof to the above case. Unfortunately, we have to use some results from algebraic topology, for instance, the simplests results of obstruction theory and of chapter 5. Thus the proof remains sketchy.

Proof. Let $p_1,\dots,p_m$ be the images of the rotation centers on $\mathbb{E}/G$. Let $D_1',\dots,D_m'$ be pairwise disjoint disks such that $p_i \in D_i'$. Now we remove the interior of each D_i' and glue the boundary of another disk D_i to the boundary of D_i' by a mapping of degree h_i. (If we parameterize the boundaries of D_i' and D_i by the complex numbers of absolute value 1 then such gluing mapping is given by $z \mapsto z^{h_i}$.) We obtain a topological space M which looks like a manifold except at the points of the ∂D_i. The fundamental group of M is isomorphic to G.

Let X_i be a $K(G_i,1)$ space, i.e. a CW-complex with $\pi_1(X_i) = G_i$ and $\pi_j(X_i) = 0$ for $j \geq 2$. The embeddings $A \hookrightarrow G_i$ can be realized by mappings $g_i\colon S^1 \to X_i$, and we obtain the mapping cylinders $X_{-1} \cup S^1 \times [-1,0]$ and $S^1 \times [0,1] \cup X_1$. We glue those together at $S^1 = S^1 \times 0$ and obtain the space $X = X_{-1} \cup S^1 \times [-1,1] \cup X_1$, see figure:

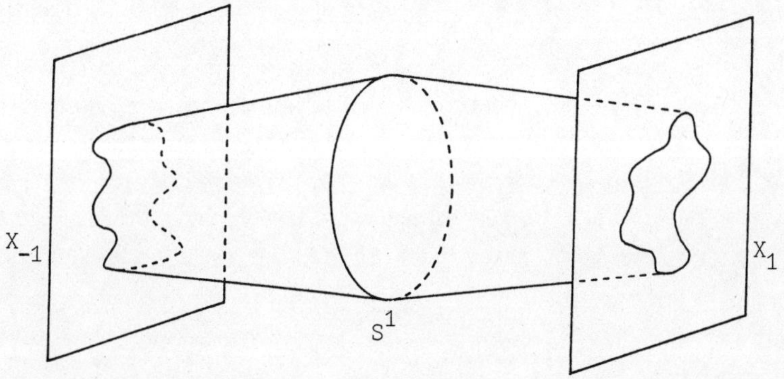

From the Seifert-van Kampen theorem it follows that $\pi_1(X) = G$ and that the fundamental groups of X_{-1}, X_1 and S^1 are embedded into G the same way as G_{-1}, G_1 and A. The space X is a $K(G,1)$-space, hence we may define a mapping $f\colon M \to X$ which induces an isomorphism between the fundamental groups $\pi_1(M)$ and $\pi_1(X)$. Since in a free product with amalgamation an element of finite order is conjugate to an element of a factor we may assume that the boundary of any disk D_j is mapped into X_{-1} or X_1, hence it does not intersect S^1.

Next we deform f to bring $f^{-1}(S^1)$ into general position; we denote the resulting mapping also by f. Then $f^{-1}(S^1)$ consists of a number of simple closed curves in M, none of which intersects the boundary of a disk D_j or is nullhomotopic. A slight generalization of the Baer theorem 5.11.1 is that two disjoint simple closed curves on M bound an annulus if they are homotopic, but not null-homotopic and do not cross the ∂D_j. By a deformation of f we can get rid of a pair of neighbouring curves from $f^{-1}(S^1)$. Finally we end with a mapping, again denoted by f, which has at most one curve in $f^{-1}(S^1)$. If $f^{-1}(S^1)$ is empty then M would be mapped into one 'side' of X, thus $\pi_1(M)$ into one of the factors $\pi_1(X_i) = G_i$. This contradicts the assumption that f induces an isomorphism of the fundamental groups, since each factor $\pi_1(X_i)$ has infinite index in $\pi_1(X)$. Next we use the fact that a simple closed curve in M that avoids the ∂D_i is not homotopic to a proper power of another curve, see exercise E 3.10. Therefore f maps $f^{-1}(S^1)$ homeomorphically onto S^1. From the Seifert-van Kampen theorem it follows that $f^{-1}(S^1)$ separates M into two parts M_{-1} and M_1 which are mapped to X_{-1} and X_1, resp., and f defines isomorphisms between the fundamental groups of M_i and X_i.

Now we have proved (b). The assertion (a) is a simple consequence of (b) and the classification theorem 4.6.3 for planar discontinuous groups.

□

The argument becomes more interesting and complicated when the quotient surface is non-orientable because of the appearance of Möbius strips; this is the main part of the proof in [Hendriks-Shastri 1978].

The algebraic decompositions of planar discontinuous groups into free products with amalgamation on one hand and the geometric decomposition of the action on the other, were considered in [Zieschang 1976] and [Lyndon 1978]. The geometric problem is solved, but for the algebraic question there are some unsolved cases. For example, the groups $\langle t,u \mid [t,u]^{2k+1} \rangle$, $k \geq 1$, have indecomposable actions, but they can be decomposed into proper free products with amalgamated subgroups, see exercise E 4.22.

4.13 PLANAR GROUP PRESENTATIONS AND DIAGRAMS

In combinatorial group theory it is sometimes convenient to represent a group with given generators by a 1-complex, the Cayley diagram or Dehn Gruppenbild. In fact we have already used these diagrams, without saying so, in the solution of the word problem in 4.9. The groups we are interested in have nice presentations which correspond to useful 2-complexes. We have used these complexes in the proof of the existence theorem 4.7.1.

The treatment would be much simpler if we were to consider only planar discontinuous groups without reflections. This case was considered in [Zieschang 1966], following an idea of K. Reidemeister, see also [Reidemeister 1932]. The more general situations have been dealt with in [Lyndon - Schupp 1977] and [Gramberg-Zieschang 1979]. We repeat the approach of the latter here, applying the refined Reidemeister-Schreier process of 2.2, and continuing to use the notation of that section.

4.13.1 Definition. Let $G = \langle S \mid R \rangle$. Two letters $v, w \in S \cup S^{-1}$ are called *algebraic neighbours* iff v^{-1} precedes w in a word R^{ε} where $R \in R$ and $\varepsilon \in \{1,-1\}$. Here R^{ε} is considered as a 'cyclic' word.

For instance, if $R = s^k$, $k \geq 1$, then s^{-1} is an algebraic neighbour of s. If v is an algebraic neighbour of w then w is an algebraic neighbour of v, too . In the study of crystallographic groups of the plane it is convenient to characterize the reflections already in the presentations as follows, cf. [Hoare - Karrass - Solitar 1973], [Lyndon - Schupp 1977, III. 7, 8].

4.13.2 Let $G = \langle S' \cup S'' \mid R' \cup R'' \rangle$ where

$S' = \{x_i \mid i \in I'\}$, $S'' = \{y_i \mid i \in I''\}$, $S' \cap S'' = \emptyset$,
$R' = \{R_j \mid j \in J\}$, where each R_j is a word in the letters of $S' \cup S'^{-1} \cup S''$,

$R'' = \{y_i^2 \mid i \in I''\}$.

The sets I', I" and J are finite or countable. The letters y_i and y_i^{-1}, $i \in I''$, are identified.

4.13.3 For each R_j, $j \in J$, the *root* R_j^* is defined by: $R_j \triangleq R_j^{*h_j}$ where $h_j \in \mathbb{N}$ is maximal, and $\triangleq$ denotes equality in the free group.

Let R^* consist of the roots R_j^* and their inverses R_j^{*-1} if these do not only differ by a cyclic permutation. Otherwise only R_j^* belongs to R^*. (At this place we use essentially that y_i and y_i^{-1} ($i \in I''$) are identified; for example, for $xy_1x^{-1}y_2$ the inverse $y_2^{-1}xy_1^{-1}x^{-1} = y_2xy_1x^{-1}$ is a cyclic permutation of the original word and only one of these two words would belong to R^*.) The multiplicity of a neighbour v of w where $v, w \in S' \cup S'^{-1} \cup S''$ is the number of occurrences of $w^{-1}v$ in the collection R^* of (cyclic) words. The sum of the multiplicities of the algebraic neighbours is called the *number of neighbours*.

4.13.4 Assume that the presentation $\langle S' \cup S \mid R' \cup R'' \rangle$ of the group G has the following properties:

(a) The number of neighbours of any element $v \in S' \cup S'^{-1} \cup S''$ is at most 2. (This number measures how often v appears in the roots of the defining relations.)

(b) For $y \in S''$ the number of neighbours is 2. Iff the relation $R \in R'$ contains y, then R^{-1} is a cyclic permutation of R. (Remember that $y_i^{-1} = y_i$ for $i \in I''$. By this condition we characterize 'reflections' in G.)

(c) The order of $R_j^*(s)$ is h_j, $j \in J$. We have $y_i \neq 1$ for $i \in I''$.

Using (a-c) we can divide the set $S' \cup S'^{-1} \cup S''$ into a system of sequences in which consecutive letters are algebraic neighbours. We take a system of maximal sequences or - what amounts to the same - a system with a minimal number of sequences. Thus we obtain the following lemma:

4.13.5 Lemma. *Let the presentation* $\langle S' \cup S'' | R' \cup R'' \rangle$ *have the properties 4.13.4. Then the system* $S' \cup S'^{-1} \cup S''$ *can be divided into a system* L *of pairwise disjoint sequences* $L_k = (\ldots, z_{k1}, z_{k2}, \ldots)$ *such that two letters are consecutive in some sequence if and only if they are algebraic neighbours. Each letter occurs exactly once in the sequences* $L_k \in L$*; the first and the last letter in a sequence may be algebraic neighbours or not.*

□

Each L_k from 4.13.5 is called a sequence of neighboured elements from $S' \cup S'^{-1} \cup S''$, or simply a sequence of neighbours. If the first and the last letters of a finite sequence L_k are algebraic neighbours the sequence is called *closed*.

4.13.6 Definition. (a) The presentation $\langle S' \cup S'' | R' \cup R'' \rangle$ of the group G is called *planar* if it has the properties 4.13.4 (a-c) and in addition the following properties i) - iii),

i) If $|L| > 1$ each sequence is finite and the first and the last letter of L_k have at most one algebraic neighbour.

ii) If L consists only of one infinite sequence L then L does not have an initial or final member, i.e. all members have two algebraic neighbours.

iii) If L consists of one finite sequence L then L may be closed or not. In the second case the first and the last terms in L do have only one algebraic neighbour.

(b) Let $\mu(\langle S' \cup S'' | R' \cup R'' \rangle) := 2 \cdot (|S'| + |S''| - \sum_{R \in R' \cup R''} \frac{1}{|\mathrm{Stab}_R|} - 1)$.

Here we use the order R" < R' for R' ∈ R', R" ∈ R''. (For the definition of Stab cf. 2.2.4.)

In the definition of the sequence of neighbours we avoid sequences with periods and assume that each letter occurs only once in the set of sequences. Planar presentations appear in the study of discontinuous groups of motions of the euclidean or non-euclidean plane and these are the main examples:

4.13.7 Theorem. *If G is a discontinuous group of the plane then any fundamental domain defines pairs of generators (the transformations that move the fundamental domain to a neighbour with a common side) and defining relations (which correspond to the inequivalent vertices of the fundamental domain). If the generators that correspond to reflections are put into* S''*, and from each other pair one member into* S'*, the result is a planar presentation.*

This can easily be checked on the canonical presentations of finitely presented discontinuous groups, see [Macbeath - Hoare 1976], [Zieschang 1976] ; the generalization to other 'geometrical presentations' can be proved using 4.14. The fundamental domain is compact iff there is only one sequence where the first and the last letter are neighbours. Next we will show that each planar presentation can be obtained that way, except when the group G is finite. (Then the presentation belongs to a discontinuous group of the sphere.) For the proof we use *modified Cayley diagrams*, see [Lyndon - Schupp 1977, p. 134], and we will first repeat their construction:

4.13.8 Construction of modified Cayley diagrams. Let G be presented as in 4.13.2. Now we construct a 2-complex as follows:
(a) Let C^o consist of vertices which are in a 1-1 correspondence with the elements of G and we assume that the vertices are labeled by the group elements.
(b) To each pair (g,z), where $g \in G$ and $z \in S' \cup S'^{-1} \cup S''$ corresponds exactly one directed edge which starts at the vertex g and ends at the vertex gz. The inverse edge $(g,x_i^\varepsilon)^{-1}$ equals $(gx_i^\varepsilon, x_i^{-\varepsilon})$ where $i \in I'$, $\varepsilon \in \{1,-1\}$ and the inverse edge $(g,y_i)^{-1}$ equals (gy_i,y_i) for $i \in I''$.

C^1 consists of C^o and the edges described above. Note that (g,y_i) and $(gy_i,y_i) = (g,y_i)^{-1}$ are equal as undirected edges. (With this exception C^1 is the usual group diagram corresponding to the generators $S' \cup S''$.) We label the (directed) edge (g,z) with the symbol z. To each pair (g,w) where $g \in G$ and w is a word in the generators now corresponds a uniquely determined path $\hat{w}$ in C^1 with initial point g such that the edgepath is labeled by w; the final point is the vertex gw. We call $\hat{w}$ the *realization of* w *at* g.
(c) To each pair (g, R_j^ε) where $g \in G$, $\varepsilon \in \{1,-1\}$ and $j \in J$ corresponds an oriented face $D(g,R_j^\varepsilon)$. The boundary of $D(g,R_j^\varepsilon)$ is the realization of R_j^ε at g and the face with the inverse orientation equals $D(g,R_j^{-\varepsilon})$. Two faces $D(g,R_j^\varepsilon)$ and $D(g',R_j^\eta)$ which have the same boundary are considered to be identical. (Here $j \in J$ is fixed, but the direction and the initial point may be altered.) The complex $C(S, S'; R)$ consisting of C^1 and the faces described is called a *modified Cayley (or group) diagram.*

4.13.9 Remark. (a) The group G acts on the 2-complex $C(S', S''; R)$ as a group of automorphisms in the usual way: for a fixed $g \in G$ the transformation t_g corresponding to g maps the vertex g' to $g'g$, the edge (g',z) to $(g'g,z)$ and the face $D(g',R_j^\varepsilon)$ to $D(g'g,R_j^\varepsilon)$. The boundary of a face is mapped to the boundary of the image face etc.
(b) $\pi_1 C(S',S'';R') = 1$.

4.13.10 Definition. The modified group diagram $C(S',S'';R')$ is called *planar* if the complex can be embedded into the plane or 2-sphere (if G is finite) in such a way that the group $\{t_g \mid g \in G\}$ is induced by a discontinuous group of homeomorphisms of the plane or sphere. If $|S' \cup S''| = \infty$ we embed only the 'complex' $C(S', S''; R')$ without the vertices. (These can be considered as 'ends' on the 'boundary' of the non-euclidean plane.)

Our next aim is the following theorem:

4.13.11 Theorem. *If* $\langle S' \cup S'' | R' \cup R'' \rangle$ *is a planar presentation of the group* G *then the modified Cayley diagram* $C(S', S''; R')$ *is planar.*

Before proving 4.13.11 we will get a criterion for planarity:

4.13.12 Definition. Let C be a 2-dimensional complex and p a vertex of C^0. Two (directed) edges $v, w \in$ star p, i.e., both starting at p, are called *geometrical neighbours* iff $v^{-1}w$ is part of the boundary of a face. If the edge v is the full boundary of a face then v and v^{-1} are considered to be neighbours. An ordered subset $L \subset$ star p is called a *geometrical sequence* iff each pair of consecutive elements are geometric neighbours and L is maximal with this property. If an edge v is not contained in the boundary of any face then v alone forms a geometrical sequence. A *geometrical sequence* is called *closed* iff it is finite and the first and the last elements are neighbours, otherwise it is called *open*.

4.13.13 Lemma. *Let* C *be a countable 2-complex with the following properties:*
(a) Each (oriented) edge of C *occurs at most twice in the boundaries of the (oriented) faces of* C.
(b) If the star at a vertex of C *is not a closed geometrical sequence then each sequence in this star is finite and the first, as well as the last, letter in each sequence has only one geometrical neighbour.*
(c) $\pi_1 C = 1$.
(d) From (a-c) it follows that C *is orientable. We fix an orientation, i.e. from each pair of inverse faces we choose one such that each (directed) edge belongs to the positive boundaries of the selected faces at most once.*

Let the 'complex' C^* *be obtained from* C *deleting all vertices which have infinite stars. Then there exists an embedding of* C^* *into the plane or 2-sphere such that the chosen orientation extends to an orientation of the plane or 2-sphere, resp.*

Proof. On a 2-complex C in which each edge occurs at most twice in the boundaries of faces we can introduce the notion of an orientation as follows: From each pair of inverse faces we take one. It determines a 'positive' boundary. If now each oriented edge occurs at most once in the positive boundaries of the chosen faces, then we say that we have an orientation on C.

The complex C consists of 2-connected components. By definition 4.1.1, two disks belong to the same 2-component iff they can be connected by a sequence of faces from C in which consecutive faces have a common edge in the boundary. If we fix the orientation of one face in a component C_i then the orientation of all faces of this component is uniquely determined to make C_i oriented. The orientations on different components can be chosen independently. If an edge e does not belong to the boundary of a face then e alone is considered as a 2-component of C.

If a 2-connected complex C_o has trivial fundamental group it is orientable.

From the Seifert-van Kampen theorem, see 2.8,we obtain the fundamental group of C as follows: Let $C_1, C_2, \ldots$ be the 2-connected components of C, ordered in such a way that C_i has a point p_i in common with $C_1 \cup \ldots \cup C_{i-1}$. Let $C_1', C_2', \ldots$ be disjoint copies of the C_i and let p_i' denote the image of p_i in C_i'. Now we glue these copies together as follows: Let $C_1'' = C_1'$. Next we identify the point p_2' with the corresponding point in C_1'' and obtain a complex C_2''. By induction we construct $C_1'', C_2'', C_3'', \ldots$ Let C_i'' be constructed. Then we identify the point p_{i+1}' with that point of C_i'' that corresponds to p_{i+1}. The fundamental group C_{i+1}'' is isomorphic to the free product of the fundamental groups of C_i'' and C_{i+1}', see 2.8. By induction we obtain a complex C" which contains copies of all C_i and whose fundamental group is the free product of the fundamental groups of the C_i. Finally we have to identify some vertices of C" to get to C. For each pair of identified vertices we have to add an infinite cyclic group as a free factor to the fundamental group, see 2.8. From $\pi_1 C = 1$ it follows that all 2-connected components have trivial fundamental groups and that we obtain a tree if we represent each component by a vertex and connect two of them by an edge if the corresponding components have a vertex in common.

Next we construct an embedding of C in the plane or 2-sphere. In each 2-component of C we fix an orientation. This is possible as noted in (d). Hence, at each vertex of the component, we can define a 'positive' star of the edges which start there, by the following rule: the edge s follows s' iff there is a positive face such that $s^{-1}s'$ is part of a positive boundary. Next we order the sequences ω_i so obtained in some order $\ldots, \omega_1, \ldots, \omega_m, \ldots$ and define the last term of ω_i to be a neighbour of the first term of ω_{i+1}; here i+1 mod m if m is the total number of sequences. This now defines cycles or infinite sequences at the vertices of C.Finally we introduce semifaces with infinite boundaries by saying that the first edge of the sequence ω_{i+1} follows the inverse of the last edge of ω_i. Next we drop all vertices which have an infinite star. Then the modified complex obtained has trivial fundamental group and represents a 2-manifold. (Obviously there are possibilities to decompose the 'infinite' disks into infinite systems of ordinary disks.) Hence, the modified complex represents either a plane or a 2-sphere, see Theorem 4.1.7.

□

4.13.14 *Proof of* 4.13.11. $C(S', S''; R')$ is a countable 2-complex. Now let $z \in S' \cup S'^{-1} \cup S''$ occur in the word $R_j^{*\varepsilon} = uzv$, $\varepsilon \in \{1,-1\}$, $j \in J$. There are at most two possibilities for the position of z in the words of R^*, by 4.13.4 (a). Now we consider one of them. Let $g \in G$ and $z \in S' \cup S'^{-1}$. The edge $(g,z) \in C^1$ occurs in the boundary of a face $D(g',R_j^\varepsilon)$ at a place corresponding to the chosen z in R_j^* iff $g' = gu^{-1}R_j^{*\varepsilon\alpha}$ where $0 \le \alpha \le h_j-1$. By 4.13.8 (c) the faces for the different α are identified, hence, (g,z) occurs at most twice in the boundaries of faces of $C(S', S''; R')$.

If $z \in S''$ then the edge $(g,z) = (gz,z)^{-1}$ also occurs in the boundary of the faces $D(gvR_j^{*\alpha}, R_j^{-\varepsilon})$, $0 \le \alpha \le h_j-1$. Different α determine the same boundary path up to the choice of the beginning, hence, by 4.13.8 (c) they correspond to the same face. If the number of algebraic neighbours of z is 1 then the edge (g,z) occurs in the boundary of at most two (oriented) faces. If the number of algebraic neighbours is 2 then R_j^{*-1} is a cyclic permutation of R_j^*, as postulated in 4.13.4 (b). After a cyclic permutation we may assume that $R_j^* = uzu^{-1}z'$ where $z' \in S''$ and u is a word in the generators $S' \cup S''$, i.e. we assume $v = u^{-1}z'$. The boundary of $D(gv,R_j^{-\varepsilon})$ is the path $(gv,R_j^{-\varepsilon}) = (gv,(z'uzu^{-1})^{h_j})$ which is a cyclic permutation of the path $(gu^{-1},(uzu^{-1}z')^{h_j}) = (gu^{-1},R_j)$, as $v(z'uzu^{-1})^{h_j}v^{-1} = u^{-1}(uzu^{-1}z')^{h_j}u$.

Thus we have proved that $C(S', S''; R')$ is a 2-complex that has the property 4.13.13 (a). The star condition 4.13.13 (b) follows immediately from the definition of geometric neighbours and the conditions 4.13.6 i-iii). As any closed edge-path in $C(S', S''; S')$ corresponds to a relation which is product of conjugates of the defining relations it follows that $\pi_1 C(S', S''; R') = 1$. Finally we have to fix an orientation on the complex such that a mapping t_g, $g \in G$, either preserves or reverses the orientation everywhere. Let $L = \{L_i \mid i = 1,2,\ldots\}$ be the system of sequences of generators from $S' \cup S'^{-1} \cup S''$, see 4.13.5. We fix an order on L; without loss of generality we may assume that it is given by the enumeration:

$$L = \begin{cases} \{L_i \mid 1 \le i \le k\} & \text{if } |L| < \infty \\ \{L_i \mid i \in \mathbb{Z}\} & \text{if } |L| = \infty. \end{cases}$$

In the second case we assume that the sequence is infinite to both sides. Now we order all elements of $S' \cup S'^{-1} \cup S''$ into a cycle $Z^{+1} = (L_1,\ldots,L_k)$ if $|L| < \infty$ or into a sequence to both sides infinite $Z^{+1} = (\ldots,L_{-1},L_0,L_1,\ldots)$. Here each L_i is by definition a sequence of elements from $S' \cup S'^{-1} \cup S''$ and finite if $|L| > 1$. By Z^{-1} we understand the inversely directed sequence to Z^{+1}.

Now let u,v be consecutive letters in some L_i^ε. Then there is a R_j^η such that

$u^{-1}v$ is part of it. Let w denote the following letter in R_j^{η}, considered as cyclic word. If v^{-1}, w are consecutive letters in some L_r^{μ} then we define $o(v) = \varepsilon\mu$. It follows immediately: if o is defined for v then it is defined for v^{-1}, too, and $o(v) = o(v^{-1})$. Furthermore, a consequence of 4.13.4 (b) is that $o(y_i) = -1$ for $i \in I''$. On the remaining generators from S we give o an arbitrary value in $\{1,-1\}$. Then o can be extended to a homomorphism $o\colon G \to \mathbb{Z}_2 = \{1,-1\}$; because if we follow a relation R_j each pair of consecutive letters decides whether the direction of the star remains unchanged or has to be reversed, and when we are back to the beginning we get the original cycle.

Now we define the cycle (g,Z^{ε}) as positive iff $o(g) = \varepsilon$. This defines an orientation on the modified group diagram. As proved in 4.13.13 $C(S', S''; R')$ can be embedded into the plane (or sphere) so that the chosen orientation extends to an orientation of the plane (or sphere). Moreover, the transformation $t_g\colon C(S', S''; R') \to C(S', S''; R')$ (see 4.13.9 (a)) is a homeomorphism that extends to the plane. It preserves the orientation iff $o(g) = 1$. This is a consequence of the assumption that the edge cycles at the vertices equal Z^{+1} or Z^{-1}. This finishes the proof of 4.13.11.

□

4.13.15 Corollary. *Let the group G have a planar presentation $\langle S' \cup S'' \mid R' \cup R''\rangle$ where $S' \cup S''$ is finite. Then:*

(a) $R' \cup R''$ is finite, too.

(b) G is isomorphic to a discontinuous group of motions of the non-euclidean or euclidean plane or the sphere and has a canonical presentation as given in [Macbeath - Hoare 1976], [Zieschang 1976, 1.3,4].

(c) The given presentation $\langle S' \cup S'' \mid R' \cup R''\rangle$ corresponds to a fundamental domain of G and can be transformed into a canonical one by a finite number of elementary steps that correspond to bifurcations of the fundamental domain.

Proof. Dualizing the planar group diagram we obtain a planar complex C^* and an action of the group G on C^*. Then G operates simply transitively on the system of faces of C^*, thus each face is a fundamental domain for G. It is well known that the action of a finitely generated group on a planar complex can be realized by a discontinuous group of motions of the euclidean or non-euclidean plane or the sphere and that the fundamental domain can be altered by bifurcations to get a fundamental domain of canonical form, see 4.3.

□

4.13.16 Remark. That a planar presentation can be transformed into a canonical presentation of a discontinuous group of the plane can be proved without constructing a planar complex, by doing the corresponding processes in the presentation as in the proof that a discontinuous group has a canonical presentation.

4.14 SUBGROUPS AND THE RIEMANN-HURWITZ FORMULA

4.14.1 Theorem. *Let* $\langle S' \cup S'' \mid R' \cup R'' \rangle$ *be a planar presentation of the group* G *and* $U < G$ *a subgroup. (The notation is the same as in 4.13.2) On* $R' \cup R''$ *we define the order again by*

4.14.2 $R'' < R'$ *for all* $R'' \in R''$, $R' \in R'$.

Let K *be a Schreier system of coset representatives,* $\{\Theta_R \mid R \in R' \cup R''\}$ *a system of choice functions and* τ *the corresponding Reidemeister-Schreier rewriting process. Let*

4.14.3 (a) $S'(U,K) := \{x(i,K) := Kx_i(\overline{Kx_i})^{-1} \mid i \in I',\ K \in K,\ \overline{Kx_i} \neq Kx_i\}$

$\cup\ \{y(i,K) := Ky_i(\overline{Ky_i})^{-1} \mid i \in I'',\ K \in K,\ Ky_i \neq \overline{Ky_i} \neq K\}$,

(b) $S''(U,K) := \{y(i,K) := Ky_i(\overline{Ky_i})^{-1} \mid i \in I'',\ K \in K,\ \overline{Ky_i} = K\}$,

4.14.4 (a) $R'(U,K) := \{R(j,K) := (\Theta_{R_j}(K) R_j\, \Theta_{R_j}(K)^{-1})_\tau \mid j \in J,\ K \in K\}$

$\cup\ \{Q(i,K) := (\Theta_{y_i^2}(K) y_i^2\, \Theta_{y_i^2}(K)^{-1})_\tau \mid i \in I'',\ K \in K,\ \overline{\Theta_{y_i^2}(K) y_i} \neq \Theta_{y_i^2}(K)\}$,

(b) $R''(U,K) := \{Q(i,K) := (\Theta_{y_i^2}(K) y_i^2\, \Theta_{y_i^2}(K)^{-1})_\tau \mid i \in I'',\ K \in K,\ \overline{\Theta_{y_i^2}(K) y_i} = \Theta_{y_i^2}(K)\}$.

Then

4.14.5 $\langle S'(U,K) \cup S''(U,K) \mid R'(U,K) \cup R''(U,K) \rangle$ *is planar presentation of* U.

4.14.6 Corollary. *A subgroup of a discontinuous group of motions of the plane is also a discontinuous group of motions of the plane.*

This is a consequence of 4.14.1 and 4.13.15. Of course, the statement 4.14.6 is obvious by the definition of discontinuity of a group. But here we have a purely algebraic proof that subgroups of groups with a planar presentation, especially with a canonical one, have again a planar presentation. Moreover, we give a general method to find a planar presentation for the subgroup, see remark 4.13.16. Another proof without geometry is in [Hoare, Karrass, Solitar 1971, 1972, 1973].

4.14.7 Examples. For $\langle s,t \mid s^4, t^4, (st)^2\rangle \to \mathbb{Z}_4$ and $\langle s,t \mid s^6, t^3, (st)^2\rangle \to \mathbb{Z}_6$ the kernel is $\mathbb{Z} \oplus \mathbb{Z}$ and one obtains planar presentation when using the modified Reidemeister-Schreier method, as in 2.2.3. Another more intuitive example has been given in 2.2.13.

4.14.8 *Proof of theorem 4.14.1.* It is $G_{y_i^2} = \hat{G}$ and $\mathrm{Stab}_{y_i^2} = \{Y_i^\ell \mid \ell \in \mathbb{Z}\}$. If $K \sim_{y_i^2} K'$ where $K, K' \in \mathcal{K}$ then by definition 2.2.4

$$K'Y_i^\ell K^{-1} \in \varphi^{-1}(\mathcal{U}).$$

If $2 \mid \ell$ then $K' = K$. Otherwise $K' = uKy_i$ where $u \in \mathcal{U}$, hence,

14.14.9 $K' = \overline{K'} = \overline{uKy_i} = \overline{Ky_i}$.

$$y(i,K) \in \mathcal{S}''(\mathcal{U},K) \Leftrightarrow \overline{Ky_i} = K$$
$$\Leftrightarrow \overline{Ky_i} = K,\ [K' \sim_{y_i^2} K,\ K' \in \mathcal{K} \Rightarrow K' = K]$$
$$\Leftrightarrow \overline{Ky_i} = K,\ \Theta_{y_i^2}(K) = K \Leftrightarrow Q(i,K) = y(i,K)^2$$
$$\Leftrightarrow y(i,K)^2 \in \mathcal{R}''(\mathcal{U},K).$$

This proves:

4.14.10 $\mathcal{S}''(\mathcal{U},K)$ and $\mathcal{R}''(\mathcal{U},K)$ are in the correspondence as postulated in 4.13.2.

Again we identify $y(i,K)$ and $y(i,K)^{-1}$ in this case.
Let $R^*(j,K)$ denote the root of $R(j,K)$, cf. 4.13.3:
$R(j,K) = R^*(j,K)^{h(j,K)}$ where $1 \le h(j,K)$ is maximal.

For the following argument we may assume that $K = \Theta_{R_j}(K)$. Then

$$R^*(j,K)^{h(j,K)} = R(j,K) = KR_jK^{-1} = KR_j^{*h_j}K^{-1} = (KR_j^*K^{-1})^{h_j}.$$

It follows that $h(j,K) \mid h_j$ as h_j is maximal for G, but $h(j,K)$ is maximal only for $\mathcal{U} \subset G$. Then

4.14.11 $((KR_j^*K^{-1})^{m(j,K)})_\tau \triangleq R^*(j,K)$ where $h_j = m(j,K)\cdot h(j,K)$.

We assume now that $R^*(j,K)^{-1}$ is a cyclic permutation of $R^*(j,K)$. (Remember that $y(i,K) \in \mathcal{S}''(\mathcal{U},K)$ is identified with its inverse; but this is also a consequence of the identification of the y with y^{-1}.) Hence

$$K\, R_j^{*-m(j,K)} K^{-1} = U\, R_j^{*m(j,K)} U^{-1} \qquad \text{where } U \in \varphi^{-1}(\mathcal{U}).$$

This implies that R_j^{*-1} is a cyclic permutation of R_j^*. Hence:

4.14.12 For the determination of the number of neighbours of a generator $x(i,K)$ or $y(i,K)$, see 4.13.3, the relations R_j and $R(j,K)$ behave the same way.

Now we consider a fixed position in some R_j^*, $j \in J$, for some $z \in S' \cup S''$. Then $R_j^* = vzw$. Let $z(K)$ denote the generator $Kz(\overline{Kz})^{-1}$, $\overline{Kz} \neq Kz$, of $\mathcal{U}$. Then we have the following equivalences:

$z(K)$ occurs in $R^*(j,K')$ at the corresponding position; here $K' \in \mathcal{K}$ and $K' = \Theta_{R_j}(K')$.

$$\iff (R^*(j,K))_\tau = (K'R_j^{*a}v\, \overline{K'R_j^{*a}v}^{-1})_\tau \cdot z(K) \cdot (\overline{Kz}\, w\, R_j^{*m(j,K')-a-1}K)^{-1})_\tau$$

$$\iff \overline{K'R_j^*v} = K$$

$$\iff K'R_j^{*a}vK^{-1} \in \mathcal{U}$$

$$\iff K' \sim_{R_j} \overline{Kv^{-1}} \quad (\text{as } R_j^{*a} \in \mathrm{Stab}_{R_j})$$

$$\iff K' = \Theta_{R_j}(\overline{Kv^{-1}}).$$

This proves:

4.14.13 There is exactly one root $R_j^*(j,K')$ and in it one appropriate position that corresponds to a fixed position in R_j and the generator $z(K)$.

Similar arguments can be applied for $z \in S'^{-1}$ or the inverse of a root R_j^*. Now condition 4.13.4 (a) for the presentation of G implies that the presentation 4.14.5 of $\mathcal{U}$ has the property 4.13.4 (a).

As the presentation $G = \langle S' \cup S'' \mid R' \cup R'' \rangle$ has the property 4.13.4 (b) and as the elements von $S''(\mathcal{U},K)$ have a 'kernel' from S'' it follows from 4.13.12 and 4.14.13 that the presentation 4.14.5 has the property 4.13.4 (b) too. That $R^*(j,K)$ has the order $h(j,K)$ follows from $h_j = h(j,K)m(j,K)$, 4.14.11 and the assumption that R_j^* has order h_j. Finally, if $y(i,K) \in S''(\mathcal{U},K)$ then $y(i,K) = Ky_iK^{-1} \neq 1$. Now the proof of 4.14.1 is completed.

□

Next we will give some refinements of theorem 4.14.1 and prove the combinatorial Riemann-Hurwitz formula. Here we will restrict ourselves to the case of groups with compact fundamental domain, although most of the steps can be extended to the general case. The difficulty then arising is the fact that generators can be dropped to get an obvious presentation of a free product, and that generators may lose their geometric properties.

4.14.14 Definition. Let $\langle S' \cup S'' | R' \cup R'' \rangle$ be a planar presentation of the group G that determines a closed sequence of neighboured elements, see 4.13.6. Let $R \in R'$ and $x \in S'$ be such that $R = vx^{\varepsilon}w$, where v,w are words in the generators different from x and $\varepsilon \in \{1,-1\}$. Now let $\overline{S'} = S' \setminus \{x\}$ and replace x in all relations of $R' \setminus \{R\}$ by $(wv)^{-\varepsilon}$; let $\overline{R'}$ denote the system of relations obtained for the generators $\overline{S'} \cup S''$. The replacement of the presentation $\langle S' \cup S'' | R' \cup R'' \rangle$ by $\langle \overline{S'} \cup S'' | \overline{R'} \cup R'' \rangle$ is called a *geometrical reduction*, the inverse process a *geometrical extension*. Two presentations are called *geometrically equivalent* if one results from the other by a finite number of geometrical reductions and extensions.

The definition of geometrical equivalence makes sense because of the following lemma which is a direct consequence of the definitions:

4.14.15 Lemma. *Let two presentations* $\langle S' \cup S'' | R' \cup R'' \rangle$ *and* $\langle \overline{S'} \cup S'' | \overline{R'} \cup R'' \rangle$ *be connected as in 4.14.14. If one of the presentations is planar and determines a closed sequence of neighbours then the other does also.* □

4.14.16 Lemma. *Assume that the presentations* $\langle S' \cup S'' | R' \cup R'' \rangle$ *and* $\langle S'_o \cup S''_o | R'_o \cup R''_o \rangle$ *are planar, determine closed sequences of neighbours and are geometrically equivalent. Then:*

(a) $|S'| - |R'| = |S'_0| - |R'_0|$.

(b) $\mu(\langle S' \cup S'' | R' \cup R'' \rangle) = \mu(\langle S'_o \cup S''_o | R'_o \cup R''_o \rangle)$.

(For the definition of μ *see 4.13.6 (b).)*

Proof. We consider an elementary step 4.14.14. Here we lose one relation from R' and one generator from S'. The replacement of x by $(vw)^{-\varepsilon}$ cannot turn any relation of $R' \setminus \{R\}$ into a trivial relation, as follows from 4.13.4 and the assumption that $\langle S' \cup S'' | R' \cup R'' \rangle$ involves a closed sequence of neighbours. Now (a) follows by induction. Since a replaced element x belongs to S', and x, x^{-1} together occur only once in the relation R it follows that $|\varphi(\mathrm{Stab}_R)| = 1$, and this implies (b). □

4.14.17 Corollary. $\mu(\langle S' \cup S'' | R' \cup R'' \rangle)$ *is an invariant of the group G. We define* $\mu(G) := \mu(\langle S' \cup S'' | R' \cup R'' \rangle)$, *if* $\langle S' \cup S'' | R' \cup R'' \rangle$ *is a planar presentation of G.* □

4.14.18 Corollary. *Let $\langle S' \cup S'' \mid R' \cup R'' \rangle$ be a planar presentation which determines a closed sequence of neighbours. We assume that the presented group F is torsion-free. Then:*

(a) $S'' = R'' = \emptyset$ and F is isomorphic to the fundamental group of a closed surface F_g of genus g where

$$g = \begin{cases} \frac{1}{2}(|S'| - (|R'| - 1)) & \text{if } F_g \text{ is orientable,} \\ |S'| - (|R'| - 1) & \text{if } F_g \text{ non-orientable.} \end{cases}$$

(b) $\mu(F) = 2\alpha g - 3$ *where* $\alpha = 2$ *if* F_g *is orientable and* $\alpha = 1$ *if* F_g *is non-orientable.*

Proof. By 4.13.15 (b), F is isomorphic to a torsionfree group that acts discontinuously on the plane, hence, F is isomorphic to the fundamental group of a surface. From the classification theory of discontinuous groups of the plane it is well known that any planar presentation is geometrically equivalent to the canonical presentation, see 4.6. Thus 4.14.18 is a consequence of 4.14.16.

□

A discontinuous group G of motions of the euclidean or non-euclidean plane with compact fundamental region has a planar presentation of the following type, see theorem 4.5.6.

4.14.19 Generators:

(a) s_i $\quad 1 \le i \le m$,

(b,2) t_j, u_j $\quad 1 \le j \le g$ (or (b,1) v_j $\quad 1 \le i \le g$),

(c) e_ℓ $\quad 1 \le \ell \le q$,

(d) $c_{\ell k}$ $\quad 1 \le \ell \le q$, $1 \le k \le m_\ell + 1$, $m_\ell \ge 1$,

4.14.20 Defining relations:

(a) $s_i^{-h_i}$ $\quad$ for $1 \le i \le m$ where $h_i \ge 2$,

(b) $c_{\ell k}^2$ $\quad$ for $1 \le \ell \le q$, $1 \le k \le m_\ell + 1$,

(c) $(c_{\ell,k} c_{\ell,k+1})^{h_{\ell k}}$ for $1 \le \ell \le q$, $1 \le k \le m_\ell$ where $h_{\ell k} \ge 2$,

$c_{\ell 1} e_\ell c_{\ell, m_\ell + 1} e_\ell^{-1}$ for $1 \le \ell \le q$,

(d,2) $\prod_{i=1}^{m} s_i \prod_{j=1}^{g} [t_j, g_j] \prod_{\ell=1}^{q} e_\ell$ (or (d,1) $\prod_{i=1}^{m} s_i \prod_{j=1}^{g} v_j^2 \prod_{\ell=1}^{q} e_\ell$).

We have:

4.14.21 $$\mu(G) := 2\cdot \sum_{i=1}^{m} \left(1-\frac{1}{h_i}\right) + \sum_{\ell=1}^{q} \sum_{k=1}^{m_\ell} \left(1-\frac{1}{h_{\ell k}}\right) + 2q + 2\alpha g - 4$$

Here $\alpha = 2$ in the case 4.14.19 (b,2) and 4.14.20 (d,2), otherwise $\alpha = 1$. (If G is a group of motions of the non-euclidean plane then $\mu(G)\cdot\pi$ is the measure of the fundamental domain of G. A group G given by a presentation <4.14.19|4.14.20> acts as a group of motions on the non-euclidean plane if $\mu(G) > 0$, on the euclidean plane if $\mu(G) = 0$ and on the sphere if $\mu(G) < 0$.)

The number $\mu(G)$ can be determined purely algebraically: A group that contains orientation reversing elements is not isomorphic to a group consisting only of orientation preserving mappings, hence, the subgroup of orientation preserving elements is characteristic, see Corollary 4.8.5. Now the numbers m, h_i, $h_{\ell k}$, m_ℓ and q can be determined and finally g and the type α.

4.14.22 Theorem of the Riemann-Hurwitz-formula. *Let G be a discontinuous group of motions of the non-euclidean or euclidean plane with compact fundamental domain and H a subgroup of G of finite index. Then*

$$\mu(H) = [G:H]\cdot\mu(G).$$

The proof of the theorem is a consequence of the following proposition and the fact that each discontinuous group with compact fundamental domain contains a subgroup of finite index which is isomorphic to the fundamental group of a closed orientable 2-manifold, see Theorem 4.10.1 (The last statement can also be proved purely algebraically.)

4.14.23 Proposition. *Let G be a discontinuous group of motions of the non-euclidean or euclidean plane and let U be a subgroup of finite index that is isomorphic to the fundamental group of a closed orientable surface of genus γ. Then*

$$4\gamma - 4 = \mu(U) = [G:U]\cdot\mu(G).$$

Proof. We apply theorem 4.14.1 and determine the numbers of generators and relations in the presentation 4.14.5.

As U is torsionfree, $S''(U,K) = \emptyset = R''(U,K)$ for any system of coset representatives. Now, by 2.2.

4.14.24 $$\begin{aligned} |S'(U,K)| &= [G:U]\cdot(|S'| + |S''| - 1) + 1 \\ &= [G:U]\cdot\left(m + \alpha g + \sum_{\ell=1}^{g} (m_\ell+1) + q - 1\right) + 1, \end{aligned}$$

where we used the presentation <4.14.19|4.14.20> of G and put the elements $c_{\ell,k}$ into S'', the other generators into S'.

The partial order in the set of relations is given in 4.14.2. Now we have to determine how many relations in $R'(U,K)$ are obtained from a given relation $R \in R' \cup R''$. We will prove:

4.14.25

relation	$c_{\ell k}^2$	$s_i^{-h_i}$	$(c_{\ell k}c_{\ell,k+1})^{-h_{\ell k}}$	$c_{\ell 1}e_\ell c_{\ell,m_\ell+1}e_1^{-1}$	Π
$\lambda(R,K)$	2	h_i	$2h_{\ell k}$	2	1

where Π denotes the long product relation 4.14.20 (d,α).

For the proof we will apply lemma 2.2.5 and we have to determine the orders of the stabilizers. By simple calculations it follows that

$$\varphi(\mathrm{Stab}_{c_{\ell k}^2}) = \{1,c_{\ell k}\},\ \varphi(\mathrm{Stab}_{s_i^{h_i}}) = \{s_i^a \mid 0 \le a < h_i\},$$

$$\varphi(\mathrm{Stab}_{(c_{\ell k}c_{\ell,k+1})^{h_{k\ell}}}) = \{(c_{\ell k}c_{\ell,k+1})^a c_{\ell k}^b \mid 0 \le a < h_{\ell k},\ 0 \le b \le 1\},$$

$$\varphi(\mathrm{Stab}_{c_{\ell 1}e_\ell c_{\ell,m_\ell+1}e_\ell^{-1}}) = \{1,c_{\ell 1}\}$$

and that the stabilizer of the product relation Π consists of the powers of this element only which are mapped to the unit element by φ. Hence $\varphi(\mathrm{Stab}_R)$ is finite for all relations, thus $\lambda(R,K)$ does not depend on K.
As U is torsionfree it follows from 2.2.5 that

$$\lambda(R,K) = \lambda(R,1) = |\varphi(\mathrm{Stab}_R)| \quad \text{for } R \in R' \cup R'',\ K \in K.$$

This finishes the proof of 4.14.25.

From 4.141, 2.2.11 and 4.14.25 it follows that $R'(U,K)$ consists of

$$[G:U]\cdot\left(\frac{1}{2}\sum_{\ell=1}^{q}(m_\ell+1) + \sum_{i=1}^{m}\frac{1}{h_i} + \sum_{\ell=1}^{q}\sum_{k=1}^{m_\ell}\frac{1}{2h_{\ell k}} + \frac{g}{2} + 1\right)$$

relations. As U is the fundamental group of an orientable closed surface of genus γ it has a planar presentation with 2γ generators and one defining relation; the canonical presentation is an example. Using 4.14.18,24 and 4.14.21 we conclude that

$$\frac{1}{2}(\mu(U) + 2) = 2\gamma - 1 = |S'(U,K)| - |R'(U,K)|$$

$$= [G:U] \cdot \left[m + \alpha g + \sum_{\ell=1}^{q} (m_\ell + 1) + q - 1 \right] + 1$$

$$- [G:U] \cdot \left[\frac{1}{2} \cdot \sum_{\ell=1}^{q} (m_\ell + 1) + \sum_{i=1}^{m} \frac{1}{h_i} + \sum_{\ell=1}^{q} \sum_{k=1}^{m_\ell} \frac{1}{2h_{\ell k}} + \frac{q}{2} + 1 \right]$$

$$= 1 + \frac{1}{2}[G:U] \cdot 2 \cdot \left[\sum_{i=1}^{m} (1 - \frac{1}{h_i}) + \sum_{\ell=1}^{q} \sum_{k=1}^{m_\ell} (1 - \frac{1}{h_{\ell k}}) + 2q + 2\alpha g - 4 \right]$$

$$= 1 + \frac{1}{2}[G:U] \cdot \mu(G)$$

This finishes the proof of 4.14.23.

□

The formula in 4.14.22 or the more special one in 4.14.23 is known as Riemann-Hurwitz formula in the theory of Fuchsian groups, in fact, both sides multiplied by π. Here we have given a combinatorial, algebraic proof of the formula without using results of the theory of functions in a complex variable. Our proof is similar to that in [Hoare-Karrass-Solitar 1972, 1972, 1973]. Another combinatorial proof is given in [Lyndon 1976] where an (combinatorial) angle measure for presentations is used. This angle measure corresponds to the measure of polygons in the non-euclidean plane, as does our $\mu(G)$. See also [Shepardson 1973], [Zimmermann 1977].

As a first application we give a bound for the number of 'automorphisms' of a closed surface.

<u>4.14.24 Corollary ([Hurwitz 1893] [Siegel 1954]).</u> *Let* F *be a closed orientable surface of genus* $\gamma \geq 2$ *and let* A *be a discontinuous group of orientation preserving homeomorphisms of* F. *Then* $|A| \leq 84(\gamma-1)$.

Proof. We lift A to the universal cover $\mathbb{E}$ of F and obtain a discontinuous group G on $\mathbb{E}$ with compact fundamental domain, hence, G has a presentation <4.5.7|4.5.8> with $q = 0$. The Riemann-Hurwitz formula 4.14.22 and the definition 4.14.21 now read:

$$0 < 2\gamma - 2 = |A| \cdot \left[\sum_{i=1}^{m} (1 - \frac{1}{h_i}) + 2g - 2 \right] =: |A| \cdot I.$$

We have $I > 0$. We are looking for a lower bound for I. Let $h_1 \leq h_2 \leq \ldots \leq h_m$. Now the following is easily checked:

$$\begin{array}{lll}
g > 0 & \Rightarrow & I > \frac{1}{2} \\
g = 0,\ m \geq 5 & \Rightarrow & I \geq \frac{1}{2} \\
\quad m = 4 \Rightarrow h_3 \geq 3 & \Rightarrow & I \geq \frac{1}{3} \\
\quad m = 3,\ h_1 \geq 4 & \Rightarrow & I \geq \frac{1}{4} \\
\qquad h_1 = 3 \Rightarrow h_3 \geq 4 & \Rightarrow & I \geq \frac{1}{12} \\
\qquad h_1 = 2,\ h_2 \geq 4 \Rightarrow h_3 \geq 5 & \Rightarrow & I \geq \frac{1}{20} \\
\qquad\quad h_2 = 3 \Rightarrow h_3 \geq 7 & \Rightarrow & I \geq \frac{1}{42}\,.
\end{array}$$

Hence, $42\cdot 2\cdot(\gamma - 1) \geq |A|$.

□

For more results of the type above see [Hurwitz 1893, pg. 423]. The first case where the bound $84(g-1)$ is obtained is in [Klein 1879].

4.14.25 Corollary. *(a) If* F *is a torsionfree normal subgroup of the triangle group* $\mathcal{D} = \langle s_1, s_2, s_3 \mid s_1^2, s_2^3, s_3^7, s_1 s_2 s_3\rangle$ *and* $[\mathcal{D}:F]$ *is finite, then* F *is the fundamental group of an orientable closed surfaces of genus* γ *where*

$$[\mathcal{D}:F] = 84\cdot(\gamma - 1).$$

This way infinitely many surfaces can be obtained ([Macbeath 1965']).

(b) Let F *be a closed orientable surface of genus* $\gamma \geq 2$ *and let* $\pi_1(F)$ *not be isomorphic to a normal subgroup of* $\mathcal{D}$*. Then any discontinuous group of orientation preserving self homeomorphisms of* F *has an order* $\leq 48\cdot(\gamma - 1)$.

Proof. The assertions are easy consequences of 4.14.22 and 4.10.1 and the fact that every closed orientable surface of genus $\gamma \geq 2$ is covered by infinitely many other closed surfaces, see E 4.22.

□

The question about upper bounds for abelian or cyclic groups of orientation preserving self mappings of a closed surface, or for lower bounds for a given genus, is more complicated. From E 2.23 we extract some partial answers:

4.14.26 Proposition. *Let* F *be a closed orientable surface of genus* γ*. Then* F *has the following groups of orientation preserving self homeomorphisms:*

(a) The group C *of order* $8(\gamma + 1)$*.*

(b) The abelian group A *of order* $4(\gamma + 1)$*.*

(c) A cyclic group of order $2(\gamma+1)$.

Proof. Take $k := g+1$, $g \geq 2$. Using the Reidemeister-Schreier method we show that (in the notation of E 2.22) $A \cong \mathbb{Z}_2 \oplus \mathbb{Z}_{2(g+1)}$. It follows that the elements a,b, ab from C have orders 4, $2(g+1)$, and 2, resp. The group $\mathcal{D}$ from E 2.23 is a triangle group, hence the elements of finite order are conjugate to powers of s_1, s_2 and s_1s_2, see 4.8.1. Thus Kern ρ is torsionfree, hence, isomorphic to the fundamental group of a closed orientable surface of some genus γ, see 4.8.1 (d). Using the Riemann-Hurwitz formula we conclude:

$$2\gamma - 2 = 8(g+1)\left(1 - \frac{1}{2} - \frac{1}{4} - \frac{1}{2(g+1)}\right),$$

hence $\gamma = g$. This proves (a) (and E 2.23 (a,b)). Now (b) and (c) are direct consequences from $A \cong \mathbb{Z}_2 \oplus \mathbb{Z}_{2(g+1)}$.

□

By a detailed study of the Riemann-Hurwitz formula one can obtain the following

4.14.27 Results. *Let* $N(\gamma)$ *and* $N_A(\gamma)$ *denote the maximal order of a group or an abelian group of orientation preserving self homeomorphisms of the closed orientable surface of genus* γ. *We have seen:*

$$8(\gamma+1) \leq N(\gamma) \leq 84\cdot(\gamma-1), \quad 4(\gamma+1) \leq N_A(\gamma).$$

Further results:

(a) $N(\gamma) = 8(\gamma+1)$ *for infinitely many* γ ([Accola 1968], [Maclachlan 1969]). The simplest γ to mention are in [Accola 1968]:

(b) If p *is a prime such that* $(\frac{p-1}{2}, 42) = 1$ *and* $p > 863$, *then* $N(2p+1) = 8(2p+2)$.

(c) If p *is a prime such that* $p \equiv 59 \bmod 60$ *and* $p > 214$, *then* $N(p+1) = 8(p+4)$, ([Accola 1968]). Other cases in [Kiley 1970].

(d) $N_A(\gamma) = 4(\gamma+1)$. This result can be extracted from [Maclachlan 1965] according to [Maclachlan 1969, proof of Theorem 3]. See also [Accola 1968 (5)].

(e) The maximal order of a cyclic group acting on the closed orientable surface of genus γ *ist* $2(2\gamma+1)$ [Wiman 1895/96], [Harvey 1966]. From the second paper we take the example that shows that $\mathbb{Z}_{2(2\gamma+1)}$ acts on the surface of genus γ. *There is an epimorphism*

$$\mathcal{D}_0 = \langle s_1, s_2, s_3 \mid s_1^2, s_2^{2\gamma+1}, s_3^{2(2\gamma+1)}, s_1s_2s_3 \rangle \to \mathbb{Z}_{2(2\gamma+1)}$$

such that the kernel is isomorphic to the fundamental group of a closed orientable surface of genus γ. *(Proof* as Exercise E 4.23.)

Automorphisms of Riemann surfaces are the subject of many papers. We mention here some more, especially those which contain results that can be proved without using the complex analytic structure, but only combinatorial arguments based on the Riemann-Hurwitz formula 4.14.22: [Accola 1970, 1971], [Greenberg 1960, 1963], [Jones - Singerman 1978], [Macbeath 1961', 1965'], [Maclachlan 1971], [Natanson 1978] [Sah 1969], [Singerman 1974', 1976], [Timmann 1976]. See also the paper mentioned in 4.15.

Exercises: E 4.18-23.

4.15 FINITE GROUPS ACTING ON CLOSED SURFACES

As we have seen in 4.10, each planar discontinuous group G contains a normal subgroup F of finite index that is isomorphic to the fundamental group of an orientable surface F. If G has compact fundamental domain then the surface F is closed. We can interpret F as factor space $\mathbb{E}/F$ and the factor group $A := G/F$ acts on F.

Conversely, if A is a finite group acting on the closed orientable surface F then, by lifting the action to the universal cover, we obtain a planar discontinuous group G; here we assume that F is not a sphere. In 4.14.24-26 we have given relations between the genus of F and the order of A. Let us now consider the effect to the action to the groups associated with F. Since the basepoint is not fixed by the action, the effect on the fundamental group is not properly defined. This is better with the homology group, which is the abelianized fundamental group. Our basic result ist the following:

4.15.2 Proposition. *Let G be a planar discontinuous group with compact fundamental domain and let $F \triangleleft G$ be a normal subgroup which is isomorphic to the fundamental group of a closed orientable surface* S *of genus* $\gamma > 1$. *Then* $[G:F] < \infty$ *and the action of* G/F *on* $F/[F,F]$:

$$y \cdot [F,F] \mapsto x^{-1} y x \cdot [F,F], \; x \in G, \; y \in F,$$

is effective, i.e.

$$x^{-1}yx[F,F] = y\cdot[F,F] \quad \textit{for all } y \in F \quad \Rightarrow \quad x \in F.$$

Let us first prove that $[G:F]$ is finite:

4.15.2 Lemma. Let G be a planar discontinuous group and $N \neq 1$ a finitely generated normal subgroup of G. Then $[G:N] < \infty$.

Proof. Because of 4.10.8 we may restrict ourselves to the case where G is the fundamental group of a closed orientable surface. If $[G:N] = \infty$ then N acts on $\mathbb{E}$ with non-compact fundamental domain and is a free group as follows from the presentation 4.11.3, since $m = 0$ and $r = r_1 > 0$. But the fundamental group of a closed surface, different from S^2, is not free, see E 2.18.

□

Proof of 4.15.1. Assume that the action of G/F on $F/[F,F]$ is not effective. Then there is a $x \in G$, $x \notin F$ such that $x^{-1}yx \in y\cdot[F,F]$ for all $y \in F$. We may assume that the relative order of x is a prime p: $x^p \in F$. Let H be the subgroup of G generated by x and F and let $\varphi: H \to \mathbb{Z}_p$ be the homomorphism that maps F to 0 and x to 1. Then Kern $\varphi = F$ and the sequence

$$1 \to F \hookrightarrow H \to \mathbb{Z}_p \to 1$$

is exact. Since $[F,F] \subset$ Kern φ, the sequence

$$1 \to F/[F,F] \to H/[F,F] \to \mathbb{Z}_p \to 1$$

is also exakt. Since x operates trivially on F it follows (E 4.23) that

(1) $\quad H/[F,F] = \mathbb{Z}^{2\gamma} \oplus B \qquad$ where $B = \begin{cases} 1 & \text{or} \\ \mathbb{Z}_p & \end{cases}$.

Hence, $[H,H] \subset [F,F]$, thus
$[F,F] = [H,H]$.

The group H is a planar discontinuous group with compact fundamental domain. Therefore it has a presentation as in 4.5.6.

First, let us assume that x preserves orientation. Then H has a presentation <4.5.7|4.5.8> with $q = 0$ and $h_i = p$ $(1 \le i \le m)$. By abelianizing H we deduce from <4.5.7|4.5.8> and (1) that $g = \gamma$. From the Riemann-Hurwitz formula 4.12.23 and the definition 4.12.21 we obtain

$$4\gamma - 4 = p\cdot [2m(1 - \frac{1}{p}) + 4g - 4] \quad \Longleftrightarrow$$
$$2 - 2\gamma = m, \text{ since } p \neq 1.$$

This contradicts the assumptions $\gamma \geq 2$ and $m \geq 0$.

Now let x be orientation reversing. Then $p = 2$. If H has the presentation <4.5.7 (a,b,c,d)|4.58 (a,b,c,d)> then $q \geq 1$, since the v_j and $c_{i,j}$ are the only orientation reversing elements. By abelianizing it follows that

(2) $\qquad q = 1,\ m = 0 \text{ and } \gamma = g.$

If H has a presentation <4.5.7 (a,b',c,d)|4.5.8 (a,b,c,d')> then there are the following two possibilities:

(3) $\qquad q = 1,\ m = 0,\ 2\gamma = g,$

(4) $\qquad q = 0,\ m = 1,\ 2\gamma = g - 1.$

For the cases (3) and (4) we obtain from the Riemann-Hurwitz formula 4.12.23 that

$$4\gamma - 4 = 2\cdot[\sum_{k=1}^{m_1} (1- \frac{1}{h_{1k}}) + 2 + 2\alpha g - 4] \text{ where } \alpha = \begin{cases} 2 & \text{for the case (2)} \\ 1 & \text{for case (3).} \end{cases}$$

In both cases, $\alpha g = 2\gamma$, hence the above equation is equivalent to

$$\sum_{k=1}^{m_1} (1 - \frac{1}{h_{1k}}) + 2\gamma = 0.$$

This equation contradicts $h_{1k} \geq 2$, $\gamma \geq 2$. In case (4) we have

$$4\gamma - 4 = 2\cdot[2(1 - \frac{1}{h_1}) + 2g - 4],$$

which is equivalent to $0 = 2\gamma + 2 - \frac{2}{h_1}$, a contradiction.

□

Remark: The case where x reverses orientation can easily be excluded, since an orientation reversing transformation changes the sign of intersection numbers which are homological invariants.

By projecting to the factor surface $F = \mathbb{E}/F$ we obtain:

4.15.3 Corollary ([Hurwitz 1893]). *Let* F *be a closed orientable surface of genus* $\gamma \geq 2$ *and let* A *be a finite group acting on* F. *Then the induced action of* A *on the homology group* $H_1(F)$ *is effective.*

□

See exercise E 4.25.

Next we deal with another approach to finite groups of mappings on a surface and we will prove the even stronger results 4.15.12. We make the general

4.15.4 Assumption. Let F be a closed orientable surface of genus $\gamma \geq 2$ and $\psi: F \to F$ an orientation-preserving homeomorphism of prime order p. Let F_o denote the factor surface $F/\langle\psi\rangle$. Then the projection $F \to F_o$ is a branched covering, see 3.3.1. Let m denote the number of branch points in F_o and g the genus of F_o.

If we lift the action of $\langle\psi\rangle$ to the universal cover $\mathbb{E}$ of F we obtain a planar discontinuous group G which has a presentation as in 4.5.6. But now $h_1 = \ldots = h_m = p$ and $q = 0$. Moreover, we have an epimorphism $\varphi: G \to \mathbb{Z}_p$ with Kern $\varphi = \pi_1(F)$, hence, $\varphi(s_i) = k_i$ with $(p,k_i) = 1$. Next we normalize the generators of G with respect to φ and obtain:

4.15.5 Lemma. *G has a presentation*

$$G = \langle s_1,\ldots,s_m,t_1,u_1,\ldots,t_g,u_g \mid s_1^p,\ldots, s_m^p, \prod_{i=1}^{m} s_i \prod_{j=1}^{g} [t_j,u_j]\rangle$$

where

$$\varphi(s_i) = k_i \textit{ with } (p,k_i) = 1 \textit{ for } 1 \leq i \leq m,$$
$$\varphi(t_1) = \varphi(u_1) = \ldots = \varphi(t_g) = \varphi(u_g) = 0$$

if $m > 0$ *and*

$$\varphi(u_1) = \varphi(t_2) = \varphi(u_2) = \ldots = \varphi(t_g) = \varphi(u_g) = 0$$
$$(\varphi(t_1),p) = 1$$

if $m = 0$. *We call such a* presentation adapted to φ.

Proof. Starting with any canonical presentation we obtain an adapted one by bifurcations, as have been used in 3.2, see also 5.2. For instance, by

$$\ldots xyx^{-1}y^{-1} \ldots \;\to\; \begin{cases} \ldots (x\underline{y})x^{-1}(\underline{y}^{-1}x^{-1})x \ldots \\ \ldots (\underline{x}y)y(y^{-1}\underline{x}^{-1})y^{-1} \ldots \end{cases}$$

obtain a generator with $\varphi(x) + \varphi(y)$ as φ-image. After a finite number of similar steps we find generators x', y' for the free group generated by x,y with

$x'y'x'^{-1}y'^{-1} = xyx^{-1}y^{-1}$ and $\varphi(x') = (\varphi(x),\varphi(y))$ and $\varphi(y') = 0$. That there are processes to reduce the φ-image of elements of different commutators can also be proved by bifurcations, we pose it as exercise E 4.26. (The Lemma can also be deduced from Theorem 3.6.7 which has been proved by the same steps as describe above.)

□

Next we apply the Reidemeister-Schreier 2.2.1 method to get a presentation of $\pi_1(F)$; here it is convenient to use the improved form described in 2.2.8. As coset representatives we choose

$$\{s_1^{\ell} \mid 0 \le \ell < p\} \quad \text{if } m > 0,$$

$$\{t_1^{\ell} \mid 0 \le \ell < p\} \quad \text{if } m = 0.$$

Then we obtain in the first case:

4.15.6 Generators. $t_{j\ell} = s_1^{\ell} t_j s_1^{-\ell}$, $u_{j\ell} = s_1^{\ell} u_i s_1^{-\ell}$ $(1 \le j \le g,\ 0 \le \ell < p)$

$$s_i^{\ell} = s_i^{\ell} s_i s_1^{-\ell-k_i} \quad (2 \le i \le m,\ 0 \le \ell < p).$$

4.15.7 Defining relations:

$$\prod_{\ell=0}^{p-1} s_{i,\ell k_i} \qquad (2 \le i \le m,\ \ell k_i \text{ considered mod } p)$$

$$\Big(\prod_{i=2}^{m} s_{i,\ell+k_1+\ldots k_{i-1}}\Big) \prod_{j=1}^{g} [t_{j\ell},u_{j\ell}] \qquad (0 \le \ell < p).$$

In the second case:

4.15.6' Generators: $t = t_1^p$, $u_{1\ell} = t_1^{\ell} u_1 t_1^{-\ell}$,

$$t_{j\ell} = t_1^{\ell} t_j t_1^{-\ell},\ u_{j\ell} = t_1^{\ell} u_j t_1^{-\ell} \quad (2 \le j \le g, 0 \le \ell < p).$$

4.15.7' Defining relations:

$$u_{1,\ell+1} u_{1\ell}^{-1} \prod_{j=2}^{g} [t_{j\ell},u_{j\ell}] \qquad (0 \le \ell \le p-2)$$

$$t u_{10} t^{-1} u_{1,p-1} \prod_{j=2}^{g} [t_{j,p-1},u_{j,p-1}].$$

Let $\chi\colon \pi_1(F) \to \pi_1(F)$ be the automorphism $x \mapsto s_1 x s_2^{-1}$ or $x \mapsto t_1 x t_1^{-1}$, resp. Then $s_{j\ell} = \chi^{\ell}(s_{j0}) = \chi^{\ell}(s_j)$ etc. Now we obtain:

4.15.8 Theorem ([Nielsen 1937], [Gilman 1977]). *Let the situation be as in 4.16.4. Then there is an automorphism* $\chi: \pi_1(F) \to \pi_1(F)$ *"obtained from* ψ*"* [1] *such that the following is true:*

(a) If $m > 0$, $\pi_1(F)$ *is generated by elements*

$$\chi^{\ell}(t_j),\ \chi^{\ell}(u_j) \qquad 1 \le j \le g,\ 0 \le \ell \le p-1$$

$$\chi^{\ell}(s_i) \qquad 3 \le i \le m,\ 0 \le \ell \le p-2$$

such that $\pi_1(F)$ *has a single defining relation in which each generator and its inverse occurs exactly once. Hence,*

$$2\gamma = p\cdot 2g + (m-2)(p-1).$$

Further, for each $i \in \{3,\ldots,m\}$ *we have*

$$\prod_{\ell=0}^{p-1} \chi^{\ell k_i}(s_i) = 1,$$

or in homology:

$$s_i + \psi_*(s_i) + \ldots + \psi_*^{p-1}(s_i) \sim 0.$$

(b) If $m = 0$, $\pi_1(F)$ *is generated by*

$$\chi^{\ell}(t_j),\ \chi^{\ell}(u_j) \qquad 2 \le j \le g,\ 0 \le \ell \le p-1$$

$$t = t_1^p,\ u_1.$$

Again, $\pi_1(F)$ *has a single defining relation in which each generator and its inverse occur exactly once. We have* $\chi(t) = t$ *and* $\chi(u_1) \sim u_1$.

(c) In either case, the generators for $\pi_1(F)$ *have the following properties with respect to intersection numbers (for their definition see 3.*

$$\nu(\chi^{\ell}(t_j), \chi^{k}(u_i)) = \delta_{ij}\delta_{\ell k}$$

$$\nu(\chi^{\ell}(t_j), \chi^{k}(t_i)) = 0$$

$$\nu(\chi^{\ell}(u_j), \chi^{k}(u_i)) = 0$$

for $0 \le \ell, k \le p-1$, $1 \le i,j \le g$. *However, when case (ii) occurs,*

$$\nu(\chi^{\ell}(t), \chi^{k}(u_1)) = 1 \text{ for } 0 \le \ell, k \le p-1.$$

---

[1] In general, ψ does not induce an endomorphism ψ_* of $\pi_1(F)$ since ψ may be fixed point free, but ψ_* ist defined for homology.

Proof. (a,b) have been proved before the formulation of the theorem. (c) is a direct consequence since the commutators from 4.15.7 or 7' are brought to commutators of the single defining relation of $\pi_1(F)$.

□

Remark: The assertions (a) and (b) are from [Nielsen 1937], (c) is from [Gilman 1977].

Now we study the action of $\langle\psi\rangle$ on the homology $H_1(F)$, the abelianized $\pi_1(F)$. For $x \in \pi_1(F)$ we will denote the image in $H_1(F)$ also by x, but we write the operation additively. As free basis for $H_1(F)$ we use the basis obtained from the generators from 4.15.8:

$$\chi^{\ell}(t_j),\ \chi^{\ell}(u_j) \quad (1 \le j \le g,\ 0 \le \ell \le p-1),\ \chi^{\ell}(s_i)\ (3 \le i \le m,\ 0 \le \ell \le p-2)$$

or $$\chi^{\ell}(t_j),\ \chi^{\ell}(u_j) \quad (2 \le j \le g,\ 0 \le \ell \le p-1),\ t,\ u_1, \text{ resp.}$$

4.15.9 Definition. (a) $nH := \{c \in H_1(F) \mid c = nd \text{ for some } d \in H_1(F)\}$,
$C(n,\psi) := \{c \in H_1(F) \mid c - \psi_*(c) \in nH\}$.
(b) For $c \in H_1(F)$ let

$$\{c\} := \begin{cases} c & \text{if } \psi_*(c) = c \\ c + \psi_*(c) + \ldots + \psi_*^{p-1}(c) & \text{otherwise and} \end{cases}$$

$$[c] := \sum_{j=0}^{p-2} (j+1)\, \psi_*^j(c).$$

Note that both nH and $C(n,\psi)$ are subgroups of $H_1(F)$ and that $C(n,\psi) \subset nH$.

4.15.10 Corollary. *In the situation 4.16.4* $H_1(F)$ *has a basis that contains no element of* $C(n,\psi)$ *for* $n \ge 3$.

Proof. If $m \neq 0$, this is a direct consequence of 4.15.8. If $m = 0$, replace t by $t + t_2$ and u by $u + u_2$. Here $g \ge 2$, as follows, for instance, from the Riemann-Hurwitz formula 4.14.22.

□

4.15.11 Lemma. *Let* $c \in H_1(F)$. *Then* $\psi_*(c) \sim c$ *is and only if*

$$c = \sum_{i=1}^{g} (a_i\{t_i\} + b_i\{u_i\}) \text{ with } a_i, b_i \in \mathbb{Z}.$$

□

4.15.12 Corollary ([Accola 1967]). *Let F be a closed orientable surface of genus* $\gamma \geq 2$ *and let* $\psi: F \to F$ *be an orientation preserving homeomorphism of finite order. Assume that there are four linearily independend elements* $c_1, c_2, c_3, c_4 \in H_1(F)$ *such that*

$$\nu(c_1,c_3) = \nu(c_2,c_4) = 1,$$
$$\nu(c_1,c_2) = \nu(c_1,c_4) = \nu(c_2,c_3) = \nu(c_3,c_4) = 0, \text{ and}$$
$$\psi_*(c_i) = c_i \text{ for } 1 \leq i \leq 4.$$

Then ψ *is the identity.*

Proof. If $\psi \neq id_F$, we may assume that the order of ψ is a prime p. We choose a basis for $H_1(F)$ which is obtained from a system of generators of $\pi_1(F)$ adapted to ψ, as given in 4.15.8.

If $m \neq 0$, Lemma 4.15.11 and 4.15.8 (c) imply that for any $c, d \in H_1(F)$

$$\psi_*(c) = c, \ \psi_*(d) = d$$

may happen only if $\nu(c,d) \equiv 0 \bmod p$. Since by assumption, $\nu(c_1,c_3) = 1$ and $\psi_*(c_i) = c_i$, we see that $m = 0$.

Now let $c_i = w_i + a_i t + b_i u_1$ for $1 \leq i \leq 4$ where the w_i are linear combinations of the $\{t_j\}$, $\{u_j\}$ $(2 \leq j \leq g)$. From 4.15.8 (c) again it follows that

$$\nu(w_i,w_k) \equiv 0 \bmod p \text{ for } 1 \leq i, k \leq 4.$$

Now we compute:

$$1 = (c_1,c_3) = a_1b_3 - a_3b_1 + (w_1,w_3) \equiv a_1b_3 - a_3b_1 \bmod p$$
$$1 \equiv a_2b_4 - a_4b_2 \bmod p$$
$$0 \equiv a_1b_2 - a_2b_1 \equiv a_1b_4 - a_4b_1 \equiv a_2b_3 - a_3b_2 \equiv a_3b_4 - a_4b_3 \bmod p,$$

hence,

$$0 \equiv a_2(a_3b_4-a_4b_3) + a_4(a_2b_3-a_3b_2) = a_3(a_2b_4-a_4b_2) \equiv a_3 \bmod p,$$
$$0 \equiv b_4(a_2b_3-a_3b_2) + b_2(a_3b_4-a_4b_3) = b_3(a_2b_4-b_2a_4) \equiv b_3 \bmod p.$$

This contradicts $1 \equiv a_1b_3 - a_3b_1 \bmod p$.

Thus the order of ψ is 1.

□

4.15.13 Lemma. *Let* p *be the order of* ψ *and* $c \in H_1(F)$.
(a) Assume that p *and* n *are relatively prime or that* $m = 0$. *Then*

$$c \in C(n,\psi) \iff c = \sum_{j=1}^{g} (a_j\{t_j\} + b_j\{u_j\}) + nd$$

where $a_j, b_j \in \mathbb{Z}$ and $d \in H_1(F)$.

(b) Assume that $p = n$ *and* $m > 0$. *Then*

$$c \in C(n,\psi) \iff c = \sum_{j=1}^{g} (a_j\{t_j\} + b_j\{u_j\}) + \sum_{i=3}^{m} e_i[s_i] + pd$$

where $a_i, b_i, e_i \in \mathbb{Z}$ and $d \in H_1(F)$.

Proof. Write c as an integral linear combination of the basis adapted to ψ and equate coefficients mod n. □

4.15.14 Proposition. *The notation is as in 4.15.4.*
(a) If the homeomorphism $\psi: F \to F$ *has prime order* p *then* $C(n,\psi)/nH \cong (\mathbb{Z}_n)^q$ *where*

(i) $q = 2g$ *if* p *and* n *are relatively prime or* $m = 0$, *and*
(ii) $q = 2g + m - 2$ *if* $p = n$ *and* $m > 0$.

(b) Any basis of $H_1(F)$ *can obtain at most* q *elements from* $C(n,\psi)$.

Proof. (a) is a direct consequence of Lemma 4.15.13.

(b) Elements of a basis of $H_1(F)$ are not proper powers of other elements. Hence, if some basis contains r elements from $C(n,\psi)$ then the order of $C(n,\psi)/nH$ is at least n^r. □

4.15.15 Corollary ([Serre 1960/61]). *Let* F *be an orientable closed surface of genus* $\gamma \geq 2$, *and* $\psi: F \to F$ *a homeomorphism of finiter order. If* ψ *induces the identity on* $H_1(F, \mathbb{Z}_n)$ *for some* $n \geq 3$, *then* ψ *is the identity.*

Proof. It suffices to prove the assertion if the order of ψ is a prime $p > 1$. Let us assume that n' is a prime dividing n. Since ψ induces the identity on $H_1(F, \mathbb{Z}_{n'})$ we conclude (e.g. from the universal coefficient theorem [Spanier 1966, 5.2]) that

$$C(n',\psi) = H_1(F) \cong \mathbb{Z}^{2f}.$$

From 4.15.14 we obtain:

$$2\gamma = 2g \geq 4 \quad \text{if } (p,n') = 1 \text{ or } m = 0$$

$$2\gamma = 2g + m - 2 \quad \text{if } p = n' \text{ and } m > 0.$$

From the Riemann-Hurwitz formula 4.14.22:

$$4 \leq 4\gamma - 4 = p\ [2 \cdot \sum_{i=1}^{m} (1 - \frac{1}{p}) + 4g - 4],$$

we conclude that in the first case $p = 1$, hence $\psi = id_F$, and in the second case $g = 0$, hence $m \geq 6$, and $p = n' = 2$.

Thus only the case where $p = 2$ and n is a power of 2 remains: Let $n = 2^\alpha$. As before, $g = 0$, hence:

$$G = \langle s_1,\ldots,s_m \mid s_1^2,\ldots,s_m^2,\ s_1\cdot\ldots\cdot s_m\rangle$$

and $\pi_1(F)$ = Kern φ where $\varphi: G \to \mathbb{Z}_2$, $s_i \mapsto 1$. Using the Reidemeister-Schreier method 2.2.1 we obtain

$$\pi_1(F) = \langle x_3,y_3,\ldots,x_m,y_m \mid R\rangle \text{ with } x_i = s_i s_1^{-1},\ y_i = s_1 s_i$$

where each generator and its inverse occurs in R exactly once. Hence they form a basis for $H_1(F)$. The action of $G/\pi_1(F)$ on $\pi_1(F)$ is given by $x_i \mapsto s_1 x_i s_1^{-1} = y_i$, $y_i \mapsto s_1 y_i s_1^{-1} = x_i$, hence, with respect to the above basis the action of ψ_* in $H_1(F)$ gives a non-trivial permutation of the basis vectors. Thus the action does not induce the identity on $H_1(F, \mathbb{Z}_n)$.

□

4.15.16 Corollary. *Let F be a closed orientable surface of genus $\gamma \geq 2$ and $\psi: F \to F$ a homeomorphism of finite order.*

(a) Let the order of ψ divide n and let p be the smallest prime dividing n. Assume that some homology basis of $H_1(F)$ contains more than $\frac{2\gamma-1}{p} + 2$ elements c with $\psi_*(c) = c$. Then ψ is the identity.

(b) If there is a basis of $H_1(F)$ that contains more than $\gamma + 1$ elements fixed by ψ_*, then ψ is the identity.

Proof as exercise E 4.27.

Finite groups of symmetries of surfaces have mainly been studied as groups of conformal self mappings of Riemann surfaces. So Theorem 4.15.3 is from [Hurwitz 1893]. The complex analytic concept is also basic for [Accola 1967] and [Gilman 1977]. In this chapter we have developed a combinatorial theory of planar discontinuous groups, including the Riemann-Hurwitz formula, hence, we could translate [Gilman 1977] into the language of the combinatorial topology and prove the results in 4.15.4-16 without using complex analytic theorems.

The proof of proposition 4.15.1 is obtained in collaboration with M. Mellis and U. Schadowski (Bochum) in the study of (euclidean) crystallograpic groups, see [Mellis, Schadowski, Zieschang, preprint]. For more literature on the theory of automorphisms of Riemann surfaces see the end of section 4.14 and chap. 6. In addition:
[Gilman 1976, preprint], [Macbeath 1973], [Moore 1970, 1972].

Exercises: E 4.24-27.

4.16 ON THE RANK OF PLANAR DISCONTINUOUS GROUPS

The (free) rank of a group is the minimal number of elements necessary to generate the group (2.1.11). The Grushko theorem 2.9.1 states that the rank behaves additively with free products. This is obviously not the case with free products with amalgamation. J. Nielsen posed the problem: decide whether the planar discontinous group $\langle s_1,s_2,s_3,s_4,s_5 \mid s_1^2,s_2^3,s_3^5,s_4^7,s_5^{11},s_1s_2s_3s_4s_5\rangle$ in fact needs at least 4 generators. In a more general version, we may ask for the rank of an arbitrary planar discontinuous group. Using the Grushko theorem this problem can be reduced to simpler ones if the group does not have a compact fundamental region. The problem is also solved for groups with compact fundamental region, if they do not contain reflections, by the following theorem:

<u>4.16.1 Theorem</u>. *The group*

$$G = \langle s_1,\dots,s_m,t_1,u_1,\dots,t_g,u_g \mid s_1^{h_1},\dots,s_m^{h_m}, \prod_{i=1}^{m} s_i \prod_{j=1}^{g} [t_j,u_j]\rangle$$

where $h_i \geq 2$, *has the rank*

(a) 2g *if* m = 0,

(b) m-2 *if* g = 0,m *even and all* h_i *except one equal* 2 *and one is odd,*

(c) 2g+m-1 *in all other cases of infinite groups, i.e. for*

g > 0, m > 0

g = 0, m ≥ 4 *even and at least two* h_i *different from* 2

g = 0, m ≥ 3 *odd,*

g = 0, m ≥ 3 *and all* h_i *even.*

The proof of this theorem is a long combinatorial one, based on the generalized Nielsen method which we have described in 2.10; it can be found in [Zieschang 1970], [Peczynski, Rosenberger, Zieschang 1975]. The exceptional case has been found by Burns, Karrass, Pietrowski, and Purzitzky in 1973, and was overlooked in [Zieschang 1970]; we have given it in exercise E 4.28. From 4.14.1 we easily conclude the following corollary:

4.16.2 Corollary. *The group*

$$G = \langle s_1,\dots,s_m,v_1,\dots,v_g \mid s_1^{h_1},\dots,s_m^{h_m},\ \prod_{i=1}^{m} s_i \prod_{j=1}^{g} v_j^2\rangle$$

has rank g *for* $g > 1$, $m = 0$ *and rank* $m+g-1$ *for* $g \geq 1$, $m \geq 1$.

Proof. For $m = 0$ this follows by abelianizing. If $m = 1$ we introduce the relation $s_1 = 1$ and the assertion follows from the preceding. Now suppose that $m \geq 2$. The group G contains orientation reversing elements, namely the v_j. The subgroup U of index two of orientation preserving elements has the presentation

$$U = \langle s_1,s_1',\dots,s_m,s_m',x_1,y_1,\dots,x_{g-1},y_{g-1} \mid s_1^{h_1},\ s_1'^{h_1},\ \dots,s_m^{h_m},s_m'^{h_m},$$

$$\prod_{i=1}^{m} s_i \prod_{j=1}^{g-1} x_j y_j \left(\prod_{i=1}^{m} s_i'^{-1} \prod_{j=1}^{g-1} y_j x_j\right)^{-1}\rangle$$

see E 4.14 where this was an example for the Reidemeister-Schreier method. To U we may apply 4.16.1 with 2m and g-1 instead of m and g. For $g > 1$ the rank is $2(g-1) + 2m - 1$; for $g = 1$ U has an even number of rotation generators with odd order, hence the exceptional case 4.16.1 (b) cannot occur and the rank is $2m-1$.

From the Reidemeister-Schreier theorem 2.2.1 on the presentation of subgroups it follows: if r is the rank of G, U a subgroup of G of index c, then

$$\operatorname{rank} U \leq c(r-1) + 1.$$

Hence,

$$2(g-1) + 2m - 1 \leq 2(r-1) + 1 \quad \text{or}$$
$$r \geq m+g-1.$$

□

The determination of the rank for groups that contain reflections cannot be done the same way, and this question is still open. Another minimal number of generators is of interest in connection with the action of a group on the plane, namely the minimal number of generators obtained from fundamental regions, see the construction in 4.5.

4.16.3 Definition. Let G be a group of automorphisms of a complex of $\mathbb{E}$ and F a fundamental domain for G. Let r_F denote the number of pairs $\{x,x^{-1}\} \subset G$ such that $F \cap Fx$ is 1-dimensional. (With the notation of 4.11, F and Fx are neighbours.) The geometrical rank of G is by definition the minimum of the r_F where F varies over all fundamental domains. (We recall that the complex on $\mathbb{E}$ may be changed, see 4.3, and G altered appropriately.)

4.16.4 Theorem. *The group*

$$G = \langle s_1,\ldots,s_m,t_1,u_1,\ldots,t_g,u_g \mid s_1^{h_1},\ldots,s_m^{h_m}, \prod_{i=1}^{m} s_i \prod_{j=1}^{g} [t_j,u_j]\rangle$$

of automorphisms of $\mathbb{E}$ *has the geometrical rank*

$2g$ *if* $m = 0$,

$2g+m-1$ *if* $g \geq 0$, $m > 0$ *or* $g = 0$, $m \geq 3$.

The group $G = \langle s_1,\ldots,s_m,v_1,\ldots,v_g \mid s_1^{h_1},\ldots,s_m^{h_m}, \prod_{i=1}^{m} s_i \prod_{j=1}^{g} v_j^2 \rangle$

has the geometrical rank

g *if* $m = 0$,

$g+m-1$ *if* $m > 0$, $g > 0$.

Proof. The given numbers are upper bounds on the geometrical rank because of the fact that we do not need a special basepoint on the quotient surface $\mathbb{E}/G$, but can choose some rotation center for it, see figure. If we do this with the center of s_m then we get the presentation

$$G = \langle s_1,\ldots,s_{m-1},t_1,u_1,\ldots,u_g \mid s_1^{h_1},\ldots,s_{m-1}^{h_{m-1}}, ([t_1,u_1]\ldots[t_g,u_g]s_1\ldots s_{m-1})^{h_m}\rangle$$ with

$2g+m-1$ generators.

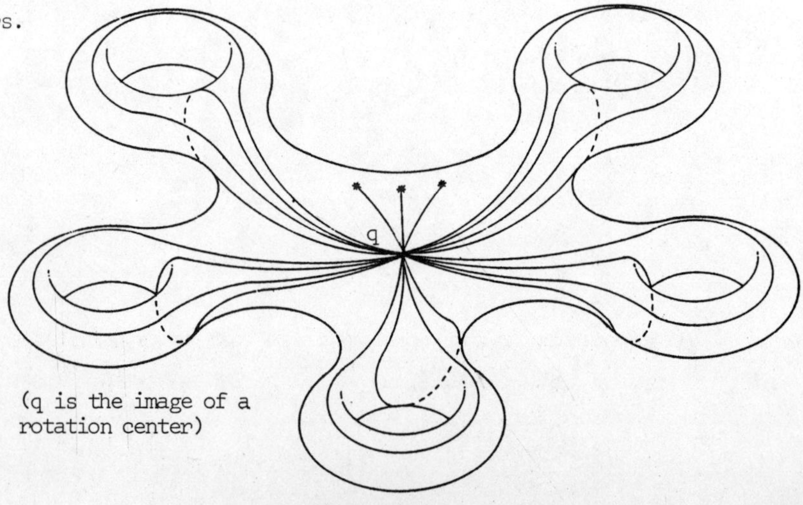

(q is the image of a rotation center)

To see that the number given in the theorem is a lower bound for the geometrical rank we look at the Euler characteristic of the quotient surface $\mathbb{E}/G$. Let F be a fundamental region. Without change of the Euler characteristic we may combine the faces of F into one. Then all edges and vertices of F are in the boundary of F. Moreover we may assume that ∂F has exactly r_F pairs of edges. Hence $\mathbb{E}/G$ carries a complex of one face and r_F edges. The images of the rotation centers are vertices (this needs some more consideration for rotations of order 2). Hence the Euler characteristic is bigger than $m-r_F+1$ in the orientable case, thus

$$2 - 2g \geq m - r_F + 1 \text{ and } r_F \geq 2g + m - 1.$$

For the non-orientable case we obtain

$$2 - g \geq m - r_F + 1 \text{ and } r_F \geq g + m - 1.$$

This gives the desired formulas for $m > 0$. If $m = 0$ then the complex on $\mathbb{E}/G$ contains at least one vertex. Now it follows that

$$2 - 2g \geq 1 - r_F + 1 \text{ and } 2 - g \geq 1 - r_F + 1, \text{ resp.}$$

□

The geometrical rank does not change if we leave the combinatorial theory to consider groups of homeomorphisms acting on the plane, and allow quite general fundamental domains, see [Peczynski- Rosenberger- Zieschang 1975, § 4].

Exercises: E 4.28-29.

Additional literature to chapter 4:
[Baumslag 1962], [Appell-Goursat 1930], [Armstrong 1968], [Bailey 1961], [Best 1973], [Cohen 1974], [Jonsson 1970], [Maskit 1965], [Mednyk 1979], [Purzitsky 1974], [Purzitsky-Rosenberger 1972], [Rosenberger 1973], [Rosenberger 1972], [Schattschneider 1978], [Scherrer 1929], [Stothers 1977].

EXERCISES

E 4.1 Prove Lemma 4.1.9.

E 4.2 If x is an element of infinite order, P a point and R a compact region, prove that Px^n lies outside R for all sufficiently large n.

E 4.3 Prove that the universal coverings (i.e. the coverings corresponding to the subgroup 1) of orientable surfaces of genus > 0 and non-orientable surfaces of genus > 1 are planar nets.

E 4.4 Let $\mathbb{E}$ be a planar net and G a planar discontinuous group acting on $\mathbb{E}$, without reflections and with compact fundamental domain. Prove that the canonical mapping $\mathbb{E} \to \mathbb{E}/G$ is a branched covering , see 3.3.1.

E 4.5 Prove Theorem 4.4.3.

E 4.6 Consider the square net in the euclidean plane, whose vertices have integer coordinates. Prove that the group generated by the rotations of order 4 around (0,0) and (0,1) has compact fundamental domain, and find its canonical presentation.

E 4.7 Determine fundamental domain and canonical presentation for the group of all automorphisms of the square net.

E 4.8 (a) Determine all groups of automorphisms of the square net with compact fundamental domain.
(b) Same for the triangular and hexagonal nets.

E 4.9 Prove Theorem 4.6.3 (c).

E 4.10 Prove that the maximal finite subgroups of a planar discontinuous group with a presentation <4.5.7|4.5.8> or <4.5.7'|4.5.8'> are
(a) subgroups of order 2 generated by reflections.
(b) cyclic subgroups of order h_j , and
(c) dihedral groups of order $2h_{k,j}$

E 4.11 Let G,G' have presentations <4.5.7|4.5.8> and let G and G' be isomorphic. Prove that $m = m'$ and that the h_i and the h_i' coincide up to a permutation.

E 4.12 Prove that the normal subgroup generated by the elements s_i is characteristic.

E 4.13 Prove Lemma 4.7.5 for groups with orientation reversing elements.

E 4.14 Give the numbers g', m' and the rotation orders $h_1',\ldots,h_m'$ for the subgroup G of "orientation-preserving" elements (for the definition of these numbers see the presentations of groups in Theorem 4.5.6). E.g., use the Reidemeister-Schreier method of 2.2.

E 4.15 Find the exceptional cases to the claim that, when no reflections are present, two elements commute only when they are powers of the same element and prove the claim in the other cases. What happens if reflections are admitted?

E 4.16 Determine which of the groups 4.11.3 are not proper free products.

E 4.17 Prove that the triangle group $\langle s_1,s_2,s_3 \mid s_1^a,s_2^b,s_3^c,s_1s_2s_3\rangle$, contains $\mathbf{Z}\oplus\mathbf{Z}$ as a subgroup of index

4 for $a = b = 4$, $c = 2$
3 for $a = b = c = 3$
6 for $a = b$, $b = 3$, $c = 2$.

E 4.18 Prove that $\langle s_1,s_2,s_3 \mid s_1^n,s_2^n,s_3^n,s_1s_2s_3\rangle$ n odd, contains the fundamental group of a closed orientable surface of genus $(n-1)/2$ as a normal subgroup of index n. What happens when n is even? (In particular, prove that the group with $n = 8$ does not contain the fundamental group of the closed orientable surface of genus 2, 3, 4 or 5.)

E 4.19 Prove that the orientable surface of genus 6 does not cover that of genus 3.

E 4.20 Let G be a planar discontinuous group of orientation-preserving automorphisms, which contains a triangle group $\mathcal{D}$. Prove that $\mathcal{D}$ is of finite index in G and that G is also a triangle group.

E 4.21 Determine the finite groups of orientation preserving homeomorphisms of the torus T, i.e. the closed orientable surface of genus 1, and compare with 4.14.24.

E 4.22 Prove that the fundamental group of a closed orientable surface F of genus $\gamma \geq 2$ contains infinitely many normal (even infinitely many characteristic) subgroups of finite index which are pairwise non-isomorphic. Prove that they are fundamental groups of closed orientable surfaces of higher genus than γ.

E 4.23 Proof of the final assertion in 4.14.27 (e).

E 4.24 Let A be a finite abelian group which acts trivially on $\mathcal{U} \cong \mathbb{Z}^n$. Prove that each extension G of $\mathcal{U}$ by G is isomorphic to $\mathbb{Z}^n \oplus \mathcal{B}$, where $\mathcal{B}$ is a subgroup of A.

E 4.25 Discuss the situation corresponding to 4.15.3 for surfaces of genus $\gamma \leq 1$.

E 4.26 Prove Lemma 4.15.5.

E 4.27 Proof of 4.15.16.

E 4.28 Prove that $\langle s_1,s_2,s_3,s_4 \mid s_1^2,s_2^2,s_3^2,s_4^2,s_1s_2s_3s_4\rangle$ has rank 2.

E 4.29 Prove that $\langle t,u \mid [t,u]^{2n+1}\rangle$ $(n \geq 2)$ is decomposable, using the fact that it has the group in E 4.28 as a quotient.

E 4.30 Let p,q,r be different primes and $\mathcal{D} = \langle s_1,s_2,s_3 \mid s_1^p, s\, s_2^q, s_3^r, s_1s_2s_3\rangle$. Then each proper normal subgroup $N \triangleleft \mathcal{D}$ is torsionfree.

E 4.31 G be a discontinuous group of the plane all elements of which preserve orientation. If G has torsion and is not a triangle group as in E 4.30 then G contains a proper normal subgroup N that has torsion.

E 4.32 Let $n \geq 3$, $n \geq m$, $a_i,b_i \geq 2$, $G = \langle s_1,\dots,s_m \mid s_1^{a_1},\dots,s_m^{a_m},s_1\dots s_m\rangle$, $H = \langle x_1,\dots,x_n \mid x_1^{b_1},\dots,x_n^{b_n},x_1\dots x_n\rangle$ and $f\colon G \to H$ an epimorphism. Then:

(a) $f(s_i) = v_i\, x_{k_i}^{\alpha_i}\, v_1^{-1}$ for $1 \leq i \leq m$,

(b) $m = n, (\alpha_i,b_{k_i}) = 1$, and

(c) if $n \geq 4$, then $\begin{pmatrix} 1 & \dots & m \\ k_1 & \dots & k_m \end{pmatrix}$ is a permutation.

(For the proof use the rank theorem 4.16.1.)

E 4.33 Let G and H be as in E 4.33, and let $f: G \to H$ be the epimorphism with $f(s_i) = v_i x_{k_i}^{\alpha_i} v_i^{-1}$. If Kern f is torsionfree and if $\begin{pmatrix} 1 & \dots & m \\ k_1 & \dots & k_m \end{pmatrix}$ is a permutation, then f is an isomorphism. Furthermore, for each i we have $a_i = b_{k_i}$ and $\alpha_i \equiv \varepsilon \bmod a_i$ where $\varepsilon = \pm 1$.

E 4.34 Let $a_i, b_i \geq 2, \frac{1}{a_1} + \frac{1}{a_2} + \frac{1}{a_3} \leq 1$,

$$G = \langle s_1, s_2, s_3 \mid s_1^{a_1}, s_2^{a_2}, s_3^{a_3}, s_1 s_2 s_3 \rangle,$$

$$H = \langle x_1, x_2, x_3 \mid x_1^{b_1}, x_2^{b_2}, x_3^{b_3}, x_1 x_2 x_3 \rangle,$$

and let $f: G \to H$ be an epimorphism with torsionfree kernel. Prove:

(a) If the numbers a_1, a_2, a_3 are pairwise distinct, then f is an isomorphism.

(b) If $b_1, b_2, b_3 \geq 6$, then f is an isomorphism.

E 4.35 Let $G = \langle s_1, s_2 \mid s_1^a, s_2^b, (s_1 s_2)^a \rangle$ and $H = \langle x_1, x_2 \mid x_1^a, x_2^b, (x_1 x_2)^2 \rangle$. Prove:

(a) $s_1 \mapsto x_1$, $s_2 \mapsto x_1^{-1} x_2^2 x_1$ defines a homomorphism $\alpha: G \to H$.

(b) If $a > 2$ and $2 \nmid b$ then α is an epimorphism and Kern α is non-trivial and torsionfree.

E 4.36 Let G and H be as in E 4.32 and suppose that there is an epimorphism $f: G \to H$ with non-trivial kernel. If Kern f is torsionfree, then $2 \nmid b$ and $a > 2$.

E 4.37 (a) Let F be the fundamental group of a closed surface F. Suppose that F contains a finitely generated non-trivial normal subgroup N. Then $[F:N] < \infty$ or F is a torus or a Klein bottle.

(b) Discuss the same question for arbitrary surfaces, see E 2.9, and for normal subgroups of planar discontinuous groups.

(The exercises E 4.30-36 are from [Zieschang 1976']; E 4.35 can also be obtained from [Knapp 1968] where further examples of this type can be found.)

5. AUTOMORPHISMS OF PLANAR GROUPS

In this chapter we prove that each automorphism of the fundamental group of a surface is induced by a homeomorphism (Dehn-Nielsen theorem) and we characterize those homeomorphisms which induce the identical automorphism in the fundamental group (Baer theorem). In this chapter we use mainly combinatorial group theoretic arguments, in contrast to the proof of the Dehn-Nielsen theorem in 3.3. The advantage of our approach is that we can also prove a similar result for non-orientable surfaces and planar discontinuous groups.

5.1 PRELIMINARY CONSIDERATIONS

If $\mathbb{E}$ is a planar net, G a group of automorphisms on $\mathbb{E}$ and α an automorphism of G, then we say that "α can be realized by a homeomorphism" when there are two subdivisions $\mathbb{E}'$ and $\mathbb{E}''$ of $\mathbb{E}$ and an isomorphism $\eta: \mathbb{E}' \to \mathbb{E}''$ with $\alpha(x) = \eta^{-1}x\eta$, $x \in G$. The operation of G on $\mathbb{E}$ extends in a natural way to $\mathbb{E}'$ and $\mathbb{E}''$.

We call η a *homeomorphism* and carry over this concept to surfaces. If one takes, say, the piecewise linear theory as basis, then one can obviously choose η to be a piecewise linear homeomorphism.

We shall prove that each automorphism of a planar discontinuous group without reflections can be realized by a homeomorphism (Theorem 5.8.2). If G is a surface group (i.e. there are no generators s_i) then a subdivision of $\mathbb{E}$ induces one of $\mathbb{E}/G$ and η goes over to a homeomorphism of $\mathbb{E}/G$. Each automorphism of the fundamental group of a surface F can therefore be realized by a homeomorphism of F. This result is what we have called the Dehn-Nielsen theorem 3.3.11 [Nielsen 1927].

Of course, when we speak of automorphisms of the fundamental group we must keep in mind the basepoint of the group; we must therefore insist that η leaves the initial point of G fixed in $\mathbb{E}/G$. If we also bear in mind the rotation centers on $\mathbb{E}/G$, then the general theorem may also be expressed as a theorem about surfaces. Namely, a homeomorphism of $\mathbb{E}/G$ which permutes the rotation centers and leaves the basepoint fixed induces an automorphism of G. If we remove a "small" disk from $\mathbb{E}/G$ around each rotation center and around the basepoint then the fundamental group of the perforated surface is the free group $\hat{G}$ in the generators $S_1,S_2,\dots,S_m$ and $T_1,U_1,\dots,T_g,U_g$ (resp. $V_1,\dots,V_g$) "of G". An automorphism $\hat{\alpha}: \hat{G} \to \hat{G}$ which permutes the path classes of boundary curves corresponding to rotations of the same order (or maps them into their inverses) and carries the cuts of a canonical dissection into each other, may be realized by a homeomorphism. One can extend it to $\mathbb{E}/G$ when one maps the associated disks correspondingly. If the automorphism $\hat{\alpha}$ induces α, then the homeomorphism of the plane which covers the homeomorphism of the surface is a realization of α. The conditions that $\hat{\alpha}$ permutes the boundary curves of rotations of the same order and leaves fixed the curve around the base point are expressed algebraically as

$$\hat{\alpha}(S_i) = L_i S_{\alpha_i}^{\varepsilon_i} L_i^{-1}$$

$$\hat{\alpha}(\Pi S_i \Pi [T_i, U_i]) = L(\Pi S_i \Pi [T_i, U_i])^{\varepsilon} L^{-1} \text{ for } \varepsilon, \varepsilon_i = \pm 1.$$

We shall attain an $\hat{\alpha}$ to a given α.

5.2 BINARY PRODUCTS

For the proof of the Dehn-Nielsen theorem we use some results on special equations between elements of a free group. This can be treated by a combinatorial group theoretic method which corresponds to changing 'geometrical' generators of fundamental groups of surfaces by bifurcations: the method of binary products.

5.2.1 Definition. Let S be a free group on the generators $S_1, S_2, \ldots$. Let $X_1, X_2, \ldots, X_n$ be elements of S with

$$X_i = \tilde{X}_i S_i^{\varepsilon_i} \tilde{X}_i^{-1}, \ \varepsilon_i = \pm 1, \ i = 1, \ldots, m \le n.$$

Let $X_1, \ldots, X_n$ be symbols and let $\Pi_X = \Pi_X(X_1, \ldots, X_n)$ be a word in the $X_i^{\pm 1}$ such that each symbol $X_1, \ldots, X_m$ appears exactly once (either with exponent + 1 or - 1) and each of $X_{m+1}, \ldots, X_n$ exactly twice (hence at most twice with exponent + 1). We call $\{X_1, \ldots, X_n; \Pi_X\}$ a *binary product*. $\Pi_X(X) = \Pi_X(X_1, \ldots, X_n)$ is the element of S obtained when X_i is substituted for X_i, and is called the *value* of the binary product in S. $\{X_1, \ldots, X_n; \Pi_X\}$ is called *alternating* when $X_{m+1}, \ldots, X_n$ appear once with exponent + 1 and once with exponent - 1. The X_i (sometimes also the X_i) are called *factors* of the binary product.

We define bifurcations of binary products in analogy with bifurcations of surfaces.

5.2.2 Definition. The following processes are called *bifurcations*.

(a) If Π_X has a place $\ldots X_i X \ldots$ $1 \le i \le m$ then we define symbols Y_j, $j = 1, \ldots, n$ by $Y_i = X^{-1} X_i X$, $Y_j = X_j$, $j \ne i$ and corresponding new elements $Y_i = X^{-1} X_i X$, $Y_j = X_j$, $j \ne i$. Π_Y results from Π_X when one writes the word YY_i in place of $X_i X$ and elsewhere replaces X_j by Y_j.

(b) The process "inverse" to (a). We then have a place $\ldots XX_i \ldots$ to replace using $y_i = XX_iX^{-1}$, $y_j = X_j$, $j \neq i$ and we write $\Pi_y = \ldots y_i y \ldots$.
$Y_i = XX_iX^{-1}$, $Y_j = X_j$, $j \neq i$.

(c) Let $\Pi_X = \ldots X_i X \ldots X_i^{\varepsilon} \ldots$ $i > m$, $\varepsilon \in \{1,-1\}$. Then let $y_i = X_iX$, $y_j = X_j$, $j \neq i$, $Y_i = X_iX$, $Y_j = X_j$, $j \neq i$ and $\Pi_y = \ldots y_i \ldots (y_i y^{-1})^{\varepsilon} \ldots$.

(d) If $\Pi_X = \ldots XX_i \ldots X_i^{\varepsilon} \ldots$ $i > m$, $\varepsilon \in \{1,-1\}$ let $y_i = XX_i$, $y_j = X_j$, $j \neq i$, $Y_i = XX_i$, $Y_j = X_j$, $j \neq i$, $\Pi_y = \ldots y_i \ldots (y^{-1}y_i)^{\varepsilon} \ldots$.

5.2.3 Definition. Two binary products are called *related* if one may be converted into the other by finitely many of the following processes:

(a) Renumbering the first m and the last n-m generators.

(b) Replacement of a factor by its inverse.

(c) Bifurcations 5.2.2 (a-d).

The following properties are immediate.

5.2.4 Lemma. *(a) The factors of related products generate the same subgroup of S.*
(b) Related binary products have the same value $\Pi_X(X)$ *in S.*
(c) Alternating products remain alternating.

□

In order to describe the bifurcation process (and the processes 5.2.3 (a,b)) in geometric terms, we consider a surface with m perforations. Then the boundary curve of the disk which results from a dissection defines a binary product. It is alternating when the surface is orientable, but not in the non-orientable case. If X_i is a closed curve of the dissection which traverses the ith boundary once, and if X is the curve of the dissection which follows X_i in the boundary path, then $X^{-1}X_iX$ is also a curve which traverses the ith boundary once, and $\ldots X(X^{-1}X_iX) \ldots$ is again a dissection, resulting from the first by a bifurcation of the surface. The other processes may be visualized analogously (cf. 3.2). Thus:

5.2.5 Lemma. *If a binary product stems from cutting a surface (with boundaries), then all related binary products likewise stem from cutting this surface.*

□

Each factor of a binary product has a length as a word in the generators S_i, and we can speak of halves of factors. A *factor* X is called *inessential* when it stands next to X^{-1} in Π_X or is the identity of S. The process for finding generators of a subgroup of a free group with the Nielsen property can also be used to prove the following theorem (cf. [Zieschang 1964]).

5.2.6 Theorem. *For each binary product there is a related binary product with the properties:*

(a) No more than half of any essential factor is cancelled by a neighbour.
(b) No essential factor has halves cancelled by both neighbours.
(c) The initial factor, if essential, loses less than a half by cancellation. □

One obtains 5.2.6 (c) by treating an initial factor which loses half by cancellation as an factor which has halves cancelled on both sides.

5.2.7 Definition. A binary product which satisfies 5.2.6 (a) is called *reduced*, if it satisfies 5.2.6 (b,c) also, then it has the *Nielsen property*.

Let the group S have free generators $S_1,\dots,S_m$ and either $T_1,U_1,\dots,T_g,U_g$ or $V_1,\dots,V_g$ respectively, and let $\Pi_* = S_1\dots S_m\,\Pi[T_i,U_i]$ or $\Pi_* = S_1\dots S_m\,V_1^2\dots V_g^2$ respectively. In order to avoid distinguishing the two cases continually, we write S in the generators $H_1,\dots,H_n$. $\{H_1,\dots,H_n;\Pi_*\}$ is a binary product.

5.2.8 Theorem. *If* $\{X_1,\dots,X_{n'};\Pi_X\}$ *is a binary product in* S *with* $n' \le n$, $X_i = \tilde{X}_i H_{r_i}^{\varepsilon_i} \tilde{X}_i^{-1}$, $1 \le r_i \le m$, $i \le m'$, $\varepsilon_i = \pm 1$ *and* $\Pi_X(X) = \Pi_*(H)$ *in* S *then* $m' = m$, $n' = n$ *and* $\{X_1,\dots,X_n;\Pi_X\}$ *is related to* $\{H_1,\dots,H_n;\Pi_*\}$.

Proof. We remove all inessential elements of $\{X_1,\dots,X_{n'};\Pi_X\}$ and obtain a binary product $\{X_1,\dots,X_{n''};\Pi_X'\}$ with $n'' \le n$ and $m'' = m'$ factors which appear once. This is converted into a related product with the Nielsen property, which we shall denote by the same symbols. We remove all inessential elements from this product and again work to achieve the Nielsen property. This process terminates after a finite number of steps in a binary product without inessential elements, and with the Nielsen property. In it, at least one letter remains of each factor. Let $X_i^{\pm1}$ be the first factor of the form $L_i S_{r_i}^{\varepsilon} L_i^{-1}$ in Π_X'. We now write simply r in place of r_i. S_r is not cancelled, so $\varepsilon = +1$. If K is the part of Π_X' which stands before $X_i^{\pm1}$, then we can convert Π_X' into $(KL_iS_rL_i^{-1}K^{-1})\ \dots = \Pi_X''$ by finitely many of the processes 5.2.2 (b), as a result of which X_i goes into $X_i' = (KL_iS_rL_i^{-1}K^{-1})$. It is clear that S_r is also not cancelled in Π_X''. Since $\Pi_X''(X) = \Pi_X(X)$, then $\Pi_X''(X) = (KL_iS_rL_i^{-1}K^{-1})\ \dots$ has the form $(S_1\dots S_{r-1}S_rS_{r-1}\dots S_1^{-1})\ S_1\dots S_{r-1}S_{r+1}\dots S_m\dots$. If we leave X_i out of Π_X' and S_r out of Π_* then we obtain binary products $\{X_1,\dots,X_{i-1},X_{i+1},\dots,X_{n''};\underline{\Pi}_X\}$ and $\{S_1,\dots,S_{r-1}, S_{r+1},\dots,H_{m+1},\dots H_n;\underline{\Pi}_*\}$ which have the same value in S, namely $S_1\dots S_{r-1}S_{r+1}\dots S_m\Pi[T_i,U_i]$ or $S_1\dots S_{r-1}S_{r+1}\dots S_nV_1^2\dots V_g^2$ respectively. The hypotheses of the theorem are still satisfied, though m and m' are reduced by one. By a series of steps we can remove all factors from Π_X which appear once. If one abelianizes and computes mod 2 then one sees that no factor appears only once in Π_* either (i.e. $m'' = m$). Because of the Nielsen property we can also conclude that the first factor equals S_1. By induction, we can go to a related binary product in which the first m factors equal $S_1,\dots,S_m$.

We have therefore reduced Theorem 5.2.8 to the following situation: $\{X_1,\ldots,X_{n'};\Pi_X\}$ is a binary product in which each factor appears twice, and Π_* is a product of commutators $[T_i,U_i]$ or squares V_i^2 respectively. To be sure, generators S_i may still appear in the words X_j. We suppose that $\{X_1,\ldots,X_{n'};\Pi_X\}$ has the Nielsen property and no inessential elements appear, and first deal with the "non-orientable" case $\Pi_X(X) = \Pi_*(V) = V_1^2\ldots V_n^2$. By renumbering, let $X_1^{\pm 1}$ be the first factor in Π_X, so that $X_1^{\pm 1} = V_1^2\ldots V_{i-1}^2 V_i L$ or $X_1^{\pm 1} = V_1^2\ldots V_i^2 L$ for $i \leq g$, where L is the part of $X_1^{\pm 1}$ cancelled in Π_X. Here $\ell(L) \leq 2(i-1)$ or $2(i-1)+1$. Since $X_1^{\pm 1}$ appears again in Π_X, the second case cannot occur. In the other case at most V_i can remain from the greater front half $V_1^2\ldots V_{i-1}^2 V_i$, and this is so because of the Nielsen property also. Therefore, the factor following $X_1^{\pm 1}$ is again $X_1^{\pm 1}$, and we have $X_1^{\pm 1} = V_1^2\ldots V_{i-1}^2 V_i (V_1^2\ldots V_{i-1}^2)^{-1}$. As above it follows from $i-1>0$, because of the Nielsen property, that $n' < n$. We now have $\Pi_X = X_1^{\pm 2}\Pi'_X$ with $\Pi'_X(X) = V_1^2\ldots V_{i-1}^2 V_{i+1}^2\ldots V_n^2$. If we now apply the same process to $\{X_2,\ldots,X_{n'};\Pi'_X\}$, then $n' < n$ gives a contradiction. But it follows by induction from $n' = n$ that the original binary product is related to $\{V_1,\ldots,V_n;\Pi_*\}$.

In the "orientable" case $\Pi_X(X) = T_1U_1T_1^{-1}U_1^{-1}\ldots T_gU_gT_g^{-1}U_g^{-1}$ with $g = n/2$. Let X_1 be the first factor in Π_X. Because of 5.2.6 (c), less than half of X_1 is cancelled. If L denotes the part of X_1 which is cancelled, we have the following four cases.

5.2.9 $\quad X_1 = \prod_{j=1}^{i-1}[T_j,U_j]T_iL \qquad \ell(L) \leq 4(i-1)$

5.2.10 $\quad X_1 = \prod_{j=1}^{i-1}[T_j,U_j]T_iU_iL \qquad \ell(L) \leq 4(i-1)+1$

5.2.11 $\quad X_1 = \prod_{j=1}^{i-1}[T_j,U_j]T_iU_iT_i^{-1}L \qquad \ell(L) \leq 4(i-1)+2$

5.2.12 $\quad X_1 = \prod_{j=1}^{i}[T_j,U_j]L \qquad \ell(L) \leq 4(i-1)+3.$

We shall show that only 5.2.9 is possible.

Since X_1 appears once again and L is less than half of X_i some of $\prod_{j=1}^{i}[T_j,U_j]$ remains at the second appearance of X_1 in case 5.2.12. But that cannot happen, since Π_* no longer contains any of these letters. For 5.2.11, at most U_i can remain from $\prod_{j=1}^{i-1}[T_j,U_j]T_iU_iT_i^{-1}$ at the second appearance of $X_1^{\pm 1}$. To do this for X_1^{-1}, X_1 and X_1^{-1} must be neighbours, hence inessential. In case 5.2.10 the first (i-1) commutators will be cancelled out by the second appearance of $X_1^{\pm 1}$. But $T_i^{\pm 1}$ or $U_i^{\pm 1}$ remains, since otherwise more than half X_1^{-1} would be cancelled. For that reason, X_1 appears the second time with exponent -1, and, since it is essential, T_i^{-1} cannot remain, though U_i^{-1} can. If X_2 is the factor after X_1, then the T_i^{-1} must appear in X_2 and not be cancelled. Then $X_2 = L^{-1}T_i^{-1}M$ and X_1^{-1} follows X_2 in Π_X. But then $X_2 = L^{-1}T_i^{-1}L$ and T_i^{-1} must remain at the second appearance of X_2. This is not the case, so only 5.2.9 remains.

Let $K^{(i)} = \prod_{j=1}^{i-1} [T_j,U_j]$. Then $X_1 = K^{(i)}T_iL$. If X_2 is the second factor, then X_2 has the form $X_2 = L^{-1}U_iM$, and only U_i remains of X_2, since $K^{(i)}$ is cancelled at the second appearance of X_1, so that T_i^{-1} must remain. Thus X_1^{-1} follows X_2 in Π_X, and hence has the form $M^{-1}T_i^{-1}(K^{(i)})^{-1}$, so that $M = L$. Then because the U_i^{-1} remains at the second appearance of X_2^{-1}, X_2^{-1} must follow X_1^{-1}, and we have

$$\Pi_X = X_1X_2X_1^{-1}X_2^{-1}\Pi_X', \quad X_1X_2X_1^{-1}X_2^{-1}\ \Pi_X'(X) = \Pi_*(H).$$

Since $X_1X_2X_1^{-1}X_2^{-1} = K^{(i)}[T_i,U_i](K^{(i)})^{-1}$ we have

$$\Pi_X'(X) = K^{(i)}[T_{i+1},U_{i+1}] \cdots [T_g,U_g] = \Pi_*'(H).$$

If we continue in the same way with $\{X_3,X_4,\ldots,X_n;\Pi_X'\}$ and $\{T_1,U_1,\ldots,T_{i-1},U_{i-1},T_{i+1},U_{i+1},\ldots,T_g,U_g;\Pi_*'\}$, then we again obtain n" = n and the theorem follows.

□

5.2.13 Corollary. *If* α *is an endomorphism of S with* $\varepsilon, \varepsilon_i = \pm 1$, $\alpha(S_i) = L_iS_{r_i}^{\varepsilon_i}L_i^{-1}$, $i = 1,\ldots,m$ *and* $\alpha(\Pi_*(H)) = L\Pi_*^{\varepsilon}(H)L^{-1}$ *then* α *is an automorphism.*

Proof. Let ϕ be the inner automorphism of *S* which maps X to $L^{-1}XL$. Then $\alpha\phi$ is an endomorphism with $\alpha\phi(S_i) = L^{-1}L_iS_{r_i}^{\varepsilon_i}L_i^{-1}L$ and $\alpha\phi(\Pi_*) = \Pi_*^{\varepsilon}$. If $\varepsilon = -1$, let ψ be the automorphism which maps T_i to U_{g+1-i}, U_i to T_{g+1-i} or V_i to V_{g+1-i}^{-1} respectively, and S_i to $(S_{i+1}\ldots S_m\Pi[T_j,U_j])^{-1}S_i^{-1}(S_{i+1}\ldots S_m\Pi[T_j,U_j])$ or $(S_{i+1}\ldots S_mV_1^2\ldots V_g^2)^{-1}S_i^{-1}(S_{i+1}\ldots S_mV_1^2\ldots V_g^2)$ respectively. Then Π_* goes into Π_*^{-1}. Because of Theorem 5.2.8, $\alpha\phi\psi$ maps $\{H_1,\ldots,H_n;\Pi_*\}$ onto the related binary product $\{\alpha\phi\psi H_1,\ldots,\alpha\phi\psi H_n;\Pi_*\}$; in particular, $\alpha\phi\psi H_1,\ldots,\alpha\phi\psi H_n$ generate the same subgroup as $H_1,\ldots,H_n$, namely all of *S*. But then $\alpha\phi\psi$, and hence α, is an automorphism.

□

For g = 1 this corollary is in [Nielsen 1918]. It is also known for g = 0. Our processes 5.2.2 (a,b) are then the braid processes of [Artin 1925, 1947, 1947'].

The method of binary products, first that of alternating products, was introduced in [Zieschang 1964,1965], originally for some problems about handlebodies. The theorem 5.2.8 gave the impulse to look for a new proof of the Dehn-Nielsen theorem from [Nielsen 1927]. For the applications to handlebodies the equations from the exercises E 5.1-3 are fundamental. The equation $a^2b^2 = c^2$ in a free group had already been considered in [Lyndon 1959], but with other methods. See also [Rosenberger 1978, 1978'].

Exercises: E 5.1-4

5.3 HOMOTOPIC BINARY PRODUCTS

If we apply only bifurations to a binary product then the value does not change. This will not be enough for the proof of the Dehn-Nielsen theorem, which requires something like homotopic deformation of curves during which the basepoint is passed. We now give an algebraic formulation of this.

5.3.1 Definition. Let G be a finitely generated group and $\hat{G}$ the free group on the generators of G. The mapping which associates the free generators of $\hat{G}$ with the corresponding generators of G defines a homomorphism φ: $\hat{G} \to G$ with kernel $\hat{N}$.
(a) Let $\{X_1,\ldots,X_n;\Pi_X\}$ and $\{Y_1,\ldots,Y_n;\Pi_Y\}$ be binary products in $\hat{G}$. If $X_i = \tilde{X}_i S_{r_i}^{\varepsilon_i} \tilde{X}_i^{-1}$ and $Y_i = \tilde{Y}_i S_{r_i}^{\varepsilon_i} \tilde{Y}_i^{-1}$ for $i \leq m$ with $\tilde{Y}_i = \tilde{X}_i \tilde{N}_i$, $\tilde{N}_i \in \hat{N}$ and $Y_i = X_i N_i$, $N_i \in \hat{N}$ for $i > m$, and if the word Π_X results from Π_Y when each Y_i is replaced by X_i, then the binary products are called *homotopic*. We then have $\varphi X_i = \varphi Y_i$, $\varphi(\Pi_X(X)) = \varphi(\Pi_Y(Y))$; i.e. two homotopic binary products of $\hat{G}$ yield the same binary product in G.
(b) We say that a binary product $\{X_1,\ldots,X_n;\Pi_X\}$ *decomposes* (over $\hat{N}$) when there is a subword Π_X' of Π_X which contains a proper subset of the factors, none of these factors appears elsewhere in Π_X, and $\varphi(\Pi_X'(X)) = 1$ in G. A binary product is called *decomposable* when it is related to one which decomposes.
(c) A binary product $\{X_1,\ldots,X_n;\Pi_X\}$ is called *simple* when, if $\Pi_X(X) \notin \hat{N}$, no proper subword of $\Pi_X(X)$ (considered as a reduced word in the generators of $\hat{G}$) lies in $\hat{N}$, and if $\Pi_X(X) \in \hat{N}$, no proper subword of the cyclically reduced form of $\Pi_X(X)$ lies in $\hat{N}$.

To obtain a geometrical interpretation we construct a graph for G, vertices of which correspond to group elements, as in the existence proof for planar groups. The edges emanating from a vertex correspond to the generators and their inverses, and may be denoted by corresponding symbols (see the proof of Theorem 4.7.1). $\Pi_X(X)$ is then a path in the graph. $\{X_1,\ldots,X_n;\Pi_X\}$ is simple when the path becomes simple after removal of spurs or, if $\Pi_X(X) \in \hat{N}$, has the form ABA^{-1} where B is simple and closed.

5.3.2 Theorem. *The length of a binary product is defined to be the length of the cyclically reduced path for $\Pi_X(X)$. If a binary product $\{X_1,\ldots,X_n;\Pi_X\}$ is not decomposable with respect to $\hat{\hat{N}}$ and is not simple, then there is a shorter product homotopic to it.*

Proof. We may assume that no factor equals 1 in $\hat{G}$ and likewise none appears when a bifurcation is made. Otherwise this element can be omitted initially and reintroduced later at the appropriate position. We regard $\Pi_X(X)$ as an unreduced path. The reduced path has a simple closed subpath which contains a double point of the reduced path. By applying an inner automorphism of $\hat{G}$ to the path we can assume that it begins at the double point. At the end of the proof we apply the reverse of this transformation. The inner automorphism carries $\{X_1,\ldots,X_n;\Pi_X\}$ onto another nonsimple, indecomposable binary product of equal length. It therefore suffices to show that the theorem holds for binary products corresponding to reduced paths which begin with a double point.

Using Theorem 5.2.6 we convert $\{X_1,\ldots,X_n;\Pi_X\}$ into a reduced binary product. This has the same reduced path and equal length. It will be denoted by the same symbols. No factor X_i then has more than half of itself cancelled by either neighbour, and no factor is completely cancelled by its two neighbours. In particular, the middle element $S_{r_i}^{\varepsilon_i}$ in a factor $X_i = \tilde{X}_i S_{r_i}^{\varepsilon_i} \tilde{X}_i^{-1}$ is never cancelled. For each letter in X_i, $i = 1,\ldots, n$ which is cancelled there is a corresponding inverse letter in the neighbour which is also cancelled. We say that the two letters constitute a *cancelling pair,* or that one is the *cancelling partner* of the other. The part of X_i which remains is called the *kernel* of X_i and will be denoted $|X_i|$.

Apart from the middle elements of X_i, $i \leq m$, we assign each letter of a factor a *formal partner.* The formal partner of a letter in $\tilde{X}_i$ or $\tilde{X}_i^{-1}$ will be the corresponding inverse letter in $\tilde{X}_i^{-1}$, $\tilde{X}_i$ respectively; the formal partner of a letter in X_i^{ε}, $i > m$ will be the same letter in X_i^{ε} in case X_i^{ε} appears again with the exponent ε, otherwise it will be the corresponding inverse letter in $X_i^{-\varepsilon}$.

The simple closed path with which the reduced $\Pi_X(X)$ begins will be called the *loop* from now on. If spurs begin at the double point, then the loop is not uniquely determined by the unreduced word $\Pi_X(X)$. In order to determine whether spurs emanating from the double point belong to the unreduced loop (loop with spurs), we make a fixation of the double point. Then up to the factor which contains the double point, it is determined for each factor whether it belongs to the (unreduced) loop or not. If one of the two factors of index $i > m$ lies on the loop and the other does not, then we say that the *double point separates the pair* X_i. Since the binary product is not decomposable, the double point either lies in the interior of a factor or it separates at least one pair.

We first assume that the pair for X_i $(i > m)$ is separated by the double point, and a letter A of X_i and its formal partner do not lie on a spur. If we then replace the A in X_i by the rest of the loop, as is indicated by the dotted lines, then this is a homotopic modification, since the loop lies in $\hat{N}$; the length of Π_X is decreased by at least 2 as a result. This reduction process is the basis of the following proof.

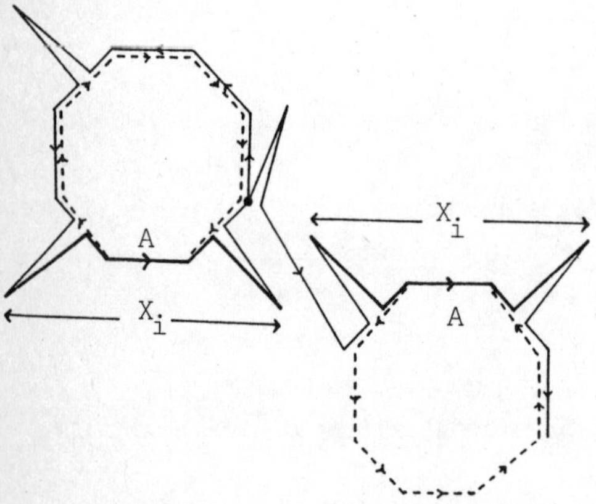

By going from a letter alternately to formal and cancelling partners we obtain a chain of symbols; we obtain another chain by beginning with the cancelling partner and then alternating. We combine these two chains into a single one for each letter, and consider all the maximal chains. Among them there are possibly chains which traverse a simple closed chain infinitely often, but in any case there will be simple open chains, the ends of which lie on the reduced path.

5.3.3 Assumption. *There is a maximal open chain, the ends of which are separated by the double point.*

It is clear that this chain contains no middle letter of an X_i $(i \le m)$.

Let the chain be $K_1,\dots,K_r$ and let K_1 lie on the loop, K_r on the rest. Let the generators of $\hat{G}$ be denoted by $H_1, H_2, \dots$. The reduced loop has the form PK_1Q where P is the path from the double point to K_1 and Q is the part of the loop from K_1 back to the double point. Suppose $K_i = H_j^{\eta_i}$, $\eta_i = \pm 1$ for a suitable fixed j and $1 \le i \le r$. If we replace K_i by $(QP)^{-\eta_i\eta_1}$ then we obtain a new binary product $\{Y_1,\dots,Y_n;\Pi_Y\}$ homotopic to the first (relative to $\hat{N}$). We shall compare $\ell(\Pi_Y(Y))$ with $\ell(\Pi_X(X))$. The loop PK_1Q is replaced by a spur and this reduces $\Pi_X(X)$ by $\ell(P) + \ell(Q) + 1$ symbols. Each pair (K_{2i}, K_{2i+1}^{-1}), $i = 1,\dots,\frac{r}{2} - 1$ is a cancelling pair. Thus $K_{2i} = K_{2i+1}^{-1}$. For each of these pairs we have introduced the longer spur $(QP)^{-\eta_{2i}\eta_1} \cdot (QP)^{\eta_{2i}\eta_1}$, but this does not alter the length of the *reduced* word. In place of K_r we have $(QP)^{-\eta_r\eta_1}$. This increases the length by at most $\ell(P) + \ell(Q) - 1$. It follows from these considerations that

$$\ell(\Pi_Y(Y)) \le \ell(\Pi_X(X)) + \ell(P) + \ell(Q) - 1 - \ell(P) - \ell(Q) - 1,$$

so the length has been reduced by at least two, and the theorem is also proved for this case.

We shall now show that one can always go to a related binary product which satisfies assumption 5.3.3.

5.3.4 Lemma. *When the double point separates the pair with index* i *then there is a related binary product in reduced form which satisfies one of the following conditions:*

(a) There is a maximal open chain, the ends of which are separated by the double point.

(b) The sum of the lengths of the factors has decreased.

(c) The number of pairs separated by the double point has decreased, and the sum of the lengths of the factors has not changed.

Proof. Let the pair corresponding to X_i $(i > m)$ be separated by the double point, and let $X_i \neq 1$. We let $X_i^{(\sigma)}$ denote the symbol belonging to the loop, and let $X_i^{(\alpha)}$ be the other.

We can assume that no formal partner of a letter from the kernel of $X_i^{(\sigma)}$ belongs to the kernel of $X_i^{(\alpha)}$, since otherwise (a) is already satisfied. Then $X_i = AB$, where A and B have equal length, and one of the halves $A^{(\sigma)}$ and $A^{(\alpha)}$ of $X^{(\sigma)}$ and $X^{(\alpha)}$ respectively corresponding to A is completely cancelled; likewise for $B^{(\sigma)}$ and $B^{(\alpha)}$. We may suppose $A^{(\sigma)}$, and hence $B^{(\alpha)}$, is fully cancelled. Let $X_j^!$ be the left neighbour of $X_i^{(\sigma)}$. Then X_j^ε has the form $X_j^\varepsilon = C(A^{(\sigma)})^{-1}$ and C and B do not cancel each other out, $\ell(C) \ge \ell(A^{(\sigma)})$. If $j \le m$ then we convert X_j^ε into $Y_j^\varepsilon = (X_i^{(\sigma)})^{-1} X_j X_i^{(\sigma)}$ by a bifurcation; for $j > m$ we make a bifurcation analogous to 5.2.2 (d).

We obtain a related binary product $\{Y_1,\dots,Y_n;\Pi_Y\}$ for which the sum of the lengths of the factors remains unchanged. In the case of the bifurcation 5.2.2 (d) this is evident. Likewise for the bifurcation 5.2.2 (b) it follows easily from the fact that X_j has the form $\tilde{X}_j^{-1} S_{r_j}^{\varepsilon_j} \tilde{X}_j$. If it is not reduced, we can decrease the sum of the lengths of the factors by bifurcations and (b) occurs.

Let $\{Y_1,\dots,Y_n;\Pi_Y\}$ be reduced. Under the bifurcations given above, all factors except $X_i^{(\sigma)}$ remain in their places. If $X_i^{(\sigma)}$ has left the loop, then (c) occurs. Otherwise, there is either a letter in the kernel of $Y_i^{(\sigma)}$ with formal partner in the kernel of $Y_i^{(\alpha)}$ (so that (a) occurs) or $Y_i^{(\sigma)}$ and $Y_i^{(\alpha)}$ constitute a pair like the previous $X_i^{(\sigma)}$ and $X_i^{(\alpha)}$. We proceed with $Y_i = X_i = AB$ as before and replace $A^{\pm 1}$ by $B^{\pm 1}$. After each step we either get (a), (b) or (c) or else the number of factors which begin or end with B or B^{-1} decreases. However, this can happen only finitely often.

□

Conclusion of the proof of Theorem 5.3.2:

Suppose now that 5.3.4 (a) does not occur during the above process, so that we obtain a reduced binary product $\{Z_1,\dots,Z_n;\Pi_Z\}$, related to $\{X_1,\dots,X_n;\Pi_X\}$, in which no pair is separated by the double point. We show that $\{Z_1,\dots,Z_n;\Pi_Z\}$ satisfies 5.3.4 (a).

Let Z_i contain the double point and let $Z_i = A|Z_i|B$. If $|Z_i|$ does not contain the double point in its interior, then either A or B is non-trivial, otherwise $\{Z_1,\dots,Z_n;\Pi_Z\}$ is decomposable. We can therefore choose a letter C_0 at the beginning or end of Z_i which is separated from its formal partner C_1 by the double point. Starting from C_0 and C_1 we construct chains $C_0C_{-1}\dots$ and $C_1C_2\dots$, by going first to the cancelling partner. We show that neither of the chains jumps from the loop to the other part or vice versa. Suppose for example that C_0 is on the loop. Since the double point separates no pairs, $C_0C_{-1}C_{-2}\dots$ can only leave the loop by means of a letter from Z_i. Since C_0 is the initial letter, all of $C_0, C_{-1},\dots$ are either initial or end letters. Thus $C_0C_{-1}C_{-2}\dots$ leads back to C_0 via a cancelling partner, $C_{-1}C_{-2}\dots$ back to C_{-1} via a formal partner, etc. By going further and further back we eventually find a letter which is its own formal or cancelling partner. However, this is impossible. One argues analogously when C_0 does not lie on the loop. Thus $\dots C_{-1}C_0C_1C_2C_3\dots$ constitutes a maximal open chain, the ends of which are separated by the double point.

Thus we have gone from our original binary product to a related one homotopic to a binary product of smaller length. If we now reverse all bifurcations, restore the necessary 1's and reverse the initial conjugation we finally obtain a binary product homotopic to the original (the concepts "homotopic" and "related" obviously commute).

□

5.3.5 Corollary. *For each indecomposable binary product there is a homotopic simple binary product of no greater length.* □

5.3.6 Theorem. *Let* $G = \langle s_1,\dots,s_m,t_1,u_1,\dots,t_g,u_g \mid s_1\dots s_m \Pi[t_i,u_i] = 1\rangle$ *or* $\langle s_1,\dots,s_m,v_1,\dots,v_g \mid s_1\dots s_m v_1^2\dots v_g^2 = 1\rangle$ *respectively, and let* $\hat{G}$ *be the free group on* $S_1,\dots,S_m, T_1,\dots,U_g$ *or* $S_1,\dots,S_m, V_1,\dots,V_g$ *respectively. Let* $\hat{\alpha}: \hat{G} \to \hat{G}$ *be an endomorphism with* $\alpha S_i = L_i S_{r_i}^{\varepsilon_i} L_i^{-1}$, $i \le m$ *and let* $\hat{\alpha}$ *induce an automorphism* α *of* G. *Then* $\{\hat{\alpha}S_1,\dots,\hat{\alpha}S_m,\hat{\alpha}T_1,\dots,\hat{\alpha}U_g;\Pi_*\}$, $\{\hat{\alpha}S_1,\dots,\hat{\alpha}S_m,\hat{\alpha}V_1,\dots,\hat{\alpha}V_g;\Pi_*\}$ *respectively are indecomposable (relative to the* $\hat{N}$ *of the standard homomorphism* $\varphi: \hat{G} \to G$*). As above,* Π_* *denotes the product* $S_1S_2 \dots S_m \prod_{i=1}^{g} [T_i,U_i]$ *or* $S_1S_2 \dots S_m V_1^2 \dots V_g^2$ *respectively.*

Proof. As above, we write G as $\{h_1,\dots,h_n;\Pi_*\}$ and let $\hat{G}$ be freely generated by $H_1,\dots,H_n$. Supposing $\{\hat{\alpha}H_1,\dots,\hat{\alpha}H_n;\Pi_*\}$ were decomposable, it would be related to a decomposed binary product $\{N_1,\dots,N_n;\Pi_N\}$. If we apply the same processes which convert $\{\hat{\alpha}H_1,\dots,\hat{\alpha}H_n;\Pi_*\}$ into the decomposed product to $\{H_1,\dots,H_n;\Pi_*\}$ then we obtain a related $\{N_1',\dots,N_n';\Pi_N\}$ with $\hat{\alpha}N_i' = N_i$. Let Π_N' be a subword of Π_N with $\Pi_N'(N) \in \hat{N}$, then $\Pi_N'(\varphi N) = 1$ in G. Because $\Pi_N'(\varphi N) = \Pi_N'(\hat{\alpha}\varphi N') = \hat{\alpha}\Pi_N'(\varphi N')$, $\Pi_N'(N')$ lies in $\hat{N}$, so $\{N_1',\dots,N_n';\Pi_N\}$ decomposes, i.e. $\{H_1,\dots,H_n;\Pi_*\}$ would be decomposable. It is not, e.g. because this product corresponds to the canonical dissection of a surface with m perforations, bifurcations correspond to geometric bifurcations (cf. 3.2) and no subword of the boundary path after cutting is trivial in the fundamental group. Another proof is based on planar group diagrams, and it also yields the following theorem [Zieschang 1965], Theorem 7. We set it as exercise E 5.5. □

5.3.7 Theorem. *Let* $G = \langle s_1,\dots,s_m,t_1,\dots,u_g \mid s_1^{-k_1},\dots,s_m^{-k_m}, \Pi_*\rangle$ *or* $\langle s_1,\dots,s_m,v_1,\dots,v_g \mid s_1^{-k_1},\dots,s_m^{-k_m},\Pi_*\rangle$ *respectively be infinite. For simplicity we write both groups* $G = \langle h_1,\dots,h_n \mid h_1^{-k_1},\dots,h_m^{-k_m},\Pi_*\rangle$. *Let* $\hat{G}$ *be the free group in the generators* $H_1,\dots,H_n$, $\hat{N}$ *the kernel of the standard homomorphism* $\varphi: \hat{G} \to G$ *and let* $\hat{\alpha}$ *be an endomorphism of* $\hat{G}$ *with* $\hat{\alpha}H_i = L_i H_{r_i}^{\varepsilon_i} L_i^{-1}$, $1 \le r_i \le m$, $1 \le i \le m$, $\varepsilon_i = \pm 1$ *which induces an automorphism of* G. *Then* $\{\hat{\alpha}H_1,\dots,\hat{\alpha}H_n,\Pi_*\}$ *is indecomposable (over* $\hat{N}$*).* □

The notion of homotopic binary products was introduced in [Zieschang 1964', 1965'] where the main theorem 5.3.2 was also proved. This result is the main step in direction of the Dehn-Nielsen theorem.

5.4 FREE GENERATORS OF THE GROUP OF RELATIONS

In this section we study the group of relations for discontinuous groups. This enables us, in the next section, to define something like the degree of a mapping. For continuous mappings of surfaces it does in fact correspond to the degree. Both sections are from [Zieschang 1966].

Let G be a planar group with presentation as in Theorem 5.3.7. For simplicity $G = \langle h_1,\dots,h_n \mid h_1^{-k_1},\dots,h_m^{-k_m},\Pi_*\rangle$. Let $\hat{G}$ be the free group on $H_1,\dots,H_n$ and let $\hat{N}$ be the kernel of the standard homomorphism. Capital letters denote elements of $\hat{G}$, small letters elements of G, and a capital letter is mapped to the corresponding small letter by the standard homomorphism.

5.4.1 Definition. Let $\omega(L)$ be $+1$ or -1 according as $L \in \hat{G}$ is orientation preserving or not, or (what comes to the same thing) whether the number of V_i appearing in L is even or odd.

Let $L = \{L\}$ be a system of representatives of the left cosets of $\hat{G}$ modulo $\hat{N}$ which satisfies the Schreier condition. Let L_i be a subset of L which contains only one representative for the cosets $L\hat{N}, LS_i\hat{N},\dots,LS_i^{k_i-1}\hat{N}$.

5.4.2 Theorem. *The elements* $L(\Pi_*(H))^{\omega(L)}L^{-1}$, $L \in L$ *and* $LS_i^{-\omega(L)k_i}L^{-1}$, $L \in L_i$, $i = 1,\dots,m$ *freely generate* $\hat{N}$.

Proof. As in the proof of Theorem 4.7.1 we consider the planar net for G. Let C be the graph of the net. Then we can regard $\hat{N}$ as the fundamental group of C. Because of the Schreier condition, the paths corresponding to the $L \in L$ constitute a spanning tree of C (cf. 1.6). Thus for each vertex of C there is a uniquely determined representative, corresponding to the path connecting 1 with the point. If σ is a segment which does not lie in the tree and if ν and μ are paths in the tree which run to the initial and final point of σ, then $\nu\sigma\mu^{-1}$ is a representative of the path class, and the set of these classes constitutes a free generating system for $\hat{N}$. Now, $\nu\sigma\mu^{-1}$ is "simple closed" and bounds a disk in the net for G. It may be that a face in the interior of the disk meeting σ corresponds to a relation $S_i^{-k_i}$; then we have the unique $L \in L_i$ which connects the 1 of the net to a point on the boundary of this surface piece. The path ξ corresponding to L lies in the disk bounded by $\nu\sigma\mu^{-1}$, and $\nu\sigma\mu^{-1}$ may be represented as a product as follows:

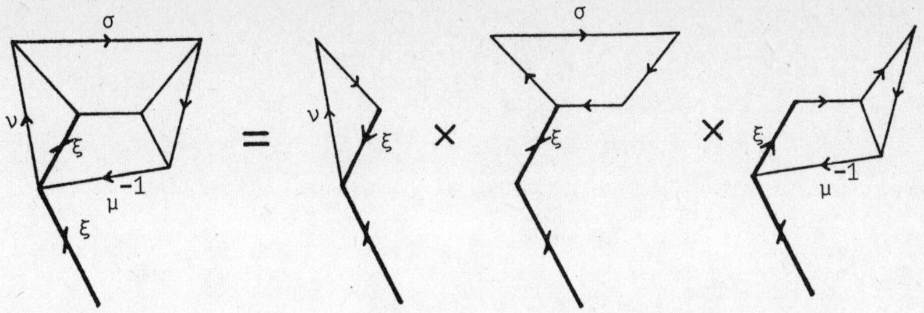

If the face meeting σ corresponds to the product relation, then we choose the $L \in \mathcal{L}$ which leads to the "initial point" of the product relation, and decompose as above. If we leave out the generator corresponding to $\nu\sigma\mu^{-1}$ and take instead $LS_i^{-\omega(L)k_i}L^{-1}$ or $L\Pi_*^{\omega(L)}L^{-1}$ respectively, then the theorem follows by induction on the number of faces in the interior of the path.

□

This theorem also follows from [Cohen-Lyndon 1963].

If we examine the proof of the Existence Theorem 4.7.1 then we see that an orientation of the plane net is defined, so that when the relations $h_1^{-k_1},\ldots,h_m^{-k_m},\Pi_*(h)$ are traced from the 1 they bound faces positively. Since positive stars which correspond to orientation-reversing elements are traversed in the opposite sense to stars which correspond to orientation-preserving elements, each generator $LS_i^{-\omega(L)k_i}L^{-1}$ or $L\Pi_*^{\omega(L)}L^{-1}$ respectively bounds its interior face positively. It follows that

5.4.3 Corollary. *For* $N \in \hat{N}$ *let the path traced from 1, apart from an approach path, be simple and closed. If we then write* N *as a reduced word in the free generators from Theorem 5.4.2 all the generators appear with the same exponent* + 1 *or* - 1.

□

Exercise: E 5.6.

5.5 THE MAPPING MATRIX

Let G, $\hat{G}$ and $\hat{N}$ have the same meaning as in 5.4. The defining relations are denoted by Π_i, $i = *, 1, \ldots, m$ where $\Pi_* = \Pi_*(H)$ and $\Pi_i = S_i^{-k_i}$ for $i = 1,\ldots,m$. Let $\mathbb{Z}^{m+1}$ be the free abelian group on the generators $e_*, e_1,\ldots,e_m$. We then have

5.5.1 Lemma. *$K\Pi_*^{\omega(K)}K^{-1} \mapsto e_*$ and $KS_i^{-\omega(K)k_i}K^{-1} \mapsto e_i$ ($i = 1,\ldots,m$ and $K \in \hat{G}$) define a homomorphism $\Theta: \hat{N} \to \mathbb{Z}^{m+1}$.*

Proof. The above mapping applied to the free generators from Theorem 5.4.2 defines a homomorphism Θ. We must show that our mapping coincides with this homomorphism.

One sees easily that all representatives of a coset modulo $\hat{N}$ have the same image under ω. Likewise ω maps all representatives of the cosets $L\hat{N}$, $LS_i^{-1}\hat{N},\ldots,LS_i^{-k_i+1}\hat{N}$ to the same integer.

□

Now let α be an automorphism of G. By Theorem 4.8.1 the elements of finite order are conjugate to powers of $s_1,\ldots,s_m$. Thus $\alpha(s_i) = \ell_i s_{r_i}^{b_i}\ell_i^{-1}$ where s_i and $s_{r_i}^{b_i}$ have the same order. We can assume that $-k_i/2 < b_i \leq +k_i/2$. Since s_i is not itself a power of an element of higher order in G, $k_i = k_{r_i}$ (so b_i is relatively prime to k_i), and $\begin{pmatrix} 1\ldots m \\ r_1\ldots r_m \end{pmatrix}$ is a permutation, as one can see, e.g. by abelianization. Thus we can induce α by an endomorphism $\hat{\alpha}$ of $\hat{G}$ for which $\hat{\alpha}(S_i) = L_i S_{r_i}^{b_i} L_i^{-1}$. The restriction of $\hat{\alpha}$ to $\hat{N}$ gives an endomorphism of $\hat{N}$.

Recalling that Π_i, for $i = *, 1, \ldots, m$, denote the defining relations, an element of $\hat{N}$ may be written as a product $N = \prod_j (K_j \Pi_{\alpha_j}^{\varepsilon_j} K_j^{-1})$, which is mapped to the element $\sum_{i=*}^{m} \left(\sum_{\alpha_j = i} \omega(K_j)\varepsilon_j \right) e_i$ by Θ. Under Θ, $\hat{\alpha}N = \prod_j ((\hat{\alpha}K_j)(\hat{\alpha}\Pi_{\alpha_j})^{\varepsilon_j}(\hat{\alpha}K_j^{-1}))$ goes into $\sum_{i=*}^{m} \left(\sum_{\alpha_j = i} \omega(\hat{\alpha}K_j)\varepsilon_j \right) \Theta(\hat{\alpha}\Pi_i)$. The group of orientation-preversing transformations is characteristic (Corollary 4.8.5), so $\omega(\hat{\alpha}K_j) = \omega(K_j)$ and $\Theta N = 0$ implies $\Theta\hat{\alpha}N = 0$.

Thus the kernel of Θ is mapped into itself by $\hat{\alpha}$. Consequently $\hat{\alpha}$ induces a homomorphism $\bar{\alpha}$ of $\mathbb{Z}^{m+1}$. $\bar{\alpha}$ can be characterised by a matrix $M(\hat{\alpha}) = (a_{ij})$, where a_{ij} is the coefficient of e_i in $\bar{\alpha}(e_j)$. Since $\hat{\alpha}(S_i) = L_i S_{r_i}^{b_i} L_i^{-1}$, all places in the

i^{th} column $(i \neq *)$ except the r_i^{th} will be zero. $\omega(L_i)b_i$ is in the r_i^{th} place. Apart from the first $(*^{th})$ place, only the ith place in the r_i^{th} row is non-zero.

$$M(\hat{\alpha}) = \begin{pmatrix} * & 0 & \cdots & 0 & \cdots & 0 \\ \vdots & \vdots & & \vdots & & \\ * & 0 & \cdots & \omega(L_i)b_i & \cdots & 0 \\ \vdots & \vdots & & \vdots & & \\ * & \omega(L_1)b_1 & \cdots & 0 & \cdots & 0 \\ \vdots & \vdots & & \vdots & & \\ * & 0 & & 0 & & \end{pmatrix} \begin{matrix} \\ \\ (r_i \\ \\ (r_1 \\ \\ \\ \end{matrix}$$

$$\qquad\qquad\quad \underset{1}{\frown} \qquad\qquad \underset{i}{\frown}$$

If $\hat{\beta}$ is an endomorphism of the same form as $\hat{\alpha}$, then $M(\hat{\beta}\hat{\alpha}) = M(\hat{\beta})M(\hat{\alpha})$. We call $M(\hat{\alpha})$ the *mapping matrix* of $\hat{\alpha}$.

<u>5.5.2 Lemma</u>. *Let* α *be an automorphism of* G, *and let* $\hat{\alpha}_1$ *and* $\hat{\alpha}_2$ *be endomorphisms of* $\hat{G}$ *which induce* α. *If*

$$\alpha(s_j) = \ell_j s_{r_j}^{b_j} \ell_j^{-1} \quad (j = 1,\ldots,m)$$

let $\hat{\alpha}_1(S_j) = L_j S_{r_j}^{b_j} L_j^{-1}$

$$\hat{\alpha}_2(S_j) = L'_j S_{r_j}^{b_j} L'^{-1}_j$$

where L_j, L'_j *are mapped to* ℓ_j *by the canonical homomorphism. Then* $M(\hat{\alpha}_1) = M(\hat{\alpha}_2)$.

Proof. Apart from the $*^{th}$ column $M(\hat{\alpha}_1)$ and $M(\hat{\alpha}_2)$ certainly coincide, since L_j and L'_j are the same modulo $\hat{N}$, hence $\omega(L_j) = \omega(L'_j)$. The rest follows easily from the fact that the images of an element under $\hat{\alpha}_1$ and $\hat{\alpha}_2$ are the same modulo $\hat{N}$. E.g. $NV_i NV_i = N(V_i N V_i^{-1})V_i^2$, so that if $N \in \hat{N}$, $(NV_i)(NV_i)$ and V_i^2 have the same image under Θ. We assign the treatment of the group with only orientation-preserving mappings as Exercise E 5.7.

□

<u>5.5.3 Corollary</u>. *Let* $\hat{\varepsilon}$ *be an endomorphism of* $\hat{G}$ *with* $\hat{\varepsilon}S_i = L_i S_i L_i^{-1}$, $L_i \in \hat{N}$, *which induces the identity. Then* $M(\hat{\varepsilon})$ *is the unit matrix.*

□

5.6 THE DEHN-NIELSEN THEOREM

Now we have developed the tools for the proof of the Dehn-Nielsen theorem. We prove it first for closed surfaces, next for surfaces with boundary, and finally in section 5.8 for planar groups which have compact fundamental domain and do not contain reflections.

Let $G = \langle t_1,u_1,\ldots,t_g,u_g \mid \Pi_* \rangle$ or $\langle v_1,\ldots,v_g \mid \Pi_* \rangle$ respectively. In order to avoid case distinctions we write $G = \langle h_1,\ldots,h_n \mid \Pi_* \rangle$. As before, let $\hat{G}$ be the free group on the generators $H_1,\ldots,H_n$. We think of the binary product $\{H_1,\ldots,H_n;\Pi_*\}$ realised by a system of segments $\eta_1,\ldots,\eta_n$ and a simple closed curve $\Pi_*(\eta)$ in the plane net for the above presentation of G (cf. the proof of Theorem 4.7.1), or by a canonical curve system on the closed surface $\mathbb{E}/G$. Cuts along the simple closed curves $\eta_1,\ldots,\eta_n$ convert $\mathbb{E}/G$ into a face with boundary $\Pi_*(\eta)$.

5.6.1 Theorem. *Each automorphism α of G is induced by an automorphism $\hat{\alpha}$ of $\hat{G}$ with* $\hat{\alpha}\Pi_* = L\Pi_*^{\pm 1}L^{-1}$.

Proof. Let $\hat{\beta}$ be an endomorphism of $\hat{G}$ which induces α. Because of Theorem 5.3.6 $\{\hat{\beta}H_1,\ldots,\hat{\beta}H_n;\Pi_*\}$ is indecomposable. There is therefore a binary product $\{K_1,\ldots,K_n;\Pi_*\}$ homotopic to $\{\hat{\beta}H_1,\ldots,\hat{\beta}H_n;\Pi_*\}$ such that the path $\Pi_*(K)$ in $\mathbb{E}$ is simple and closed. If we set $\hat{\alpha}H_i = K_i$, then this defines an endomorphism which likewise induces α; for $\{K_1,\ldots,K_n;\Pi_*\}$ and $\{\hat{\beta}H_1,\ldots,\hat{\beta}H_n;\Pi_*\}$ are homotopic. $M(\hat{\alpha})$ consists of just a single number. If an automorphism $\hat{\alpha}'$ induces α^{-1}, then $\hat{\alpha}\hat{\alpha}'$ induces the identity, and $1 = M(\hat{\alpha})M(\hat{\alpha}')$. Thus $M(\hat{\alpha}) = \pm 1$. This is the same as saying $\Theta\Pi_*(\hat{\alpha}H) = \pm e_*$; but then it follows from Corollary 5.4.3 that $\hat{\alpha}\Pi_*(H) = L\Pi_*^{\pm 1}(H)L^{-1}$. Therefore $\hat{\alpha}$ is an automorphism of $\hat{G}$ (Corollary 5.2.13).

□

Let h be a homeomorphism of a surface F. Let G be the fundamental group of F relative to a basepoint P. We require that h leaves the point P fixed. Then h maps one closed curve with initial point P to another. If γ_1 is a curve homotopic to γ (i.e. γ and γ_1 represent the same element of G), then $h(\gamma)$ and $h(\gamma_1)$ are also homotopic since the elementary processes 2.4.2 (a,b) may be extended to the image. As a result, h induces an endomorphism $h_\#$ of G, in fact an automorphism, since $(h^{-1})_\# h_\#$ and $h_\#(h^{-1})_\#$ are both the identity.

The geometric analogue of Theorem 5.6.1 is

Theorem 5.6.2 (Dehn-Nielsen). *Each automorphism of the fundamental group of a closed surface is induced by a homeomorphism.*

Proof. Let G be the fundamental group of the closed surface F and α an automorphism of G. We assume that G is infinite. Let $\eta_1,\ldots,\eta_n$ be a canonical curve system, which realizes the binary product $\{H_1,\ldots,H_n;\Pi_*\}$. The path class of η_i is then just the generator h_i of G. If $\hat{\alpha}$ is an automorphism of $\hat{G}$ which induces α, with $\hat{\alpha}\Pi_*(H) = \Pi_*(H)$, then by Theorem 5.2.8 the binary products $\{H_1,\ldots,H_n;\Pi_*\}$ and $\{\hat{\alpha}H_1,\ldots,\hat{\alpha}H_n;\Pi_*\}$ are related. In geometric terms this means that we can convert the canonical curve system $\eta_1,\ldots,\eta_n$ into a canonical curve system $\eta_1',\ldots,\eta_n'$ which realizes $\{\hat{\alpha}H_1,\ldots,\hat{\alpha}H_n;\Pi_*\}$ by surface bifurcations of F (cf. 3.2). The path class of η_i' is therefore αh_i, so the mapping $\eta_i \mapsto \eta_i'$ defines the homeomorphism sought.

Because of Theorem 5.6.1 there is an automorphism $\hat{\alpha}$ of $\hat{G}$ with $\hat{\alpha}\Pi_*(H) = L\Pi_*^{\varepsilon}(H)L^{-1}$, $\varepsilon = \pm 1$, which induces α. Now for each H_i and $\delta = \pm 1$ there is a homeomorphism which induces an automorphism β of G with $\hat{\beta}\Pi_*(H) = H_i^{\delta}\Pi_*(H)H_i^{-\delta}$. Now we consider a narrow strip around the curve η_i,

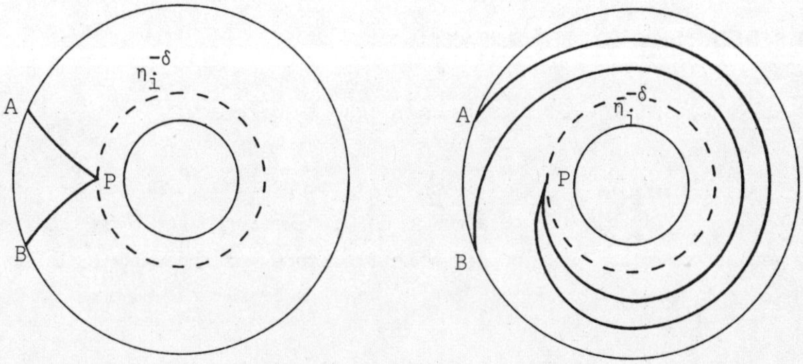

carry P once around $\eta_i^{-\delta}$ and extend this to a homeomorphism of the strip which leaves the boundary fixed. This homeomorphism may be extended to the whole of F and maps each curve η_j to one homotopic to $\eta_i^{\delta}\eta_j\eta_i^{-\delta}$. Thus $\Pi_*(\eta)$ goes into $\eta_i^{\delta}\Pi_*(\eta)\eta_i^{-\delta}$.

By carrying out the above homeomorphisms one after another one obtains, for each $L \in G$, a homeomorphism of F which induces an automorphism β of G with $\hat{\beta}\Pi_* = L^{-1}\Pi_* L$.

Finally, the homeomorphism defined by $\tau_i \mapsto \mu_{g+1-i}$, $\mu_i \mapsto \tau_{g+1-i}$ (or $\nu_i \mapsto \nu_{g-i+1}^{-1}$, respectively) defines an automorphism γ of G with $\hat{\gamma}\Pi_* = \Pi_*^{-1}$. Since $\hat{\gamma}\hat{\beta}\hat{\alpha}\Pi_* = \Pi_*$ for $\varepsilon = -1$ and $\hat{\beta}\hat{\alpha}\Pi_* = \Pi_*$ for $\varepsilon = 1$, $\gamma\beta\alpha$ and $\beta\alpha$ respectively, and hence α, may be induced by a homeomorphism.

□

For orientable surfaces theorem 5.6.3, which is the same as 3.3.11, was suggested by Dehn and first proved in [Nielsen 1927]. Another proof of it is contained implicitly in [Kneser 1930], this was carried out explicitly in [Seifert 1937]; we gave the later proof already as proof of theorem 3.3.11. For non-orientable surfaces the theorem was first proved in [Mangler 1939]. The approach as given here was announced in [Zieschang 1964'] and carried through in [Zieschang 1965, 1966]. Another approach to this problem, using the non-euclidean plane, is in the famous unpublished manuscript Fenchel-Nielsen, which has the advantage of giving directly the corresponding theorem for all finitely generated planar discontinuous groups which contain only orientation preserving transformations (Fuchsian groups).

Exercise: E 5.8

5.7 NIELSEN'S THEOREM FOR BOUNDED SURFACES

Let $G = \langle s_1,\ldots,s_m,t_1,u_1,\ldots,t_g,u_g \mid s_1\ldots s_m \prod_{i=1}^{g} [t_i,u_i]\rangle$ or

$= \langle s_1,\ldots,s_m,v_1,\ldots,v_g \mid s_1\ldots s_m v_1^2\ldots v_g^2\rangle$ respectively. As before, we write $G = \langle h_1,\ldots,h_n \mid \Pi_*(h)\rangle$ to avoid case distinctions. Let $\hat{G}$ be the free group on the generators $H_1,\ldots,H_n$ and let $\hat{N}$ be the kernel of the canonical homomorphism of $\hat{G}$ onto G.

For the groups considered here we cannot construct a planar net on which G acts as a planar group. Hence we cannot apply Corollary 5.4.3. Corollary 5.4.3 was essential in the proof of Theorem 5.6.1. Together with $\Theta(\Pi_*(\hat{\alpha}H)) = \pm e_*$ it implies that $\hat{\alpha}\Pi_*(H) = L\Pi_*^{\pm 1}(H)L^{-1}$.

In any case we can construct a 1-complex for G as in the proof of Theorem 4.7.1, so that each vertex corresponds to an element of G, edges labelled with the symbols $H_i^{\pm 1}$ emanate from each vertex and a path $W(H)$ from a fixed vertex 1 runs to the point $W(h)$: the Cayley diagram, cf. 4.13.

The groups considered here are the fundamental groups of compact surfaces with m boundary components. As in 5.6 we shall prove that we can induce automorphisms by homeomorphisms. Since homeomorphisms always map boundary components onto boundary components, the automorphisms must naturally be restricted.

<u>5.7.1 Theorem</u>. *An automorphism* α *of the fundamental group* G *of a bounded surface* F *is induced by a homeomorphism if and only if* $\alpha(s_i) = \ell_i s_{r_i}^{\varepsilon_i} \ell_i^{-1}$, $i = 1,\dots,m$ *where* $\begin{pmatrix} 1\dots m \\ r_1\dots r_m \end{pmatrix}$ *is a permutation and* $\omega(\ell_i)\varepsilon_i = \varepsilon = \pm 1$, $\ell_i \in G$.

Theorem 5.7.1 follows from the algebraic equivalent

<u>5.7.2 Theorem</u>. *Each automorphism* α *of* G *with* $\alpha(s_i) = \ell_i s_{r_i}^{\varepsilon_i} \ell_i^{-1}$, $i = 1,\dots,m$ *where*

$\begin{pmatrix} 1\dots m \\ r_1\dots r_m \end{pmatrix}$ *is a permutation,* $\varepsilon_i = \pm 1$ *and* $\ell_i \in G$ *is induced by an automorphism* $\hat{\alpha}$

of $\hat{G}$ *with* $\hat{\alpha}(S_i) = L_i S_{r_i}^{\varepsilon_i} L_i^{-1}$, $L_i \in \hat{G}$ *and* $\omega(L_i)\varepsilon_i = \varepsilon = \pm 1$.

Proof. In the proof we already use results from 5.8. Because of Theorem 5.3.6 and Corollary 5.3.5 there is an endomorphism $\hat{\alpha}$ of $\hat{G}$ with
$\hat{\alpha}(S_i) = L_i S_{r_i}^{\varepsilon_i} L_i^{-1}$, which induces α and for which $\hat{\alpha}\Pi_*(H)$ is a simple closed path in the group diagram of G. By Corollary 5.2.13 $\hat{\alpha}$ is an automorphism when $\hat{\alpha}\Pi_*(H) = L\Pi_*^{\pm 1}L^{-1}$.

Let N_f be the smallest normal subgroup of $\hat{G}$ which contains $\Pi_*(H)$ and $S_1^f,\dots,S_m^f$. We show

$$\bigcap_f N_f = \hat{N}.$$

It is clear that $\hat{N} \subset \bigcap_f N_f$. Let R be an element of $\bigcap_f N_f$. Suppose that R does not lie in $\hat{N}$. Then the element R' which results from R by cyclic reduction also lies in $\bigcap_f N_f$, but not in $\hat{N}$. If we replace a part of $\Pi_*^{\pm 1}(H)$ in R by its complement, then this element also lies in $\bigcap_f N_f \setminus \hat{N}$. Therefore we find a cyclically reduced element which contains no more than half of Π_*, in $\bigcap_f N_f \setminus \hat{N}$. But $\hat{G}/N_f$ is a planar discontinuous group, so by the Dehn solution of the word problem, Theorem 4.9.4, it follows that R contains a subword of S_i^f less at most two symbols, and this assertion must hold for all natural numbers f, which is impossible.

$\hat{\alpha}\Pi_*(H)$ is a simple closed path in the group diagram of G. Therefore no proper subword of it lies in $\hat{N}$, so there is an f such that $\hat{\alpha}\Pi_*(H)$ is a simple closed path in the planar net for $\hat{G}/N_f$. $\hat{\alpha}$ induces an automorphism of $\hat{G}/N_f$. In 5.8 we shall show that the homomorphism $\theta\colon N_f \to \mathbb{Z}^{m+1}$ described in 5.5 then maps $\hat{\alpha}\Pi_*(H)$ to $\pm e_*$. But then it follows by Corollary 5.4.3 (which we again need to apply here), that $\hat{\alpha}\Pi_*(H) = L\Pi_*^{\varepsilon'}(H)L^{-1}$. Likewise, we show in the next section that $\omega(L_i)\varepsilon_i = \omega(L)\varepsilon' = \pm 1$.

□

Proof of Theorem 5.7.1. Theorem 5.7.1 follows from Theorem 5.7.2 just as Theorem 5.6.2 follows from Theorem 5.6.1. Let F be a compact surface of genus g with m boundaries $\rho_1,\dots,\rho_m$. A *canonical curve system* is a system Σ of simple curves $\sigma_1,\dots,\sigma_m,\tau_1,\mu_1,\dots,\tau_g,\mu_g$ (when F is orientable) or $\sigma_1,\dots,\sigma_m,\nu_1,\dots,\nu_g$ (when F is non-orientable) with the following properties

5.7.3 (a) The curves have a common initial point P, and they meet only at this point.
(b) σ_i cuts an annulus from F, the other boundary of which is the ith boundary curve of F, ρ_i. No other curve of Σ meets this annulus.
(c) After removal of these annuli, the remaining curves decompose F into a disk with the boundary $\sigma_1\dots\sigma_m \prod_{i=1}^{g}[\tau_i,\mu_i]$ or $\sigma_1\dots\sigma_m\,\nu_1^2\dots\nu_g^2$ respectively.

There is a canonical curve system for an arbitrary initial point P (cf. 3.2). The fundamental group G of F with basepoint P has the presentation $\langle s_1,\dots,s_m,t_1,u_1,\dots,t_g,u_g \mid \Pi_*\rangle$ or $\langle s_1,\dots,s_m,v_1,\dots,v_g \mid \Pi_*\rangle$ respectively. In this connection, we can choose the generators of G so that $\sigma_i,\tau_j,\mu_j,\nu_k$ respectively represent the homotopy classes s_i,t_j,u_j,v_k. This follows immediately from the fact that σ_i and $x_i\rho_i x_i^{-1}$, up to the exponent + 1 or - 1, represent the same homotopy class, when x_i is a simple path from P to the ith boundary curve ρ_i which lies in the interior of the annulus cut out by σ_i. More precisely, it is a consequence of the following:

By means of elementary transformations we can convert F into a surface which consists of m annuli with boundaries $x_i\rho_i x_i^{-1}\sigma_i^{-1}$ (with suitable orientation of x_i) and a further face with boundary $\sigma_1\dots\sigma_m \prod_{i=1}^{g}[\tau_i,\mu_i]$ or $\sigma_1\dots\sigma_m\nu_1^2\dots\nu_g^2$. The tree for the fundamental group G of F then consists of the segments $x_1,\dots,x_m$. The generators corresponding to the $x_i\rho_i x_i^{-1}$ can be replaced by the generators s_i corresponding to the σ_i as $x_i\rho_i x_i^{-1}\sigma_i^{-1}$ yields a relation; and G has the presentation given, the generators of which are represented exactly by the curves of the canonical system. Now let α be an automorphism of G with $\alpha(s_i) = \ell_i s_{r_i}^{\varepsilon_i}\ell_i^{-1}$, $i = 1,\dots,m$, where $\begin{pmatrix}1\dots m\\ r_1\dots r_m\end{pmatrix}$ is a permutation and $\omega(\ell_i)\varepsilon_i = \varepsilon = \pm 1$. By Theorem 5.7.2 α may be induced by an automorphism $\hat\alpha$ of $\hat G$ with $\hat\alpha(S_i) = L_i S_{r_i}^{\varepsilon_i} L_i^{-1}$ and $\hat\alpha\Pi_* = L\Pi_*^{\varepsilon'}L^{-1}$, $\varepsilon = \omega(L)\varepsilon'$. As in the proof of Theorem 5.6.3 we can eliminate the conjugation factor and the ε' by homeomorphisms.

If on the other hand $\hat\alpha\Pi_*(H) = \Pi_*(H)$ then $\{H_1,\dots,H_n;\Pi_*\}$ and $\{\hat\alpha H_1,\dots,\hat\alpha H_n;\Pi_*\}$

are related, and the assertion follows as in the proof of Theorem 5.6.2.

The necessity of the condition follows from the fact that a homeomorphism permutes the boundary curves of F, i.e. induces an automorphism α of G with $\alpha(s_i) = \ell_i s_{r_i}^{\varepsilon_i} \ell_i^{-1}$, $\varepsilon_i = \pm 1$ and $\begin{pmatrix} 1 \ldots m \\ r_1 \ldots r_m \end{pmatrix}$ is a permutation. One sees immediately that $\omega(\ell_i)\varepsilon_i = \varepsilon' = \pm 1$ for orientable surfaces. Namely, $\omega(\ell_i) = 1$ and a homeomorphism either changes the orientation of no boundary curve or else reverses them all. For non-orientable surfaces the assertion follows from Theorem 5.7.2. Namely, we can induce α by an automorphism $\hat{\alpha}$ of $\hat{G}$ for which

$\hat{\alpha}(S_i) = L_i S_{r_i}^{\varepsilon_i} L_i^{-1}$ with $\omega(L_i)\varepsilon_i = \varepsilon = \pm 1$. Since L_i is mapped to ℓ_i by the canonical homomorphism $\hat{G} \to G$, $\omega(L_i) = \omega(\ell_i)$.

□

Theorem 5.7.1 was first proved in [Magnus 1934]. If the genus of the surface is 0, the automorphisms obtained are closely related to braids (and are often called braid automorphisms). The theory of braids is well developed, see [Artin 1926, 1947], [Burde 1963].

Exercise: E 5.9.

5.8 AUTOMORPHISMS OF PLANAR GROUPS

Let G be a planar group without reflections, and let $\hat{G}$, $\hat{N}$ be as in 5.4 and 5.5. Let Θ be the homomorphism $\hat{N} \to \mathbb{Z}^{m+1}$ from 5.5 and let $e_*, e_1, \ldots, e_m$ be the generators of $\mathbb{Z}^{m+1}$ onto which $\Pi_*, S_1^{-k_1}, \ldots, S_m^{-k_m}$ are mapped by Θ.

Let α be an automorphism of G and let the endomorphism $\hat{\alpha}$ induce α. Let $\hat{\alpha}(S_i) = L_i S_{r_i}^{b_i} L_i^{-1}$ when $\alpha(s_i) = \ell_i s_{r_i}^{b_i} \ell_i^{-1}$; L_i is then mapped onto ℓ_i by the canonical $\hat{G} \to G$. We have already seen in 5.5 that such an endomorphism always exists; for $\alpha(s_i) = \ell_i s_{r_i}^{b_i} \ell_i^{-1}$ with $k_i = k_{r_i}$, $(k_i, b_i) = 1$ and a permutation $\begin{pmatrix} 1 \ldots m \\ r_1 \ldots r_m \end{pmatrix}$.

First of all we shall show that in the mapping matrix of $\hat{\alpha}$,

$$M(\hat{\alpha}) = \begin{pmatrix} * & 0 \; . \; . \; . \; . \; . \; 0 \\ * & \\ \cdot & \\ \cdot & \\ \cdot & \omega(L_i)b_i \\ \cdot & \\ * & \end{pmatrix}$$

the first number of the first column equals + 1 or - 1, the others vanish.

By Theorem 4.10.1 we can find the fundamental group of an orientable surface as a subgroup F of finite index in G. By constructing intersections we can arrange that it is characteristic. Then α induces an automorphism of F, which we shall also call α. Let $\hat{F}$ be the free group corresponding to F, and N the kernel of the standard homomorphism $\hat{F} \to F$. We can assume that $\hat{F}$ lies in $\hat{G}$ and N in $\hat{N}$. Let Π^* be the relation of F. N is therefore "generated" by Π^*. Since Π^* lies in $\hat{N}$, we can apply Θ to Π^*. Let $\Theta(\Pi^*) = x_*e_* + x_1e_1 + \ldots + x_me_m$. Corresponding to F in $\hat{F}$ there is a binary product $\{K_1,\ldots,K_{2p};\Pi^*\}$. Although $\hat{\alpha}(K_i)$ does not necessarily lie in $\hat{F}$ (!), corresponding to $\{\hat{\alpha}K_1,\ldots,\hat{\alpha}K_{2p};\Pi^*\}$ there is a homotopic binary product $\{\hat{\beta}K_1,\ldots,\hat{\beta}K_{2p};\Pi^*\}$ relative to $\hat{N}$(!), where $\hat{\beta}$ is an endomorphism of $\hat{F}$(!) which likewise induces the automorphism α. Since Θ leaves the value $\Theta\Pi^*$ in $\mathbb{Z}^{m+1}$ fixed for homotopic binary products (cf. lemma 5.5.2), we have

$$\Theta(\hat{\alpha}\Pi^*(K)) = \Theta(\Pi^*(\hat{\alpha}K)) = \Theta(\Pi^*(\hat{\beta}K)).$$

Now if we consider just F, $\hat{F}$, N, then we have a mapping $\Theta'\colon N \to \mathbb{Z}$ and a mapping matrix for $\hat{\beta}$ relative to Θ' defined. As shown in 5.6, the matrix equals $\varepsilon = \pm 1$, and thus $\Theta'\Pi^*(\hat{\beta}K) = \varepsilon\Theta'\Pi^*(K)$. Each $N \in N$ also lies in $\hat{N}$, and

$$\Theta(N) = \Theta'(N)\Theta(\Pi^*(K));$$

for N is a product of elements conjugate to Π^* with conjugation factors which are from $\hat{F}$, and thus orientation-preserving. In summary we have:

$$\begin{aligned} \Theta(\Pi^*(\hat{\alpha}K)) &= \Theta(\Pi^*(\hat{\beta}K) \\ &= \Theta'(\Pi^*(\hat{\beta}K)\Theta(\Pi^*(K)) \\ &= \varepsilon\Theta'(\Pi^*(K))\Theta(\Pi^*(K)) \\ &= \varepsilon\Theta(\Pi^*(K)) \end{aligned}$$

so that we have proved

Lemma 5.8.1. *Under the endomorphism of* $\mathbb{Z}^{m+1}$ *induced by* $\hat{\alpha}$, $\sum_{i=*}^{m} x_i e_i = \Theta(\Pi^*(K))$ *may be fixed or mapped to* $-\sum_{i=*}^{m} x_i e_i$.

□

The correspondence which associates all generators of $\hat{G}$ with themselves and S_i with $S_i^{k_i} S_i$ defines an endomorphism on $\hat{G}$ which induces the identity in G. Further, an endomorphism is induced on $\mathbb{Z}^{m+1}$, which we can describe by its effect on the generators $e_*, e_1, \ldots, e_m$:

$$e_j \mapsto e_j,\ j \neq *, i$$

$$e_i \mapsto (k_i+1)e_i$$

for $S_i^{-k_i} \mapsto S_i^{-k_i(k_i+1)}$.

Now $\Pi_*(H) = S_1 \ldots S_i \ldots S_m \prod_{j=1}^{g} [T_j, U_j]$ is carried into $S_1 \ldots S_i^{k_i+1} S_{i+1} \ldots S_m \Pi[T_j, U_j] = S_1 \ldots S_{i-1} S_i^{k_i} S_{i-1}^{-1} \ldots S_1^{-1} S_1 S_2 \ldots S_i S_{i+1} \ldots S_m \Pi[T_j, U_j] = \Pi_*'(H)$. Thus $\Theta\Pi_*' = -e_i + e_*$. We obtain the same result in the non-orientable case. Thus e_* is mapped to $e_* - e_i$. But by lemma 5.8.1 $\sum_{i=*}^{m} s_i e_i$ remains fixed (it is not mapped to $-\sum_{i=*}^{m} x_i e_i$ since the endomorphism gives the identity on G).

We have therefore

$$x_* e_* + \ldots + x_i e_i + \ldots + x_m e_m = x_* e_* - x_* e_i + \ldots + (k_i+1) x_i e_i + \ldots + x_m e_m.$$

But that means $x_* e_i = k_i x_i e_i$. $\Pi^*(K)$ does not lie in the kernel of Θ. Namely, if we make the binary product $\{K_1, \ldots, K_{2g}; \Pi^*\}$ simple by homotopic modifications relative to $\hat{N}$, then $\Theta\Pi^*(K)$ does not change. The simple $\Pi^*(K)$ describe the perimeter of a fundamental domain of the subgroup $F \subset G$. It then follows from Corollary 5.4.3 that at least one x_i is different from zero. The x_i are thereby uniquely determined by x_*; x_* itself depends upon the embedding of $F \subset G$.

We now consider the original α again, and an endomorphism $\hat{\alpha}: \hat{G} \to \hat{G}$ with $\hat{\alpha}(S_i) = L_i S_{r_i}^{b_i} L_i^{-1}$ which induces α. We set $-\frac{k_j}{2} < b_j \leq \frac{k_j}{2}$. As before, this expression corresponds to $\ell_i s_{r_i}^{b_i} \ell_i^{-1} = \alpha(s_i)$ under the canonical homomorphism. Let $M(\hat{\alpha})$ be the mapping matrix of $\hat{\alpha}$. The endomorphism induced on $\mathbb{Z}^{m+1}$ then maps e_i onto $b_i \omega(L_i) e_{r_i}$ $(i = 1, \ldots, m)$; e_* may be mapped onto $a_* e_* + a_1 e_1 + \ldots + a_m e_m$.

Then $\Theta(\Pi^*)$ is mapped onto

$$M(\hat{\alpha})(\sum_{i=*}^{m} x_i e_i) = x_* a_* e_* + x_* a_1 e_1 + \ldots + x_* a_m e_m + x_1 b_1 \omega(L_1) e_{r_1} + \ldots + x_m b_m \omega(L_m) e_{r_m} .$$

However, by Lemma 5.8.1 this is equal to $\varepsilon \Sigma x_i e_i$ when $\varepsilon = \pm 1$. If we compare coefficients, it follows that $a_* = \varepsilon = \pm 1$. For $r_j = i$, $x_i e_i = \varepsilon(x_* a_i + x_j b_j \omega(L_j)) e_i$ and $k_i = k_j$. Because $x_* = k_i x_i$ we then have $x_i = \varepsilon k_i x_i a_i + \varepsilon x_j b_j \omega(L_j)$, and because $k_i = k_j$ we have $x_i = x_j$. Thus $1 = \varepsilon k_j a_i + \varepsilon b_j \omega(L_j)$. Considering this mod k_j, it follows from $-\frac{k_j}{2} < b_j \le \frac{k_j}{2}$, that b_j must equal $\varepsilon\omega(L_j)$ for $k_j \neq 2$ so $a_i = 0$ $(i = 1,\ldots,m)$. For $k_j = 2$ we can alter b_j to $-b_j$ if necessary so that $b_j = \varepsilon\omega(L_j)$ and then derive $a_i = 0$.

Consequently $\hat{\alpha}$ maps the S_i to $L_i S_{r_i}^{\varepsilon_i} L_i^{-1}$. Because of Theorem 5.3.7 $\{\hat{\alpha}H_1,\ldots,\hat{\alpha}H_n;\Pi_*\}$ is indecomposable and on the basis of Corollary 5.3.5, we can suppose that it is simple. It now follows from $a_* = \pm 1$ and Corollary 5.4.3 that $\hat{\alpha}\Pi_*$ runs around exactly one face and $\hat{\alpha}\Pi_* = L\Pi_*^{\varepsilon}L^{-1}$. Thus $\hat{\alpha}$ is an automorphism (Corollary 5.2.13) and we have proved

Theorem 5.8.2. *Each automorphism α of a planar group without reflections is induced by an automorphism $\hat{\alpha}$ of $\hat{G}$ with the following properties*

(a) $\hat{\alpha}(S_i) = L_i S_{r_i}^{\varepsilon_i} L_i^{-1}$ $i = 1,\ldots,m$

(b) $\hat{\alpha}\Pi_*(H) = L\Pi_*^{\varepsilon}(H)L^{-1}$,

(c) $\omega(L_i)\varepsilon_i = \omega(L)\varepsilon = \pm 1$, $k_i = k_{r_i}$ *and* $\begin{pmatrix} 1\ldots m \\ r_1\ldots r_m \end{pmatrix}$ *is a permutation.*

□

In order to obtain the geometrical analogue of Theorem 5.8.2, we consider $\mathbb{E}/G$. We remove a small disk around each rotation center. (Since no reflections appear in G, each rotation point lies in the interior of $\mathbb{E}/G$. All rotation centers are isolated, thus there is a system of disjoint disks which each contain one rotation center in the interior.) In this way we obtain a perforated surface F. Relative to a fixed basepoint A of F, the fundamental group F of F has the presentation $F = \langle S_1,\ldots,S_m,T_1,U_1,\ldots,T_g,U_g \mid \Pi_* \rangle$ or $\langle S_1,\ldots,S_m,V_1,\ldots,V_g \mid \Pi_* \rangle$ respectively. Let α be an automorphism of G and let $\hat{\alpha}$ be an automorphism of $\hat{G}$ which induces α and has the properties 5.8.2 (a-c). Because of 5.8.2 (b) $\hat{\alpha}$ induces an automorphism α' of F, which we can require to satisfy the hypotheses of Theorem 5.7.1, because of 5.8.2 (a,c). Thus α' may be induced by a homeomorphism h" of F which leaves fixed the basepoint A of F. If one removes the images of the perforations of $\mathbb{E}/G$

from the planar net $\mathbb{E}$ (i.e. removes disks around all fixed points), then one obtains a "perforated planar net" $\bar{\mathbb{E}}$, and $\bar{\mathbb{E}} \to F$ is an unbranched covering.

The homeomorphism h" of F may be lifted to a homeomorphism of $\bar{\mathbb{E}}$[1]. For this purpose one divides $\mathbb{E}$ into a net of fundamental domains which contain the vertices A_x, $x \in G$ over the basepoint A of F in the interior. A_1 lies in the fundamental domain F_1, and let F_x be the image of F_1 under $x \in G$, so that A_x lies in F_x. Now G may be defined as a discontinuous group on $\bar{\mathbb{E}}$ in the obvious way, and after removal of the disks the F_x constitute a net of fundamental domains on $\bar{\mathbb{E}}$. The images of the boundary edges of a fundamental domain constitute a system of simple curves of F which decompose F into a disk. h" maps this system onto a curve system with the same property, and we obtain a corresponding second decomposition of $\bar{\mathbb{E}}$ into fundamental domain F'_x, $x \in G$. Since h" leaves the basepoint fixed, A_x lies in the interior of F'_x, $x \in G$. We now map the boundary segments of F_x onto the boundary segments of $F'_{\alpha(x)}$ in the same way that h" maps the images of these boundary segments in F, and the face F_x onto $F'_{\alpha(x)}$ analogously.

This mapping is well-defined and is a homeomorphism of $\bar{\mathbb{E}}$. To prove this we remark that the subgroup associated with the covering $\bar{\mathbb{E}} \to F$ is the smallest normal subgroup S of F which contains $S_1^{-k_1}, \ldots, S_m^{-k_m}$, and G is the covering transformation group of the covering. Thus G is isomorphic to F/S in a canonical way.

Now if σ is an edge in the boundary of F_x and F_y, then a path ω from A_x to A_y which meets the boundary of F_x and F_y only once, and only at σ, determines a closed path $\bar{\omega}$ in F which meets no images of boundary curves of fundamental domains F_z, $z \in G$, apart from σ. The path $h''(\bar{\omega})$ therefore meets no images of boundary curves of F'_z, $z \in G$ apart from $h''(\sigma)$ (where σ is regarded as a curve in F). The path class $\{\bar{\omega}\}$ of $\bar{\omega}$ belongs to that coset of F modulo S which corresponds to $x^{-1} y \in G$. But now $\{h''(\bar{\omega})\} = \alpha'\{\bar{\omega}\}$ by definition of h", and the α' induced on $F/S = G$ is just α. This means that the path over $h''(\bar{\omega})$ beginning at $A_{\alpha(x)}$ leads to $A_{\alpha(y)}$ and meets only the boundary curve lying over $h''(\sigma)$. Thus $F'_{\alpha(x)}$ and $F'_{\alpha(y)}$ have the curve lying over $h''(\sigma)$ in common and the mappings defined on F_x and F_y coincide on σ. Since α is an automorphism and h" is a homeomorphism, the mapping defined on $\bar{\mathbb{E}}$ is a homeomorphism. We call it h'.

---

[1] In what follows we give a special proof for the lifting of homeomorphisms which map the subgroup corresponding to the covering into itself.

For $x \in G$, $x \mapsto h'xh'^{-1}$ induces the identity on F. Thus $h'xh'^{-1}$ is from G and is determined by the image of A_1. But $h'xh'^{-1}(A_1) = h'x(A_1) = h'A_x = A_{\alpha(x)}$, so $h'xh'^{-1} = \alpha(x)$. Since h' permutes the boundaries of disks in $\bar{\mathbb{E}}$ among those enclosing rotation points of the same order, h' may be completed to a homeomorphism h of E, so that h induces a homeomorphism of $\mathbb{E}/G$. Then hxh^{-1} is again an element of G and equals $\alpha(x)$, so we have proved

5.8.3 Theorem. *Each automorphism of a planar discontinuous group without reflections may be realized by a homeomorphism.* □

For the case where G has only rotations as generators this theorem was first proved in [Vollmerhaus 1963].

The homeomorphism constructed here can be regarded as semilinear in the appropriate metric (euclidean when G operates on the euclidean plane, hyperbolic when G is a plane discontinuous group of the non-euclidean plane). We can also impose arbitrary differentiability conditions on it. However, conformality cannot be attained in general.

Theorem 5.8.3 was announced in [Zieschang 1964'] and the first proof is in [Zieschang 1966]. [Macbeath 1965] contains another proof of the theorem, using Teichmüller theory and the classical Dehn-Nielsen theorem 5.6.3. This way one can also prove the corresponding theorem for planar discontinuous groups that contain reflections; we will present this type of proof in 6.6.11. If the group is generated by elements of finite order (i.e. $g = 0$) then the automorphisms are similar to braid automorphisms, by 5.5.3, 5.3.5, and the theorem can be deduced from braid theory.

5.9 COMBINATORIAL ISOTOPY

In sections 5.6 and 5.7 we saw that each automorphism of the fundamental group of a surface which preserves the boundary structure (i.e. path classes which contain boundary curves are mapped to others) is induced by a homeomorphism of the surface. Now we shall investigate how homeomorphisms which induce the same automorphism may differ from each other.

Strictly speaking, a homeomorphism can only induce an automorphism of the fundamental group of a surface when it leaves the basepoint of the fundamental group fixed. Now two fundamental groups of a surface with different basepoints are isomorphic, and the isomorphism is uniquely determined up to an inner automorphism. One can therefore say that a homeomorphism also induces an automorphism of the fundamental group when the basepoint is not fixed. We expect that homeomorphisms which differ only slightly from the identity will induce inner automorphisms. We shall now say precisely what it means to "differ slightly".

5.9.1 Definition. Let F be a surface.
(a) An *elementary combinatorial isotopy* of F is a homeomorphism H of F into itself which is the identity outside a disk S of F, and S meets the boundary of F at most in a simple non-closed path. A *combinatorial isotopy* of F is the product of elementary combinatorial isotopies.
(b) Two homeomorphisms H_0 and H_1 of F are called *combinatorially isotopic* when there is a combinatorial isotopy H with $HH_0 = H_1$.
(c) We call two simple closed paths *combinatorially isotopic* when there is a combinatorial isotopy which carries one path into the other.

Exercise: E 5.10

5.10 A SOLUTION OF THE WORD AND CONJUGACY PROBLEM IN FUNDAMENTAL GROUPS OF SURFACES

Let G be the fundamental group of the surface with a canonical presentation

$$G = \langle s_1,\dots,s_m,t_1,u_1,\dots,t_g,u_g \mid \prod_{i=1}^{m} s_i \prod_{j=1}^{g} [t_j,u_j]\rangle$$

or $$G = \langle s_1,\dots,s_m,v_1,\dots,v_g \mid \prod_{i=1}^{m} s_i \prod_{j=1}^{g} v_j^2\rangle.$$

Now we will represent each conjugacy class by a word of a special type, and it turns out that this word is well defined up to a cyclic permutation. We again denote the defining relation by Π_*.

5.10.1 Definition. A word W is called a *representative of a conjugacy class of* G if it has the following properties:
(a) W is cyclically reduced.
(b) W does not contain a subword which comprises more than half of the defining relation Π_* or its inverse; both are considered as cyclic words.
(c) If W contains a subword which comprises half of Π_*^{ε} then for g = 0 this contains s_m, and for g > 0 the t_1^{ε} or the ε-th power of the first v_1.

5.10.2 Theorem on the solution of the conjugacy problem. *The notation is as above*
(a) Each conjugacy class contains an element that can be represented by a word with

the properties above.
Let the length of the defining relation be greater than 7.
(b) If two words with the properties above represent the same conjugacy class then they coincide up to a cyclic permutation

Proof. (a) is trivial. If the defining relation has a length greater 7 then this follows from Dehn's solution of the conjugacy problem, see 4.9.3. The proof there was only for discontinuous groups with compact fundamental domain, not for the fundamental groups of surfaces with boundary. But this case can easily be reduced to the one considered as follows: For some $k \geq 9$, we introduce the additional relations s_i^k, $1 \leq i \leq m$ and obtain a group G_k. If the words W and V represent conjugate elements in G then they also define conjugates in G_k. We choose k bigger then $2(\ell(W) + \ell(V))$. By theorem 4.9.5 the words W and V coincide up to a cyclic permutation.

□

5.11 ISOTOPY OF FREE HOMOTOPIC SIMPLE CLOSED CURVES

Next we prove a deformation theorem for simple closed curves which we then will iterate on canonical systems of curves to obtain the Baer theorem for mappings.

5.11.1 Theorem. *Let* γ *and* δ *be two simple closed curves in the interior of a surface* F *which do not bound disks. If* γ *and* δ *are homotopic then they are also combinatorially isotopic.*

The restriction that γ and δ do not bound disks is necessary:

5.11.2 Proposition. *On an orientable surface* F *of genus greater than* 0 *two curves which bound disks are combinatorially isotopic if and only if the directions of the curves around the disks induce the same orientation of the surface.*

Proof. With the combinatorial methods developed as exercise E 5.11. The proposition can be generalized to non-orientable surfaces; we will give a proof of this interesting generalization, but based on the usual definition of a deformation as used in algebraic topology, see [Spanier 1966], [Massey 1967]:

5.11.3 Theorem. *Let* $f_0, f_1: (S^1,*) \to (F,P)$ *be embeddings of simple closed curves which bound disks with opposite directions, here* $*$ *and* P *are the basepoints in* S^1 *and* F. *(Note that we can define orientations in a neighbourhood of the basepoint* P, *even if* F *is non-orientable.) Then* f_0 *is isotopic to* f_1 *by an isotopy mapping* * *permanently to P if and only if* F *is a sphere.*

Proof. Suppose that $F \neq S^2$ and $f_0 \neq f_1$. If F is non-orientable, let F' be the orientable double covering, P' the basepoint of F' over P, and let $f'_0, f'_1: (S^1,*) \to (F',P')$ be liftings of f_0 and f_1. Let $h: F' \to F'$ be the covering transformation. An isotopy between f_0 and f_1, keeping the basepoint fixed, can be lifted to an isotopy between f'_0 and f'_1 in $F' \setminus h(P')$. Since $F' \setminus h(P') \neq S^2$, there is no loss of generality in assuming that F is orientable. Now the theorem follows from 5.11.2. Another proof is the following:

Let $f_t: (S^1,*) \to (F,P)$ be an isotopy between f_0 and f_1 with $f_t(*) = P$. Since f_t is nullhomotopic, $f_t(S^1)$ bounds a disk D_t. Since $F \neq S^2$, this disk is uniquely determined. The boundary path $f_t(S^1)$ assigns an orientation to D_t and hence to F (at *), for each t. Obviously, the orientation is unchanged by a small change of t. It follows that $f_0(S^1)$ and $f_1(S^1)$ bound disks and give them the same orientation, which contradicts the assumption.

□

Theorem 5.11.1 is a consequence of the following lemma which can also be used in another context, see 5.14.11 and [Zieschang 1965', 1969].

<u>5.11.4 Lemma</u>. *Let* F *be a surface,* Σ^* *a canonical dissection, and* δ *a simple closed curve on* F *which does not meet the boundary. Then, by a combinatorial isotopy,* δ *can be deformed into a curve* δ^* *that has a description with the properties 5.10.1 (a-c) relative to* Σ^*.

Proof. All the deformations of δ are of the following form:

5.11.5 *Let* δ' *be an open subpath of* δ *with initial point* A *and final point* B, *so that* $\delta = \delta_1\delta'\delta_2$. *Let* δ'' *be a simple path from* A *to* B *such that* $\delta'\delta''^{-1}$ *bounds a disk. Then* δ *is replaced by* $\bar{\delta} = \delta_1\delta''\delta_2$. The way we define the combinatorial isotopy is made clear by the following diagram:

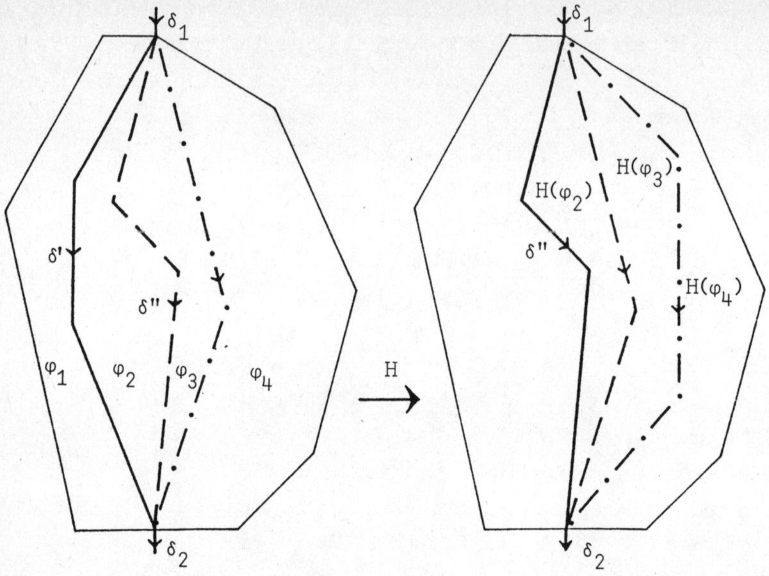

Obviously we can modify δ by a combinatorial isotopy so as to make it *properly* cut the curves of Σ^* without meeting P^*, and thus obtain a description $W(H)$ of δ, cf. 3.5. We regard $W(H)$ as a cyclic word.

(1) If $W(H)$ has a part XX^{-1}, then let ξ_1 be the subpath of the curve ξ^* corresponding to X which connects the points for the parts X and X^{-1} without meeting P^* and let δ_1 be the curve on δ connecting these points. For a suitable $\varepsilon = \pm 1$, $\delta_1^{\varepsilon}\xi_1$ is simple and closed. Since δ_1 has no points in common with Σ^* except its endpoints, the result of cutting along Σ^* outside of $\delta_1^{\varepsilon}\xi_1$ is a simple closed path in the dissected surface. Thus $\delta_1^{\varepsilon}\xi_1$ bounds a disk D on F. Since δ is simple, each part of δ entering D corresponds to another place $X^{\eta}X^{-\eta}$, $\eta = \pm 1$. However, δ can cut ξ^* at most finitely often; thus a subpath δ_o of δ together with an arc ξ_o of ξ^* bounds a disk D_o, the interior of which contains no parts of δ. Let δ_o' be a simple path with the same initial and final points as δ_o which does not lie in D_o, bounds a disk in conjunction with ξ_o, and which does not meet either δ or Σ^* except at boundary vertices. Then one replaces the δ_o in δ by δ_o', which may be done by a combinatorial

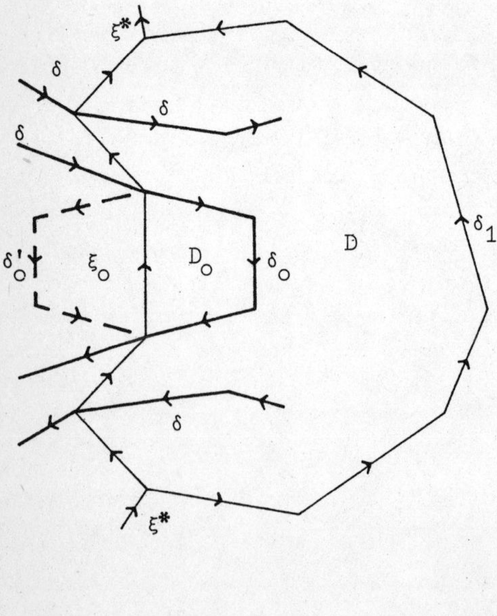

isotopy. By means of a further combinatorial isotopy we can arrange that δ no longer touches ξ^* at the initial and finalpoints of ξ_0', without any new points of intersection with Σ^* appearing. As a result, we have diminished the number of parts XX^{-1} in W(H) by one.

(2) We can now assume that the description of δ is cyclically reduced. Suppose that the description W(H) of δ contains a subword $W_1(H)$ which makes up more than half of Π_*^ε, $\varepsilon = \pm 1$. In this connection Π_*^ε and W(H) are regarded as cyclic words. Let δ_1 be the subpath of δ which corresponds to $W_1(H)$, and let η_i^*, η_j^* be the curves of Σ^* which correspond to the first and last letters of $W_1(H)$. Being a part of Π_*^ε, $W_1(H)$ defines a substar of the positive or negative star around P^* which consists of the curves of Σ^* see 3.5.1 (d). The substar begins with $\eta_i^{*\varepsilon_1\varepsilon}$ when $H_i^{\varepsilon_1}$ is the first letter, and ends with $\eta_j^{*\varepsilon_2\varepsilon}$ when $H_j^{\varepsilon_2}$ is the last letter of $W_1(H)$. Let χ_1 run along $\eta_i^{*\varepsilon_1\varepsilon}$ from P^* to the point which corresponds to the first letter of $W_1(H)$; χ_2 is the part of $\eta_j^{*\varepsilon_2\varepsilon}$ which connects P^* to the point corresponding to the last letter of $W_1(H)$. $\chi_1\delta_1\chi_2^{-1}$ bounds a disk, because $W_1(H)$ is a subword of Π_*^ε. We can now conclude, as above, that a subpath δ_o of δ with description $W_1^\eta(H)$, $\eta = \pm 1$, together with subpaths of χ_1 and χ_2, bounds a disk which contains no part of δ in its interior. We then replace δ_o by a simple path δ_o' which runs round the other side of P^*, has a description equal to the inverse of the complement of $W_1^\eta(H)$, and does not meet δ. This may also be achieved by a combinatorial isotopy.

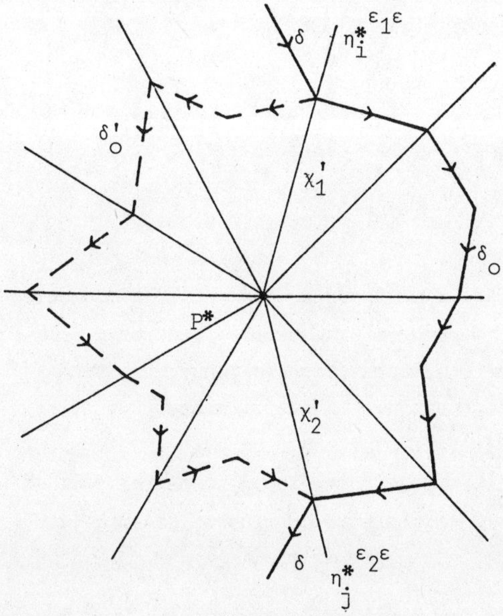

After each of these steps we cyclically reduce W(H) again and continue until the description of δ satisfies the conditions 3.5.13 and 3.5.14.

(3) If now an arc δ' on δ has a description that is half of the defining relation or its inverse then δ determines an angle at P^* and each arc of δ that enters this angle has the same description as δ' or δ'^{-1}. In particular, the innermost arc has this description; if it does not contain the one letter postulated in 3.5.15 it can be deformed by a combinatorial isotopy into an arc which fulfills 3.5.15. Hence, δ can be deformed by combinatorial isotopies into a simple curve δ^* that has the properties 5.10.1 (a-c).

□

5.11.6 *Proof of 5.11.1 for 'large' surfaces.* Assume that the defining relation for the fundamental group of F has length at least 8. We choose a dissection Σ^* relative to which γ has one of the descriptions 3.5.4. Now, by 5.11.4 we may deform δ by combinatorial isotopies till it has a description with the properties 5.10.1 (a-c). By 5.10.2 (b) we know that the descriptions of γ and δ now coincide up to a cyclic permutation. By combinatorial isotopies we may deform δ in such a way that it intersects Σ^* in the same points as γ (Proof as E 5.20). Now arcs on γ and δ which lie between the same points of intersection with Σ^* may be mapped into each other by combinatorial isotopies: for either they have no further points of intersection and therefore bound a disk between them, or else they may be divided at their further points of intersection into subarcs, and then there is at least one pair of arcs that bounds a disk. Hence, by a combinatorial isotopy as described in 5.11.5, we may either move the arc on δ to that of γ or diminish the number of points of intersection between γ and δ, of course, without altering the intersection with Σ^*.

□

5.11.7 *Proof of 5.11.1 for the surfaces not covered in 5.11.6.*

Here we have several cases. First, the fundamental group is a proper free product with cyclic amalgamated subgroup. Again choose a dissection Σ^* so that γ intersects Σ^* in the canonical form from 3.5.4. That there are the desired isotopies to move δ into γ will be shown in the second version of the proof of the Baer theorem for all cases. Here we need only some small ones, for instance, the fundamental group of the Klein bottle $G = \langle v_1, v_2 \mid v_1^2 v_2^2 \rangle = \langle v_1 \mid \rangle *_Z \langle v_2 \mid \rangle$ where $Z = \langle v_1^2 \rangle = \langle v_2^{-2} \rangle$. Now a simple closed curve is either of the form v_1, v_1^2 or $v_1 v_2$, and it is easily seen that these are the only words with the properties that represent their elements; therefore we can reach the same conclusion as in 5.11.6. This argument applies in the following cases:

Orientation	genus	number of boundary components	decomposition of the group
orient.	1	2	$\langle s_1, s_2 \mid \rangle *_{\langle s_1 s_2 \rangle} \langle t_1, u_1 \mid \rangle$
non-orient.	2	0	$\langle v_1, v_2 \mid v_1^2 v_2^2 \rangle = \langle v_1 \mid \rangle *_{\langle v_1^2 \rangle} \langle v_2 \mid \rangle$
non-orient.	1	2	$\langle s_1, s_2, v_1 \mid s_1 s_2 v_1^2 \rangle = \langle s_1, s_2 \mid \rangle *_{\langle s_1 s_2 \rangle} \langle v_1 \mid \rangle$

For 'very small' surfaces the theorem 5.11.1 is obviously true:

Orientation	genus	number of boundary components	name of surface
orient.	0	2	cylinder
orient.	0	1	disk
non-orient.	1	1	Möbius strip
non-orient.	1	0	projective plane

We are left with the torus (orientable, genus 1, closed), the torus with one hole and the sphere with 3 holes. On the torus any simple closed curve γ that is not nullhomotopic can be part of a canonical system of curves. A description with the properties of δ can consist only of the generator $[\gamma]$, hence we can again use the argument of 5.11.6. The case of a torus with a hole can be reduced to that of the closed torus; now simple closed curves are either parallel to the boundary or can be part of a system of curves. The case of a sphere with three holes also has an ad hoc solution. Proof is exercise E 5.12.

Exercises: E 5.11-13, 20

□

5.12 BASEPOINT PRESERVING ISOTOPIES OF SIMPLE CLOSED CURVES

In the proof of theorem 5.11.1 we did not care about the basepoint. But we can refine the theorem as follows:

5.12.1 Theorem. *Let* γ *and* δ *be two simple closed curves with the same initial point* P *in the interior of* F *which bound neither disks nor Möbius strips. If they are homotopic with* P *fixed (i.e. if they belong to the same element of the fundamental group of* F *with basepoint* P) *then they are isotopic under an isotopy which leaves* P *fixed.*

5.12.2 (a) Here we shall understand an isotopy to be a *semilinear isotopy*, i.e. a semi-linear homeomorphism $H: F \times I \to F \times I$ with $H(F\times t) = F \times t$ for all $t \in I$ and $H|F \times 0$ the identity on $F \times 0$, where I is the unit interval. This definition of isotopy is more useful for Theorem 5.12.1, since we are interested in the path of P during the isotopy. It is easy to show that for each elementary combinatorial isotopy H_k there is an isotopy $H: F \times I \to F \times I$ such that $H(Q,1) = (H_k(Q),1)$ for all $Q \in F$. Consequently *each combinatorial isotopy may be realized by an isotopy.* (Moreover, *each isotopy of a surface corresponds to a combinatorial isotopy.)*

(b) A point $P \in F$ is fixed *under an isotopy* H when $H(P,t) = (P,t)$ for all $t \in I$. The proof of Theorem 5.12.1 is confined to the case where the fundamental group G of F has at least four generators and $\ell(\Pi_*) \geq 8$.

Proof of Theorem 5.12.1. Because of Theorem 5.11.1 there is an isotopy $H: F \times I \to F \times I$ with $H(\gamma,1) = (\delta,1)$. If $H_t(Q)$ is given by $H(Q,t) = (H_t(Q),t)$ for all Q from F then $t \mapsto H_t(P)$ defines a semilinear mapping $I \to F$. This mapping defines a path in F, *the path* α *of the basepoint* P *during the isotopy* H. Now, H_t, $0 \leq t \leq 1$ when confined to γ, is a homotopy (in the usual sense) between γ and $\alpha\gamma\alpha^{-1}$. Then $\alpha\gamma\alpha^{-1}$ and γ are also homotopic in the usual sense when P is fixed. We again choose a canonical dissection Σ^*, relative to which γ has a description V(H) of the form 3.5.5-12. If L(H) is the description of α, then $L(h)V(h)L^{-1}(h) = V(h)$ holds in the fundamental group G of F. (Here we again consider V(h) as the product of elements of G but V(H) as a word in the symbols for the generators.) Thus L(h) and V(h) commute in G. One sees from the decomposition of G into a free product with amalgam that L(h) and V(h) must be powers of the same element. We show next that V(h) is not a power of another element, so that L(h) is a power of V(h). Now there is an isotopy which transports P around the curve γ and carries γ into itself. Therefore we can assume that the path α is null homotopic.

Suppose there is a reduced word M(H), containing at most a half of Π_*, such that $M(h)^{\ell}V(h) = 1$ in G for $\ell > 1$. There is then a cyclically reduced subword N(H) of M with $W^{-1}NW(H) = M(H)$ and $W^{-1}(h)N^{\ell}(h)W(h)V(h) = 1$ in G. We have already seen how we can apply the solution of the conjugacy problem to G, also for the case where boundaries appear. N^{ℓ} can therefore be required to have the properties 3.5.13-15, so that it is the same cyclic word as V. However, we can assume that N contains no more than a half of a relation when regarded as a cyclic word, and that it satisfies 3.5.15. But then the same is also true for N^{ℓ} because among the words 3.5.5-12, only the words V_1V_1 and $V_1 \ldots V_g V_1 \ldots V_g$, g odd, come into consideration for V(H), and these bound Möbius strips, hence are excluded by assumption.

We shall now modify the isotopy H so as to leave the point P fixed. Since α is null-homotopic there are finitely many processes $K_1,\ldots,K_r$ of the type 2.4.2 (a) or (b) which convert α into the constant path. If r = 0 there is nothing to prove. We assume then that the desired modification can be made for r < k, and next consider r = k.

Let K_1 be the removal of a spur $\sigma\sigma^{-1}$ from α, and suppose $[t_1,t_1]$ is the sime interval of I during which P traverses this spur. There is an isotopy G_t, $0 \leq t \leq 1$ with $G_t = G_o = id_F$ for $0 \leq t \leq t_1$ and $G_t = G_{t_2}$ for $t_2 \leq t \leq 1$, so that $H_tG_t(P) = H_{t_1}(P)$ for $t \in [t_1,t_2]$, i.e. for each point of time $t \in [t_1,t_2]$ we hold P back at its initial position $H_{t_1}(P)$ in the interval $[t_1,t_2]$. The existence of this isotopy can be seen intuitively by considering the part of $F \times [t_1,t_2]$ over σ and a "small" neighbourhood around it. ω_1 is the old path of P in $F \times [t_1,t_2]$, given by $t \mapsto (H_t(P),t)$, ω_2 the new path: $t \mapsto (H_{t_1}(P),t)$. The isotopy GH still carries γ into δ, but α may be carried into the constant path by $K_2,\ldots,K_r$, hence by a smaller number of combinatorial isotopies.

One follows the reverse procedure when K_1 is the insertion of a spur. Then ω_2 is carried into ω_1.

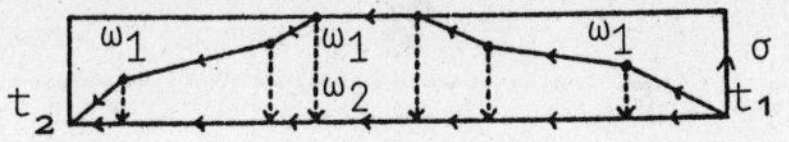

One proceeds analogously when K_1 is the replacement of part of the boundary path of a face by another which forms a triangle in conjunction with the first. Here again only points in a small neighbourhood of the triangle need to be moved. The induction hypothesis then yields an isotopy which leaves P fixed and carries γ into δ.

□

The assumption that the curves bound neither disks nor Möbius strips is necessary. For disks we have seen this in 5.11.3; for Möbius strips we will show it now. Again we use the isotopy concept from algebraic topology.

<u>5.12.4 Theorem</u>. *Let* F *be a surface and let* P *be the basepoint on* F *and* $*$ *that on* S^1 *(for instance, the complex number 1), let the image of* $f_o: (S^1,*) \to (F,P)$ *bound a Möbius strip in* F. *We assume that* $f_o(S^1)$ *is disjoint from the boundary of* F. *Then there is a isotopy* $f_t: (S^1,*) \to (F,P)$ *such that* f_1 *is an embedding, which is not isotopic to* f_o *with basepoint fixed. (The images* $f_t(S^1)$ *are disjoint from the boundary of* F.*)*

Proof. Diagram (a) shows a slightly larger Möbius strip than that bounded by $f_o(S^1)$. The embedding f_1 shown in the diagram is isotopic to f_o, by an isotopy which changes only the vertical coordinate in the diagram.

To show that there is no isotopy that leads from f_o to f_1 and keeps the base point fixed all the time, we examine two cases. Suppose first that F is a projective plane. Then f_o and f_1 bound disks with opposite orientations and we apply 5.11.3.

Suppose now that F is not a projective plane. Then every multiple of f_o is not trivial in the fundamental group of F. Lifting to the universal cover we see that each component of the inverse image of the Möbius strip is an infinite strip. Let h be the covering transformation corresponding to a generator of the fundamental group of the Möbius strip, and let $f'_o, f'_1: ([0,1],0) \to (F',P')$ be liftings of f_o, f_1; here P' denotes the basepoint of F'. Suppose we had an isotopy $f_t: (S^1,*) \to (F,P)$. Then this would lift to an isotopy

$$f'_t: ([0,1], 0,1) \to ((F' \setminus h(P')), P', h^2(P')),$$

see diagram (b). In particular, the simple closed curve obtained by going first along $f'_o([0,1])$ and then back along $f'_1([0,1])$ would be nullhomotopic, and would therefore bound a disk in $F' \setminus h(P')$; diagram (b) shows that this is not the case. □

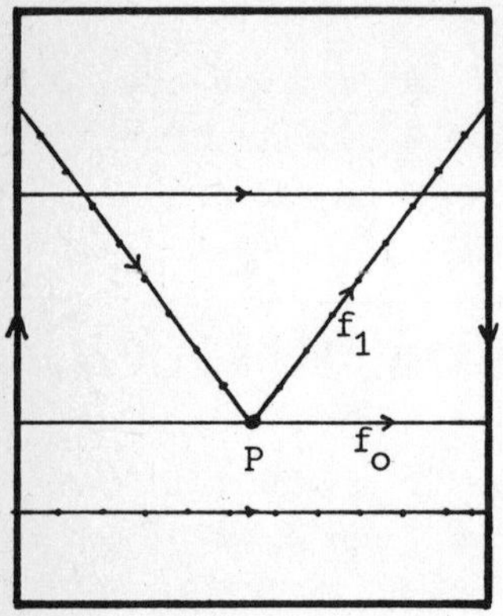

fig. (a)

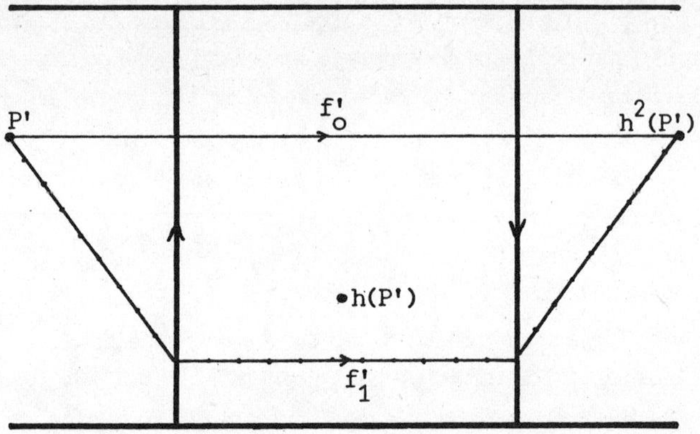

fig. (b)

[Baer 1928] has proved 5.11.2 for closed orientable surfaces; his approach is similar to that used by us in section 5.14. A combinatorial proof of the same result is in [Goeritz 1933]; it is similar to our approach in section 5.11. The refinement 5.12.1 has first been found by Epstein and was published in [Epstein 1966]. Slightly later and only for the 'most interesting' cases, but not knowing the Epstein success the result was found by Zieschang and published first in [ZVC 1970]; this proof is given here. The 'counterexample' 5.12.4 was the result of discussions of 'proofs' of a statement like 5.12.1 without the restriction about Möbius strips; it was published in [Epstein-Zieschang 1966].

5.13 INNER AUTOMORPHISMS AND ISOTOPIES

By iterated application of 5.11.1 we obtain an isotopy theorem for canonical dissections and deduce from it that homeomorphisms homotopic to the identity are isotopic to the identity. This generalizes the result from [Baer 1928] from closed orientable surfaces to arbitrary ones. Moreover, using Epstein's refinement 5.12.1 of the Baer theorem on simple closed curves we prove the refinement that a homotopy with fixed basepoint can be replaced by an isotopy which fixes the basepoint.

5.13.1 Theorem. *If a homeomorphism* h *of the surface* F *induces an inner automorphism of the fundamental group* G*, then* h *is a combinatorial isotopy of* F *(in the semi-*

linear case there is an isotopy H *of* F *from the identity to* h).

Proof. Let Σ be a canonical curve system with basepoint P. The first curve η_1 of Σ is homotopic to $h(\eta_1)$ since h induces an inner automorphism, so by Theorem 5.11.1 there is an isotopy (combinatorial isotopy) H_1 which maps $h(\eta_1)$ on to η_1 and carries h(P) into P.

By transporting P *along* $H_1h(\eta_1)$ if necessary we can assume that corresponding curves of Σ and $H_1h(\Sigma)$ are homotopic after fixing the basepoint.(Cf. the proof of Theorem 5.12.1; there it was shown that, possibly after pushing P along γ, the path α of P was null-homotopic, and this is precisely the property needed here.) If one applies the method of proof for Theorem 5.12.1 to the curve system then the result is a series of isotopies $H_1, H_2, \ldots, H_n$ with $H_i \ldots H_1h(\eta_j) = \eta_j$, $j = 1,\ldots,i$. We finally obtain an isotopy $\bar{H}$ of F which maps $h(\Sigma)$ onto Σ. $\bar{H}h$ is a homeomorphism of F which is the identity on Σ. If we remove the annuli which are bounded by the curves $\sigma_1,\ldots,\sigma_m$ of Σ and then cut F along Σ, we obtain a disk. $\bar{H}h$ induces a homeomorphism of the disk which is the identity on the boundary. By the Alexander lemma 5.13.3 below, such a homeomorphism is a combinatorial isotopy. Likewise, the homeomorphism of the annulus bounded by $\sigma_1,\ldots,\sigma_m$ which is induced by $\bar{H}h$ is a combinatorial isotopy. This also follows from Lemma 5.13.3.

□

5.13.2 Theorem. *If a homeomorphism* h *of* F *leaves the basepoint* P *of the fundamental group* G *fixed, and if it induces the identity on* G, *then* h *is isotopically deformable into the identity in such a way that* P *is fixed throughout the deformation.*

Proof. One applies Theorem 5.12.1 to η_1, which does not bound a Möbius strip, and then proceeds as above. □

5.13.3 Lemma (Alexander-Tietze deformation theorem). *A homeomorphism* h *of a disk which is the identity on the boundary may be deformed into the identity by an isotopy which does not move the points of the boundary.*

Proof. Let the disk be the unit circle $\{(x,y) \mid x^2 + y^2 \leq 1\}$ in the euclidean plane $\mathbb{R}^2$, and let h be the homeomorphism. Assume $h(0,0) = (0,0)$. Let G_t be the dilatation $(x,y) \mapsto (tx,ty)$ of $\mathbb{R}^2$ and let H be the homeomorphism which equals h on the unit disk and is the identity elsewhere. If now H_0 is the identity and $H_t = G_tHG_t^{-1}$ for $0 < t \leq 1$, then this defines an isotopy of h into the identity. - If h does not fix (0,0) we move first h(0,0) back to (0,0) by an isotopy of the disk which is the identity on the boundary and then apply the above argument.

□

5.13.4 Remark. This isotopy is semilinear if we take a triangle in place of the circle, with the same midpoint. An analogous proof may obviously be carried out in higher dimensions.

For closed orientable surfaces theorem 5.13.1 was first proved in [Baer 1928]. A combinatorial proof similar to our is in [Goeritz 1933, 1933']. For arbitrary closed surfaces it is in [Brödel 1935] and [Mangler 1939]. The general version and its proof is from [Zieschang 1966]. The refinement 5.13.2 was first proved in [Epstein 1966]. The Alexander-Tietze deformation theorem was first proved in [Tietze 1914]. The proof here is from [Alexander 1923]; it is now standard in books on algebraic or geometric topology.

In section 5.14 we give a generalization of the Baer theorem and in 5.15 we will discuss the literature of further developments connected with it.

Exercises: 5.14-16

5.14 THE BAER THEOREM FOR PLANAR DISCONTINUOUS GROUPS

In this section we generalize the Baer theorem to all finitely generated planar discontinuous groups that contain only orientation preserving transformations. (The last assumption can be dropped, but we will make it for simplicity.) Later, in 5.16, we will use the main theorem 5.14.1 to prove relative versions of the Baer theorem.

5.14.1 Baer Theorem. *Let G be a finitely generated discontinuous group of orientation preserving mappings of* $\mathbb{E}$ *which is not cyclic. Let* $\phi : \mathbb{E} \to \mathbb{E}$ *be an orientation preserving homeomorphism such that*

5.14.2 $\phi^{-1} g \phi = g$ *for all* $g \in G$

Then there is an isotopy $\phi_t : \mathbb{E} \to \mathbb{E}$ *with* $\phi_0 = \phi$, $\phi_1 = id_{\mathbb{E}}$ *and*

5.14.3 $\phi_t^{-1} g \phi_t = g$ *for all* $g \in G$ *and* $t \in [0,1]$.

It is convenient to imagine mappings and isotopies as usually done in algebraic topology. Before we make some remarks to the theorem and start the proof we introduce some useful notation.

5.14.4 Notation. For G we take a canonical fundamental domain F as described in 4.4, 4.11; the boundary of F has the form

$$5.14.5 \quad \sigma_1'\sigma_1^{-1}\ldots\sigma_m'\sigma_m^{-1}\rho_1'\square\rho_1^{-1}\ldots\rho_q'\square\rho_q^{-1}\tau_1'\mu_1^{-1}\tau_1^{-1}\mu_1'\ldots\tau_g'\mu_g^{-1}\tau_g^{-1}\mu_g'$$

where each □ denotes an 'end' of F. For simplicity, we put

$$\alpha_1 = \sigma_1,\ldots,\alpha_m = \sigma_m, \alpha_{m+1} = \rho_1,\ldots,\alpha_{m+q} = \rho_g, \alpha_{m+q+1} = \tau_1,$$

$\alpha_{m+q+2} = \mu_1,\ldots,\alpha_{m+q+2g} = \mu_g$, and similarly for the curves with a prime. To each pair of directed boundary edges α_i, α_i' there corresponds an element of G sending α_i to α_i'; we denote it by a_i. By $n = m+q+2g$ we denote the number of generators. If we use the special notation for α_i, for instance σ_i, then we use the canonical notation s_i etc. The group G has the canonical presentation

$$5.14.6 \quad G = \langle s_1,\ldots,s_m,r_1,\ldots,r_q,t_1,u_1,\ldots,t_g,u_g \mid s_1^{k_1},\ldots,s_m^{k_m}, \prod_{i=1}^{m} s_i \prod_{i=1}^{q} r_i \prod_{i=1}^{g} [t_i,u_i]\rangle.$$

By N we denote the net obtained from F by the action of G. G operates simply transitively on the faces of N. We take a point p_* from the interior of F and join it with $a_i p_*$ by a broken line γ_i $(i = 1,\ldots,n)$ which crosses α_i once; γ_i shall not pass fixed points of transformations of G. Let $C = \{\gamma_i \mid i = 1,\ldots,n\}$. GC defines a net N^* in $\mathbb{E}$, dual to N. For $g \in G$ the edge $g\gamma_i$ cuts one and only one edge or halfline from N and this piece is G-equivalent to α_i. This follows from the assumption about the γ_i for rotations of order 2.

Sometimes it is convenient to work on the surface $S = \mathbb{E}/G$. On S we mark the images $\bar{p}_1,\ldots,\bar{p}_m$ of the rotation centers and the image $\bar{p}_0$ of the other boundary points of the $\sigma_i,\rho_i,\tau_i,\mu_i$ - which are all G-equivalent. Then S is an orientable surface of genus g with q holes and m marked points. The projection $\Pi: \mathbb{E} \to S$ is a covering, branched at the points $\bar{p}_1,\ldots,\bar{p}_m$. The star at $\bar{p}_0$ has the form

$$5.14.7 \quad \Pi(\sigma_1),\ldots,\Pi(\sigma_m),\Pi(\rho_1),\ldots,\Pi(\rho_q),\Pi(\tau_1),\Pi(\mu_1),\Pi(\tau_1^{-1}), \Pi(\mu_1^{-1}),\ldots,\Pi(\tau_g^{-1}),\Pi(\mu_g^{-1}).$$

We denote these curves by $\bar{\beta}_1,\ldots,\bar{\beta}_p$, $p = m+q+4g$. (If $g > 0$, then $p \neq n$.)

Points of $\mathbb{E}$ that are mapped to the same point of S *belong to the same fiber*.

This presentation of G and this type of net N will be used, if G admits a proper decomposition as a free product with amalgamation, see 5.14.17.

5.14.8 Remarks. (a) 5.14.2 expresses that ϕ is *fiber preserving* and 5.14.3 describes a *fiber isotopy*.
(b) From 5.14.2 it follows that ϕ fixes all fixed points of non-trivial elements of G.
(c) If G does not contain elements of finite order (i.e. $m = 0$) then 5.14.1 is a reformulation of the Baer theorem 5.13.1. Our proof is quite close to the original one in [Baer 1928].

The proof of 5.14.1 will be finished in 5.14.33. First we shall prove some lemmata.

From 5.14.8 (b) it follows that $\phi(C)$ does not contain any point fixed by a non-trivial mapping. So general position arguments for the projection $\Pi\,\phi\,\Pi^{-1}$ show that ϕ can be deformed by a fiber isotopy such that

5.14.9 (a) ϕC *meets* N *only in a finite number of points.*
(b) $\phi C \cap \Pi^{-1}\,\bar{P}_o = \emptyset$.

This allows us to define for each element $\phi\gamma$ of ϕC a description, i.e. a word in the generators from 5.14.6 and their inverses as letters, which reflects how $\phi\gamma$ crosses N.

5.14.10. Let $W_i(\phi)$ denote the word corresponding to $\phi(\gamma_i)$. The length $d_i(\phi)$ of $W_i(\phi)$ is the number of letters occurring in $W_i(\phi)$. Let $d(\phi)$ be the sum of the $d_i(\phi)$. If a word $W_i(\phi)$ contains a pair $a_i^{\varepsilon} a_i^{-\varepsilon}$ ($\varepsilon = \pm 1$), we say that ϕ allows a *free cancellation.*

5.14.11 Lemma. *If a free cancellation is possible for* ϕ, *then* ϕ *can be deformed by a fiber isotopy to a mapping* ϕ_1 *which also has the properties* 5.14.2 *and* 5.14.9 *such that* $d(\phi_1) < d(\phi)$.

Proof. Repeat the argument from (1) in the proof 5.11.5 of theorem 5.11.1. See the figure.

□

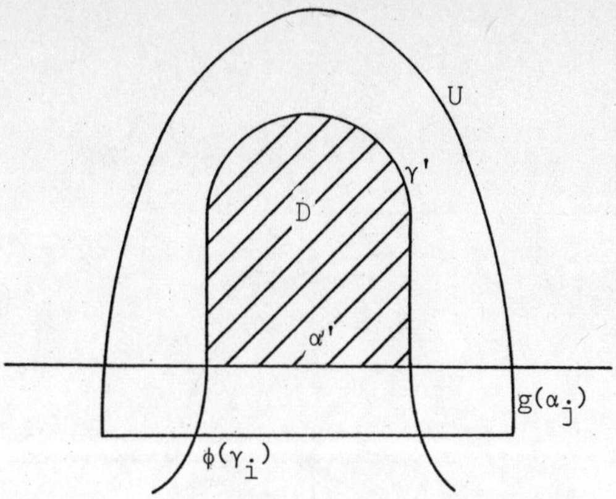

5.14.12 Lemma. *If all* $W_i(\phi)$ *start with the same letter and end with its inverse,* ϕ *can be deformed by a fiber isotopy to a mapping* ϕ_1 *with the properties* 5.14.2 *and* 5.14.9 *such that* $d(\phi_1) < d(\phi)$.

Proof. The assumption says that all curves in $\phi(GC)$ which start or end at $\phi(p_*)$ go to the same edge of N or come from it resp. It follows as in 5.14.11 that there exists a fiber isotopy which moves $\phi(p_*)$ to the other side of the segment and which diminishes $d(\phi)$.

□

5.14.13 Definition. Let us call a mapping ϕ *cyclically reduced,* if free cancellations are impossible and the situation 5.14.12 does not occur.

5.14.14 Lemma. *Let* ϕ *be cyclically reduced. Then* $W_i(\phi)$, $1 \le i \le n$, *does not contain a subword* $s_j^\varepsilon s_j^\varepsilon (j = 1,\dots,m)$ *or* $r_j^\varepsilon r_j^\varepsilon (j = 1,\dots,q)$, $\varepsilon = \pm 1$.

Proof. We assume that in $W_i(\phi)$ a letter is followed by itself and consider the image on $S = \mathbb{E}/G$, where the situation looks as in the figure. The arc $\bar{\gamma}_i'$ which goes inside produces an infinite spiral, because by assumption it cannot end in $\Pi\phi(p_*)$ and does not allow free cancellations.

□

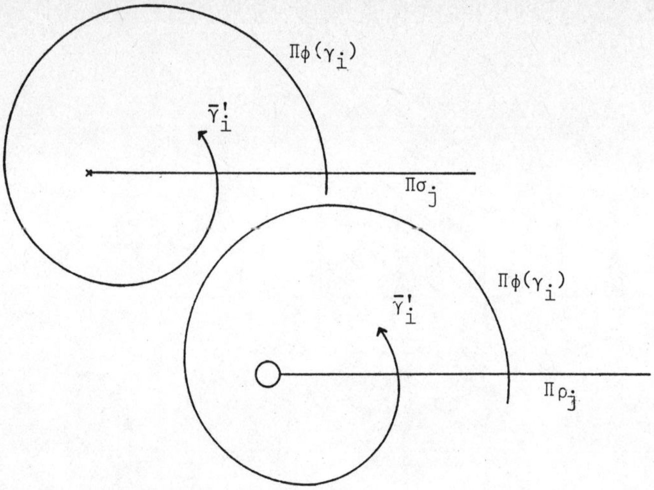

5.14.15 Lemma. *Let* $\bar{\beta}_1,\ldots,\bar{\beta}_h$ *be successive curves from the star at* $\bar{p}_0$, *and* $\bar{\beta}_{h+1},\ldots,\bar{\beta}_p$ *the rest. We assume that* ϕ *is cyclically reduced. Let* δ *be a part of one of the curves* $\Pi\phi(\gamma_i)$, $1 \le i \le n$, *such that* δ *crosses* $\bar{\beta}_1,\ldots,\bar{\beta}_h$ *in this order and is closest to* $\bar{p}_0$ *among all such parts in the* $\Pi\phi(\gamma_j)$, $1 \le j \le n$. *If the curve* $\Pi\phi(\gamma_i)$ *which contains* δ *crosses, immediately before or after* δ, *some* $\bar{\beta}_k$ *where* $1 \le k \le h$, *then all those parts of the* $(\Pi\phi(\gamma_j))^{\pm 1}$, $1 \le j \le n$, *which cross* $\bar{\beta}_1,\ldots,\bar{\beta}_h$ *in this or the inverse order also cross immediately before or after the crossing with* $\bar{\beta}_1$ *or* $\bar{\beta}_h$, *resp., some* $\bar{\beta}_\ell$, $1 \le \ell \le h$.

Proof. It is simpler to argue, if we alter the system of cuts slightly. For each handle of S we let a part of one curve shrink to the other, and we get a simple closed curve which does not contain $\bar{p}_0$. If we cut the handle at this curve, we get two disjoint holes. There are two arcs from $\bar{p}_0$ to these holes. A cut with one of them corresponds to a crossing of $\bar{\beta}_j$, a cut with the other to a sequence of crossings of $\bar{\beta}_{j+1}$, $\bar{\beta}_j$, $\bar{\beta}_{j+1}^{-1}$.

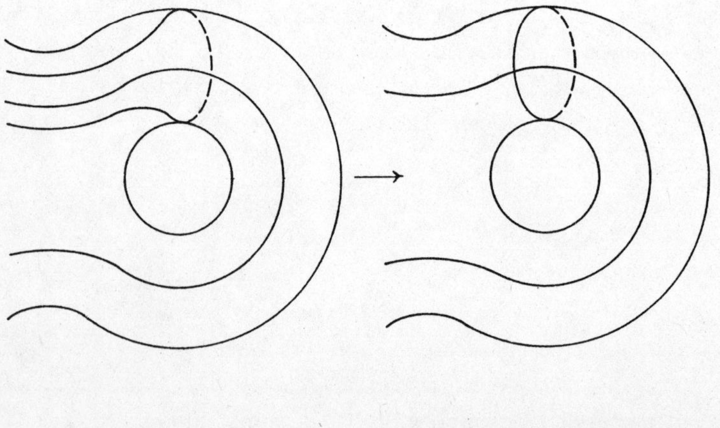

After cutting at all $\bar{\beta}_\ell$ which do not meet $\bar{p}_o$, we get a sphere with $q+2g$ holes and m marked points. From δ we obtain a simple curve δ' which goes from $\bar{\beta}_k$ back to $\bar{\beta}_k$. Let β_k' denote the arc of $\bar{\beta}_k$ between the endpoints of δ' which does not contain $\bar{p}_o$. Then $\beta_k' \cup \delta'$ bounds a disk D that may contain holes and marked points. But they belong only to the curves $\bar{\beta}_1,\ldots,\bar{\beta}_h$. If $\Pi\phi(\gamma_\mu)^{\pm 1}$ enters D, it can leave only by cutting β_k' or some of the holes. In both cases we get a cut with some $\bar{\beta}_j$, $1 \le j \le h$.

□

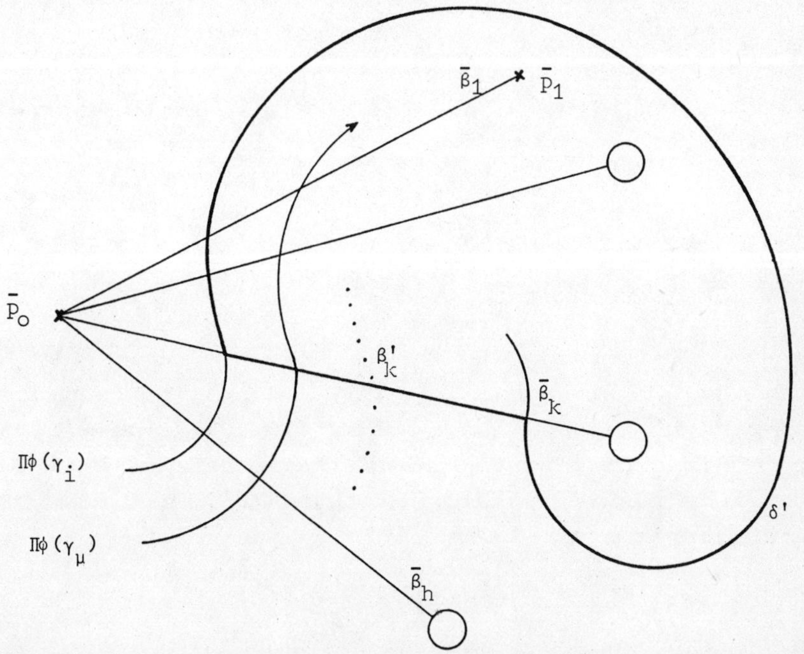

<u>5.14.16 Remark</u>. Of course, the same conclusion can be made, if δ goes from $\bar{\beta}_h$ to some of the $\bar{\beta}_j$, $1 \le j \le h$ or if δ crosses $\bar{\beta}_{h+1},\ldots,\bar{\beta}_p$ in this or the inverse order.

Because of the solution of the conjugacy problem we have to distinguish several cases. The most important is the following, and we will prove 5.14.1 for it first. The other cases are considered from 5.14.25 on.

5.14.17 *Case* $m+q+2g \ge 4$: Then the group G can be decomposed into a proper free product with amalgamated infinite cyclic subgroup, as we have done in 4.7.5. We assume that $G = G_1 *_A G_2$ and that the generators $a_1,\ldots,a_k$, $k \ge 2$, generate G_1, and $a_{k+1},\ldots,a_n$ the factor G_2. Any pair (t_i,u_i) belongs to one factor.

We now decompose the word $W_i(\phi)$ of $\phi(\gamma_i)$ into subwords

$$W_i(\phi) = W_{i1}\ldots W_{i\ell_i}$$

where all letters in one word W_{ij} are from one factor and the letters in W_{ij} and $W_{i,j+1}$ from different ones. We define $\ell(\phi) = \sum_{i=1}^{n} \ell_i$; as before $d(\phi)$ is the total number of letters in the $W_i(\phi)$, $1 \leq i \leq n$. The proof goes by induction on $(\ell(\phi), d(\phi))$ in lexicographical order. We assume that ϕ is minimal, i.e. it cannot be deformed by a fiber isotopy to a mapping with smaller $(\ell(\phi), d(\phi))$.

As the free cancellations in the proof of 5.14.11, 12 do not increase $\ell(\phi)$ but diminish $d(\phi)$, we get

5.14.18 ϕ *is cyclically reduced.*

<u>5.14.19 Lemma</u>. $W_{ij} \notin A$ *for all possible* i,j.

Proof. Assume $W_{ij} \in A$. We consider the star $\bar\beta_1,\ldots,\bar\beta_p$ at the point $\bar p_0 = \Pi(p_0)$. As $W_{ij} \in A$, the curve $\Pi\phi(\gamma_i)$ crosses $\bar\beta_1,\ldots,\bar\beta_h$ or $\bar\beta_{h+1},\ldots,\bar\beta_p$ in this or the inverse order and repeats this crossing possibly several times. Let us assume that the first system is intersected. Then the situation looks as in figure. If now a curve $\Pi\phi(\gamma_i)$ moves in the triangle determined by $\bar\beta_1$, $\bar\beta_h$ and $\Pi\phi(\gamma_i)$, then it must cut the $\bar\beta_i$, $1 \leq i \leq h$, in the same order because of 5.14.18. This implies:

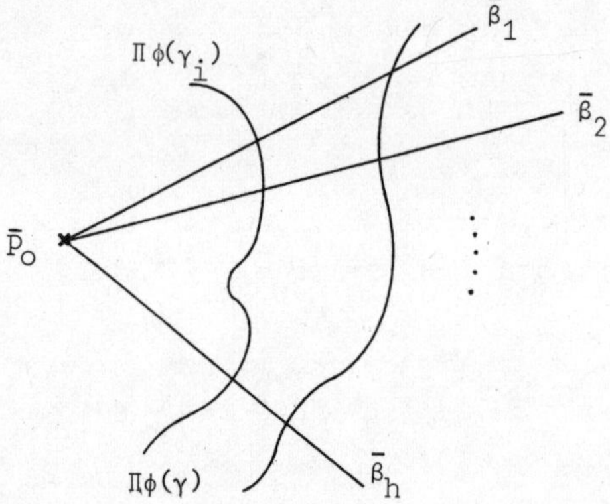

5.14.20 *The innermost arc cutting* $\bar\beta_1,\ldots,\bar\beta_h$ *in this or the inverse order corresponds to one of the words* $W_{\nu\mu}$.

Otherwise it would go before or after the cuts with $\bar\beta_1,\ldots,\bar\beta_h$, to some $\bar\beta_j$, $1 \leq j \leq h$, and so by 5.14.15 all arcs meeting $\bar\beta_1,\ldots,\bar\beta_h$ in this order would meet some $\bar\beta_j$, next to it, $1 \leq j \leq h$. As $W_{ij} \in A$, the corresponding part on γ_i meets

$\bar{\beta}_1,\ldots,\bar{\beta}_h$ several times; but the spiral argument shows that this is impossible.

The innermost arc of 5.14.20 can be deformed by an isotopy of S to the other side of the star at $\bar{p}_o$. This isotopy can be lifted to a fiber isotopy. The new mapping ϕ_1 satisfies $\ell(\phi_1) < \ell(\phi)$, if $\ell_i > 1$, contradicting the minimality of ϕ. But ℓ_i must be > 1, otherwise $W_i(\phi) \in A$ which is not possible by the solution of the word problem in groups of type 5.14.17.

□

As the basepoint moves during the deformation, there is an element $g \in G$ such that $W_i(\phi) = g a_i g^{-1}$. We write g in the normal form for elements of a free product. If the length of g is λ, then $\ell_i = 2\lambda - 1$, if a_i is from the same factor as the last part of g; $\ell_i = 2\lambda + 1$ otherwise. Because of 5.14.19 the $W_i(\phi)$ are already written in a form near to the normal form (slightly weaker, because we did not fix the representatives for the cosets with respect to the amalgamated subgroup.)

5.14.21 $\lambda \leq 1$.

Proof. Assume $\lambda \geq 2$. Then $\ell_i \geq 2$ and $W_{i1}^{-1} = AW_{i\ell_i}$ for some $A \in A$. But then $W_{i\ell_i} W_{i1} \in A$ is a subword belonging to a simple curve which contains the basepoint in the middle. Therefore either $W_{i\ell_i}$ coincides with W_{i1}^{-1} or $W_{i\ell_i} W_{i1}$ is one of the words $(b_1\ldots b_h)^{\pm1}$ or $(b_{h+1}\ldots b_p)^{\pm1}$. Here b_j corresponds to the generator or its inverse related to the cuts with $\bar{\beta}_j$. As the system $\{\phi(\gamma_i)\}$ is simple, all curves start with the same factors V or $\tilde{V}$ and end with V^{-1} or $\tilde{V}^{-1}$ where $V^{-1}\tilde{V}$ is equal to $(b_1\ldots b_h)^{\pm1}$ or $(b_{h+1}\ldots b_p)^{\pm1}$. These factors come from the first part of g. We can deform the $\tilde{V}$ isotopically to the other side of the star to get VA where A equals $(b_{h+1}\ldots b_p)^{\pm1}$ or $(b_1\ldots b_h)^{\pm1}$. This deformation does increase the ℓ_i, but now all curves start with the same V and so there is a deformation as in 5.14.12 such that all first and last terms W_{i1}, $W_{i\ell_i}$ vanish. This makes $\ell(\phi)$ smaller, contradicting the minimality of ϕ.

□

For $\lambda = 0, 1$ similar arguments can be done, and it follows that

5.14.22 $W_i(\phi) = a_i$ *or* $W_i(\phi) = a_i^{-1}$ *and* $a_i^2 = 1$.

Now move $\Pi\phi(p_o)$ back to $\Pi(p_o)$ by an isotopy of S (i.e. an ambient isotopy) without crossing $\bar{\beta}_1,\ldots,\bar{\beta}_p$ and lift this (ambient) isotopy to the plane $\mathbb{E}$.

5.14.23 Lemma. *The case* $W_i(\phi) = a_i^{-1}$ *for* $a_i^2 = 1$ *cannot occur.*

Proof. $\gamma_i a_i(\gamma_i)$ bounds a disk with the fixed point of a_i in the interior. Then $\phi(\gamma_i)$ runs through the edge that was crossed by $a_i(\gamma_i)$ and $\phi a_i(\gamma_i)$ through that of γ_i. So the disk bounded by $\gamma_i a_i(\gamma_i)$ is mapped to the disk bounded by $\phi(\gamma_i)\phi(a_i\gamma_i)$. This defines on $\mathbb{E}$ the inverse orientation, contradicting the assumption that ϕ preserves the orientation.

□

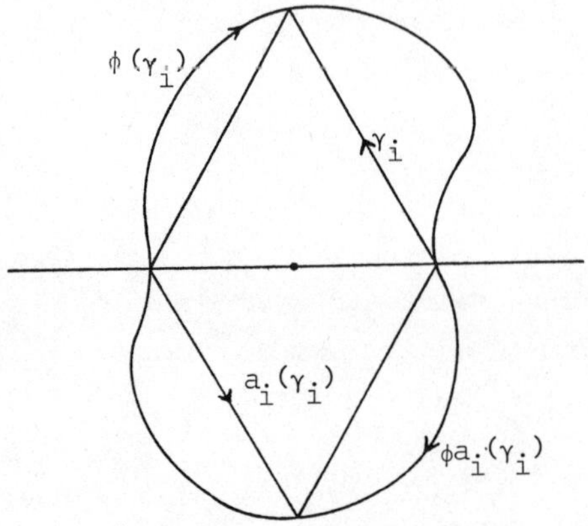

5.14.24 *Proof of 5.14.1 for the case 5.14.17.* As $W_i(\phi) = a_i$ we may isotop each curve $\Pi\phi(\gamma_i)$ to $\Pi(\gamma_i)$, and finally ϕ to the identity by a fiber isotopy.

□

5.14.25 *Case* $g = 0$, $m + q = 3$: Now the surface S is a sphere with q holes and m marked points, altogether three 'exceptions'. We take a dissection of S as described in the following figure (a) where x denotes a marked point or a hole. We lift this dissection to $\mathbb{E}$.

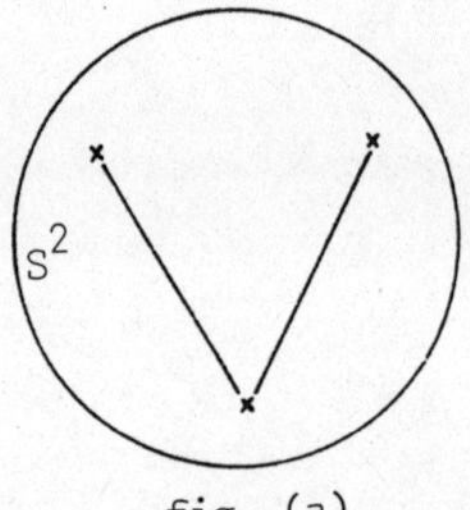

fig. (a)

From the system of cuts defined in figure (a) we get for G two generators x, y and the defining relations $x^{k_1} = y^{k_2} = (xy)^{k_3} = 1, 0 \leq k_i, k_i \neq 1$. These generators can be represented by curves ξ, η as in figure (b). A cyclically reduced simple closed curve based a $\bar{p}_o$ without free cancellations is of one of the types in figure (c), or with the roles of ξ and η interchanged. This follows by simple geometrical considerations. Such a curve can represent the generator x or y only if the word is already x or y, resp., or x^{-1} or y^{-1} for $k_1 = 2$ or $k_2 = 2$, resp. The last cases cannot occur, because ϕ is orientation preserving. This proves 5.14.1 for $g = 0$, $m+q = 3$.

□

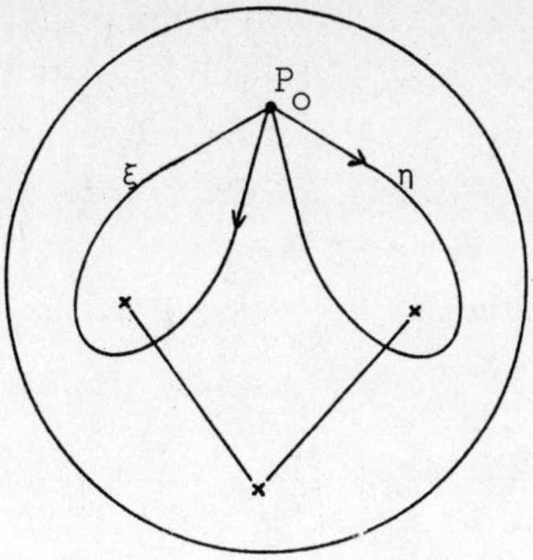

fig. (b)

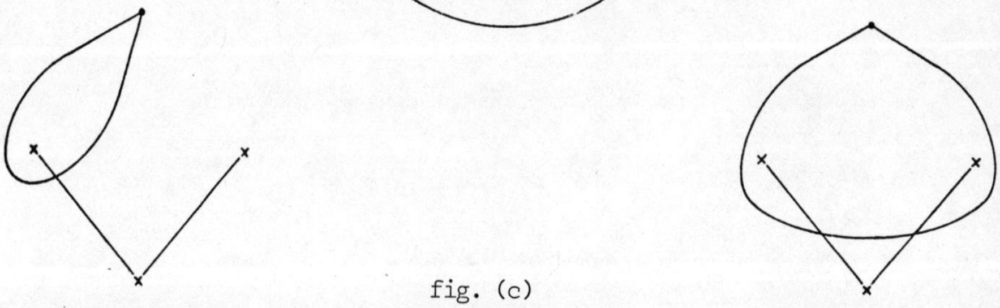

fig. (c)

The remaining cases are the following:

$m = 0$, $q = 1$, $g = 1$: a torus with a hole

$m = q = 0$, $g = 1$: a torus.

In both cases 5.14.1 is a consequence of the usual Baer theorem 5.13.1. It remains to consider

5.14.26 $m = 1$, $q = 0$, $g = 1$. Now we take the presentation $G = \langle t,u \mid [t,u]^h \rangle$, where $h > 1$.

Here t,u are the generators from a planar group diagram and we may apply Dehn's solution of the word problem, see 4.9.4, to them.

Let $d(Y)$ denote the number of letters in a word Y. Let $V = V(t,u)$ and $W = W(t,u)$ be the words of $\phi(\alpha_1)$ and $\phi(\alpha_2)$, t and u are the words for α_1 and α_2. There exists $X = X(t,u)$ such that

$$XtX^{-1} = V, \quad XuX^{-1} = W.$$

Now we will change the words etc. several times to make them 'nice'. By 5.14.11, 12 we may suppose:

5.14.27 *Free cancellations are impossible, and* V *is cyclically reduced.*

If X and V start with the same letter, we can move this letter to the end of V (which corresponds to an isotopy) and X becomes shorter. Similarly if X ends with t (or t^{-1}), we change the generators to $t, u' = tut^{-1}$ (or $t^{-1}ut$). The new generators also belong to a planar group diagram. (By these processes W perhaps will become longer, but we do not care.) We get:

5.14.28 X *and* V *do not start or end with the same letter;* X *does not end with a power of* t.

By Dehn's solution of the word problem there are the following cases, see 4.9.4.

5.14.29 (a) X = 1, V = t.

(b) *The defining relation or its inverse occurs in full length in* $XtX^{-1}V^{-1}$.

(c) *The defining relation appears once up to one letter and once up to two letters.*

(d) *The defining relation occurs three times up to two letters.*

Note that in (c) and (d) the parts of the defining relation are disjoint. Using the spiral argument we have the first statement:

5.14.30 *A subword of the defining relation which appears in* V *has length at most* 4.

5.14.31 *We can choose* X *such that it does not contain subwords which are more than a half of the defining relation.*

5.14.32 d(X) = 1 *is impossible.*

The statement 5.14.31 is trivial. For d(X) = 1 the defining relation can appear at most once because of 5.14.30. As XtX^{-1} can add only 3 letters, this again contradicts 5.14.30, 31 because the length of the defining relation is ≥ 8. This proves 5.14.32.

5.14.33 *Each of the long subwords of the defining relation quoted in 5.14.29 meets two of the parts* X, t, X^{-1}, V^{-1}.

Proof. If one part contains t and several letters from X, it can meet X^{-1} in at most one letter. If it also meets V^{-1}, then X is a subword of it and there cannot exist a second long subword meeting X^{-1} and V^{-1}: it must contain several letters of X^{-1} and this is only possible if V^{-1} has inverse letters at the front and at the end, contradicting 5.14.27. Therefore 5.14.29 (b) holds.

Let ν be the length of the subword of the defining relation belonging to X, μ the length of the part belonging to V^{-1}. Then

$$\nu + \mu + 2 \geqq 4h.$$

If we now change X to the complementary part of the defining relation, the length of X will become $4h - \nu \leqq \mu + 2$. But then there are μ free cancellations with V^{-1} possible, which will define some cyclic permutations of V^{-1}. Moreover we get one cancellation with t. After this is done we would have $V = t$ or $d(X) = 1$.

If the subword of the defining relation meets three of the terms X,t,X^{-1},V^{-1} this can only be if it meets X^{-1}, V^{-1} and X, as follows from the above. It can meet X or X^{-1} only in one letter, the other in μ letters, and V^{-1} is a subword of the defining relation of length $\nu \leqq 4$. Again it follows that 5.14.29 (b) must be true. We change the subword in X or X^{-1}, resp., by the complementary part of the defining relation. Then the part of length μ is changed to a word of length $4k - \mu = \nu + 1$. Again by cyclic permutation, things can be reduced to the case $d(X) = 0,1$.

□

Now we may assume that 5.14.33 holds, i.e. that long subwords of the defining relation will meet exactly two of the terms X,t,X^{-1},V^{-1}. But then it is impossible that t belongs to one of the parts. So there can be at most two pieces, and each of them meets X or X^{-1} in at least two letters. But these determine totally what follows or precedes in the defining relation, and V^{-1} could not be cyclically reduced. Therefore the only possiblility is that the defining relation appears once with at least four letters in X. This can be managed the same way as in the proof of 5.14.33.

This proves $V = t$ and $W = u$, and the theorem follows as before. Now the proof of 5.14.1 is finished.

□

Exercise : E 5.23

5.15 ON THE MAPPING CLASS GROUP

If we want to characterize mappings up to homotopy we are lead to the notion of a mapping class, and they can be treated using the Dehn-Nielsen and the Baer theorem as essential tools. In this section we will bring these two theorems together and prove some fundamental results on the group of mapping classes. In addition we will discuss some results from the literature.

5.15.1 Definition. (a) Two mappings $g,f\colon S \to S'$ between two surfaces are said to be from the same *mapping* or *homeotopy class* when they are homotopic. Let us denote the mapping class of f by [f].
(b) The system of all mapping classes of the homeomorphisms of a surface S onto itself forms a group and is called the *homeotopy group* or the *group of mapping classes*. The product is defined by $[f][g] = [f\cdot g]$. The group is denoted by Homeot S.

For a surface different from a disk *two homeomorphisms belong to the same mapping class if and only if they are isotopic* - this is a simple consequence of the Baer theorem 5.13.1. We denote the system of mappings isotopic to the identity by Isot S; it is also a group. Let Homeom S be the group of all homeomorphisms of S onto itself. Then we obtain from the definitions and 5.13.1:

5.15.2 Homeot S = Homeom S / Isot S.

Let $F = \pi_1(S)$ be the fundamental group of the surface S, and let $\text{Aut}_* F$ denote the group of all automorphisms of F which are induced by homeomorphisms of S. By Inn F we denote the group of all the inner automorphisms of F. Let $\lambda\colon \text{Homeot}\,(S,*) \to \text{Aut}_* F$ denote the homomorphism that adjoins to each basepoint preserving homeomorphism of S into itself the induced automorphism of the fundamental group. Mappings that can be deformed into the identity where the basepoint is not fixed during the deformation are mapped to inner automorphisms of F.

5.15.3 Theorem. *Let S be a surface different from the disk and the sphere.*
(a) $\lambda\colon \text{Homeot}\,(S,*) \to \text{Aut}_* F$ *is an isomorphism.*
(b) $\bar{\lambda}\colon \text{Homeot}\, S \to \text{Aut}_* F/\text{Inn}\, F$ *is an isomorphism.*
(c) $\text{Aut}_* F$ *consists of all automorphism of* F *that map the generators that belong to holes to elements of the same type.*

Proof. (c) is another formulation of the Dehn-Nielsen theorem 5.6.1. Now (a) is a consequence of Epstein's theorem 5.13.2, and (b) follows directly from (a). □

Now the problem arises: determine the mapping class group of a surface. For the disk and the sphere this is easily done:

5.15.4 Proposition. *The mapping class group of the sphere and the disk is isomorphic to* $\mathbb{Z}_2$ *. The unit contains all orientation preserving homeomorphisms, the other element the orientation reversing ones.*

The mapping class group of the projective plane is trivial. The homeotopy group of the Möbius strip is $\mathbb{Z}_2$*; homeomorphisms of the non-trivial class reverse the orientation of the boundary.*

Proof as exercise E 5.17.

□

In general the mapping class group is not known and is difficult to handle. For the torus we have a nice solution of the problem:

5.15.5 Theorem. *The mapping class group of the torus is isomorphic to* Aut $\mathbb{Z}^2$ *where the isomorphism is defined by* λ *from 5.15.3.* Aut $\mathbb{Z}^2$ = Gl(2,$\mathbb{Z}$) *has the following generators and defining relations:*

$$S = \begin{pmatrix} 0 & 1 \\ -1 & 1 \end{pmatrix}, \quad T = \begin{pmatrix} 0 & 1 \\ -1 & 0 \end{pmatrix}, \quad N = \begin{pmatrix} 1 & 0 \\ 0 & -1 \end{pmatrix}.$$

Defining relations: S^3T^2, S^6, T^4, N^2, $NTN^{-1}T$, $NSN^{-1}S$.

Proof see exercise E 5.18 or E 3.19. Give a geometric interpretation of the generating elements. See also [Zieschang 1980, Theorem 23.1].

□

5.15.6 *On geometric studies of the mapping class group.* For closed surfaces in general, generators of the mapping class group have been known for a long time, see [Baer 1928'], [Goeritz 1933'], [Dehn 1936], and the rediscovery of the method of the last paper in [Lickorish 1963, 1964, 1966]. See also [Birman-Chillingworth 1972]. The generators, at least most of them, correspond to twists around simple closed curves; an annulus neighbourhood of the curve is twisted such that on the boundary of the annulus we have the identity, see the figure. These twists are often called *Dehn* or *Lickorish twists*. See also [Călugăreanu 1968].

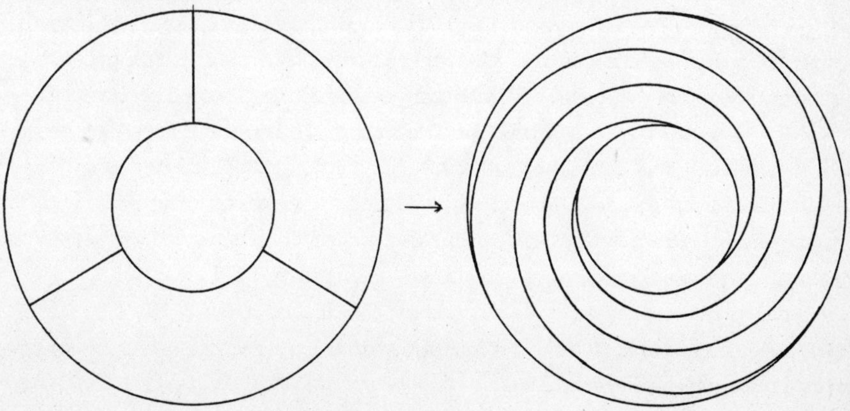

For a sphere with holes the mapping class group is closely related to the so-called *braid groups* for which generators and defining relations were already known since the original paper [Artin 1926], see also [Artin 1947]. The connection has been known since [Magnus 1934]. A picture where a system of arcs in $\mathbb{R}^3$ projected to the (y,z)-plane gives a braid and, projected to the (x,y)-plane, a dissection is the basis for [Burde 1963] and from this the relation between the theories of braids and mapping classes is obvious. Relative to the Artin presentation not only the word problem is solved, as done in [Artin 1925], but also the much more difficult conjugacy problem [Garside 1969]. Unfortunately, his solution has no direct geometric interpretation. Since 1962 a new development of braid theory has taken place, and the results can be applied to the mapping class groups of surfaces; to mention some papers: [Fadell-Neuwirth 1962], [Fadell-Van Buskirk 1962], [Fadell 1962], [Fox-Neuwirth 1962], [Van Buskirk 1966], [Arnold 1969, 1970, 1971]. This was developed further in [Birman 1969, 1969', 1969"], [Scott 1970']; a good guide to results and literature is the book [Birman 1974], see also [Magnus 1973], [Maclachlan 1978]. It is known that not all mapping classes can be obtained from braids. Already the determination of defining relations is quite difficult, but even more so the application of the presentations to solution of the word or conjugacy problem, the determination of finite subgroups, and other natural problems; the presentations are too complicated. To attack such problems, geometric methods have been used. See also [Lee 1972].

In [Hatcher-Thurston 1980] is published a general formula yielding a finite presentation for each of the mapping class groups uniformly.

5.15.7 *On an algebraic determination of a presentation of the mapping class group.*

Using the isomorphism λ or $\bar{\lambda}$ from 5.15.3 one can transform the problem of determining the mapping class group into a purely group theoretic one. We have seen that the automorphisms of the fundamental group F are induced by automorphisms of the free group $\hat{F}$ in the generators and that these automorphisms map the product Π_* of commutators or squares, resp., into a conjugate of itself or its inverse, see 5.6.1 and 5.7.2. A system of generators for the group of these automorphism can be obtained by applying the Whitehead method, see 1.9.3. [McCool 1975] considers the Whitehead method more closely and gives a method to find defining relations for the group of free automorphisms that fix an element; hence we can get a presentation for the group of automorphisms that map Π_* into a conjugate of itself or the inverse. From this we easily get a presentation of the mapping class group, see E 5.8. For details see also [Lyndon-Schupp 1977, I 4, 5]. This determination of a presentation of the mapping class group can be programmed for computers; unfortunately, even for simple cases it takes plenty of computer time. Another disadvantage of this approach is that we do not get a general presentation for the mapping class groups of all surfaces in which the generators and defining relations have some visible dependence on the genus etc.

5.15.8 *Finding simple closed curves.* From the proofs of the Dehn-Nielsen and Baer theorem it becomes clear that the simple curves are an efficient tool. Now the question arises: what elements of the fundamental group contain simple closed curves? In 3.5 we have seen that such an element may be represented by a curve that has a description with the properties 3.5.13-15, with respect to any given dissection of the surface. Now it can easily decided by drawing a diagram whether the class contains a simple closed curve or not; this has been done in [Zieschang 1963, 1965, 1969], with methods and proof belonging to the combinatorial topology. (For the corresponding homological problem see 3.6.11.)

The first algorithmic decision for this problem was given in [Reinhart 1960, 1962], he uses an interesting winding number, see also [Reinhart 1963]. His algorithm as well as those in [Čalugareanu 1966-1975] and [Chillingworth 1969, 1971, 1972, 1972'] also use facts from the geometry of the non-euclidean plane.

5.15.9 *On homotopy properties of the space of surface homeomorphisms.*

Until now we have neglected deformations. Another type of problem arises if the space of all homeomorphisms of a surface obtains a topology, for instance the compact-open, and we study this space. It is not path-connected, because under a deformation of a transformation we can never come from the trivial mapping class to a non-trivial one. It is clear that the study of the identity component is of special interest, since the knowledge of it and of the mapping class group together will yield any informations wanted. It has been proved that the identity component is contractible, hence all components are. This has been done for the space of all homeomorphisms in [Hamstrom 1966], for the space of diffeomorphisms in Earle-[Eells 1967], and for the space of PL homeomorphisms in [Scott 1970]. For spheres with holes this was proved in [Morton 1967]. Further literature on this subject [Hamstrom 1962, 1965].

5.15.10 *On finite groups of mapping classes.*

It is conjectured that any finite group of mapping classes of a surface can be realized by an isomorphic group of mappings. This was originally considered in [Nielsen 1942] for the case of one mapping class of finite order, later generalized by [Fenchel 1948, 1950]. We will give the proof of Fenchel in section 6.12. Nowadays there is a big literature on this interesting problem with partial solutions; this will be discussed in the continuation of this treatment [Zieschang 1980]. A full solution of the problem has been announced in [Kerckhoff 1980].

Exercises: E 5.17-19

5.16 COVERINGS AND THE MAPPING CLASS GROUP

One way to get informations about the mapping class group of a surface is to study its connection with the mapping class group of other surfaces which are related to the original surface by a covering. This is equivalent to comparing the automorphisms of a planar discontinuous group with those of one of its subgroups. The main result from which the known theorems can be deduced is given in theorem 5.16.2; for convenience, we postpone the proof of 5.16.2 (a), a relative Dehn-Nielsen theorem, to 6.6.11 although a proof based on the results already obtained is possible (E 5.21).

5.16.1 Definition. Let G be a planar discontinuous group. A *G-fiber* consists of all points (or vertices, edges, faces) which can be mapped one into the other by a transformation of G. As before in 4.3.1 the elements of a G-fiber are called *G-equivalent*. An isotopy is called a *G-isotopy* if at each moment the images of G-equivalent points are G-equivalent. An automorphism α of G is called *geometrical* if it is induced by a homeomorphism $\varphi\colon \mathbb{E} \to \mathbb{E}$:

$$\alpha(g) = \varphi^{-1} g\varphi.$$

5.16.2 Theorem. *Let G be a non-cyclic finitely generated planar discontinuous group consisting of orientation preserving elements only, and let F be a subgroup of finite index. Let G be of* hyperbolic *type, i.e. the centralizer of any non-trivial element is cyclic. Then:*

(a) A geometrical automorphism α of F is induced by a G-fiber preserving homeomorphism of $\mathbb{E}$ if and only if α is induced by an automorphism $\hat{\alpha}$ of G.

(b) If a G-fiber preserving homeomorphism φ of $\mathbb{E}$ is F-fiber isotopic to the identity, then φ is G-isotopic to the identity.

Proof. (a) will be proved in 6.6.11.

(b) We may assume that F is normal in G. Since φ is F-fiber isotopic to the identity the equation $\varphi^{-1}f\varphi = f$ is valid for all $f \in F$. For an arbitrary $f \in F$ we get

$$(\varphi g\varphi^{-1})\, f(\varphi g^{-1}\varphi^{-1}) = \varphi gfg^{-1}\varphi^{-1} = gfg^{-1}.$$

Since G is of hyperbolic type the center of G is trivial, see 6.4.9, hence $\varphi g\varphi^{-1} = g$. Now 5.16.2 is a consequence of 5.14.1. □

Next we apply this to coverings between surfaces.

5.16.3 Let $\pi: S_1 \to S$ be a regular covering with finitely many branch points and sheets. Furthermore let S_1 be a 'hyperbolic' surface, i.e. $2g_1 + q_1 \geq 3$ for genus g_1 and the number q_1 of holes. Let g and q denote genus and number of holes for S. The fundamental group F_1 of S_1 acts on the universal cover of S_1. The covering transformations for the covering π can be lifted to $\mathbb{E}$ and this defines a planar discontinuous group G in which F_1 has finite index. From Theorem 5.16.2 it follows that

5.16.4 <u>Corollary</u>. *If a fiber preserving mapping* $\varphi: S_1 \to S_1$ *is isotopic to the identity, then* φ *is fiber isotopic to the identity.*

□

The part (a) of 5.16.2 can be translated in a similar way, but the formulation is somewhat longer:

5.16.5 <u>Corollary</u>. *For the situation 5.16.3 let* B *be the group of geometrical automorphisms of* $\pi_1(S_1)$ *which contains the group of inner automorphisms* J *such that* B/J *is the group of outer automorphisms determined by the covering transformations. A geometrical automorphism* α *of* $\pi_1(S_1)$ *is induced by a fiber preserving mapping with respect to the covering* $\pi: S_1 \to S$ *if and only if* $\beta \mapsto \alpha\beta\alpha^{-1}$ *defines an automorphism of* B.

□

5.16.4, 5 describe the relationship between the relative homeotopy group to π and the group of automorphism classes. We remark that orientation reversing mappings of S_1 do not induce the identity on $\pi_1(S_1)$. By a small modification of the proof we get the same results, if the surfaces S_1 and S are not open, but have boundary components.

5.16.6 <u>Corollary (Baer-Nielsen-Theorem)</u>. *Let* G *be as in* 5.16.2. *Then the quotient of the group of* G*-fiber preserving homeomorphisms of* $\mathbb{E}$ *by the* G*-fiber isotopies is in the natural way isomorphic to the quotient of the group of geometrical automorphisms by the inner automorphisms.*

Proof. That all geometric automorphisms are induced by G-fiber preserving homeomorphisms of $\mathbb{E}$ is theorem 5.8.3. The rest is 5.16.2 (b).

□

The stronger form of the Baer theorem 5.13.2 has the following formulation if we lift all to the universal cover:

5.16.7 Theorem. *Let G be as in 5.16.2, but without elements of finite order. Let* $\varphi\colon \mathbb{E} \to \mathbb{E}$ *be a homeomorphism such that*

$$\varphi^{-1} g \varphi = g, \ \forall g \in G.$$

Moreover, let $\varphi(x_o) = x_o$ *hold for some* $x_o \in \mathbb{E}$. *Then there is a fiber isotopy* φ_t *of* φ *to the identity such that* $\varphi_t(x_o) = x_o$ *for* $t \in [0,1]$.

□

5.16.8 Remark. The proof of the generalized Baer Theorem 5.13.1 follows the lines of the proof in the section 5.11; the conclusions there to get Epstein's generalization are obstructed by the *rotation centers*:

Move the basepoint p_o by an ambient isotopy k times around the center of a rotation of order k. Then this isotopy induces the identity on the discontinous group. But it cannot be deformed to the identity by an isotopy which fixes the basepoint all the time, because the basepoint cannot cross a rotation center during any fiber isotopy. Therefore we need an isotopy for the complement of the fixed points,which has a free fundamental group H. A homeomorphism η can be deformed to the identity by an isotopy which does not move the basepoint, iff η induces the identity on H.

Next we consider the euclidean case:

In $\pi\colon S_1 \to S$ we assume that S_1 is a cylinder or torus. By a direct consideration the analogue to 5.16.5, 6 can be proved for unbranched coverings:

5.16.9 Proposition. *If* $\pi\colon S_1 \to S$ *is an unbranched regular covering, then a fiber preserving mapping* $\varphi\colon S_1 \to S_1$ *which is isotopic to the identity is fiber isotopic to the identity. An isotopy class of homeomorphisms of* S_1 *contains a fiber preserving homeomorphism, iff the induced automorphism of* $\pi_\#(\pi_1(S_1)) \subset \pi_1(S)$ *extends to* $\pi_1(S)$.

□

5.16.10 For branched coverings the situation is rather different. In this case S is a disk or sphere, as follows from the Euler characteristic. Let p be the branching order, r the number of branch points of orders $k_1,\dots,k_r$. From the Euler characteristic we get

5.16.11 $$p\left(\frac{1}{k_1} + \dots + \frac{1}{k_r} - r + d\right) = 0.$$

where d = 1 for S a disk and d = 2 for a sphere. There will be five cases (see Table 5.16.13). By lifting the covering transformations to the universal cover we get a group G. Now G allows a homomorphism χ to $\mathbb{Z}_2$, $\mathbb{Z}_3$, $\mathbb{Z}_4$ or $\mathbb{Z}_6$. (We put $\mathbb{Z}_m = \{0, 1, \ldots, m-1\}$.) The kernel R does not contain elements of finite order. The subgroup R is characteristic. Let F be a subgroup of finite index in G. Then $R \cap F$ is a subgroup of finite index in R and in F. So an automorphism of G induces the identity of F, iff it induces the identity on R. If an automorphism of F is induced by an automorphism of G, then on $R \cap F$ it is induced by an automorphism of R. By 5.16.9 we can answer the questions for the unbranched coverings $\mathbb{E}/R \cap F \to \mathbb{E}/R$, $\mathbb{E}/R \cap F \to \mathbb{E}/F$, and for the full answer we need only consider the branched covering $\mathbb{E}/R \to \mathbb{E}/G$.

In the following we assume $S_1 = \mathbb{E}/R$. We determine generators x, y for $R \cong \mathbb{Z}^2$ (cases a - d) or x for $R \cong \mathbb{Z}$ (case e) and the "inner" action of the s_i on these generators by the Schreier method 2.2.1. We drop the case $R \cong \mathbb{Z}$ in the following considerations, we will only give the result in the table.) Here $R = \langle x,y \mid [x,y] \rangle$.

$$T:\ R \to R,\ x \mapsto y, \qquad y \mapsto x^{-1}$$
$$S:\ R \to R,\ x \mapsto y^{-1}, \qquad y \mapsto xy$$
$$N:\ R \to R,\ x \mapsto y, \qquad y \mapsto x$$

are the generators for Aut R and Aut $R = \langle S,T,N \mid S^3T^2, S^6, T^4, N^2, NTN^{-1}T, NSN^{-1}S \rangle$, see 5.15.5.

As the elements of finite order in G are conjugate to the powers of the s_i or s_1s_2, resp. an automorphism of G has the form $s_i \mapsto v_i s_{k_i}^{\varepsilon} v_i^{-1}$, where $\begin{pmatrix} 1 \ldots r \\ k_1 \ldots k_r \end{pmatrix}$ is a permutation and $\varepsilon = \pm 1$. (The notation is from 5.16.13.)

Therefore there is in each class of automorphisms a mapping of the following form

5.16.12 (a) $\quad s_1 \mapsto v s_{k_1} v^{-1},\ s_2 \mapsto v_2 s_{k_2} v_2^{-1},\ s_3 \mapsto s_{k_3}.$

5.16.12 (b-d) $\quad s_1 \mapsto v s_{k_1}^{\varepsilon} v^{-1},\ s_2 \mapsto s_{k_2}^{\varepsilon}.$

As $\chi(s_1)$ [and $\chi(s_2)$ for the case (a)] generates G/R, we can assume $v = x^a y^b$, $v_2 = x^c y^d$.

The fact that the induced mapping is an automorphism restricts the numbers

Case No	r	k_i	S_1	S	G	$\chi: G \to \mathbb{Z}_m$	Generators x, y for $\mathcal{R}$	Induced autom. of $\mathcal{R}$	Kernel of ρ: Aut $G \to$ Aut $\mathcal{R}$, $\alpha \mapsto \alpha \mid \mathcal{R}$
(a)	4	$k_i = 2$	torus	sphere	$\langle s_1, s_2, s_3, s_4 \mid s_i^2 = s_1 s_2 s_3 s_4 = 1\rangle$	$\chi: G \to \mathbb{Z}_2$ $s_i \mapsto 1$	$x = s_1^{-1} s_2$ $y = s_1^{-1} s_3$	Aut $\mathcal{R}$	id, $(s_1 \mapsto s_1 s_2 s_1, s_2 \mapsto s_1, s_3 \mapsto s_1 s_2 s_3)$ $(s_2 \mapsto s_1 s_3 s_1, s_2 \mapsto s_1 s_3 s_2, s_3 \mapsto s_1)$ $(s_1 \mapsto s_1 s_3 s_1 s_2 s_1, s_2 \mapsto s_1 s_3 s_1, s_3 \mapsto s_1 s_2 s_1)$
(b)	3	$k_i = 3$	torus	sphere	$\langle s_1, s_2, s_3 \mid s_i^3 = s_1 s_2 s_3 = 1\rangle$	$\chi: G \to \mathbb{Z}_3$ $s_i \mapsto 1$	$x = s_1^{-1} s_2$ $y = s_1 s_2 s_1$	S^i, NU^i $i = 0, \ldots, 5$	id, $(s_1 \mapsto s_1 s_2^{-1} s_1, s_2 \mapsto s_1)$ $(s_1 \mapsto s_1 s_2^{-1} s_1 s_2^{-1} s_1, s_2 \mapsto s_1 s_2^{-1} s_1)$
(c)	3	$k_1 = 4$ $k_2 = 4$ $k_3 = 2$	torus	sphere	$\langle s_1, s_2 \mid s_1^4 = s_2^4 = (s_1 s_2)^2 = 1\rangle$	$\chi: G \to \mathbb{Z}_4$ $s_i \mapsto 1$	$x = s_1^{-1} s_2$ $y = s_1^{-1} s_1 s_2$	T^i, NT^i $i = 0, \ldots, 3$	id $s_1 \mapsto s_1 s_2^{-1} s_1, s_2 \mapsto s_1$
(d)	3	$k_1 = 6$ $k_2 = 3$ $k_3 = 2$	torus	sphere	$\langle s_1, s_2 \mid s_1^6 = s_2^3 = (s_1 s_2)^2 = 1\rangle$	$\chi: G \to \mathbb{Z}_6$ $s_1 \mapsto 1$ $s_2 \mapsto 2$	$x = s_1^{-2} s_2$ $y = s_1^{-3} s_2 s_1$	1, N	id
(e)	2	$k_i = 2$	cylinder	disc	$\langle s_1, s_2 \mid s_1^2 = s_2^2 = 1\rangle$	$\chi: G \to \mathbb{Z}_2$ $s_i \mapsto 1$	$x = s_1 s_2$	Aut $\mathcal{R} \cong \mathbb{Z}_2$	id $s_1 \mapsto s_1 s_2 s_1, s_2 \mapsto s_1$

Table 5.16.13

a, b, c, d, and it is simple to determine the automorphisms which are induced. For the calculation of the kernel of ρ: Aut $G \to$ Aut R, $\alpha \mapsto \alpha|R$ it is helpful that a mapping from the kernel with the trivial permutation $k_i = i$ is equal to the identity on G. We describe the steps in detail only for the most complicated case (a):

The permutation $\begin{pmatrix} 1 & 2 & 3 & 4 \\ 2 & 1 & 3 & 4 \end{pmatrix}$ can be induced by $s_1 \mapsto s_1 s_2 s_1$, $s_2 \mapsto s_1$, $s_3 \mapsto s_3$ and this defines on R the automorphism $NT^{-1}S^{-1}T^{-1}$, $\begin{pmatrix} 1 & 2 & 3 & 4 \\ 2 & 3 & 4 & 1 \end{pmatrix}$ is induced by $s_i \mapsto s_{i+1}$, which gives $TSTS^{-1}T^{-1}$ on R. For mappings with the trivial permutation we get on R the mappings $x \mapsto x^{2a-2c+1}y^{2b-2d}$, $y \mapsto x^{2a}y^{2b+1}$, and the condition is

$$\begin{vmatrix} 2a-2c+1 & 2a \\ 2b-2d & 2b+1 \end{vmatrix} = \pm 1$$

It is easily checked that all integer matrices $\begin{pmatrix} a_{11} & a_{12} \\ a_{21} & a_{22} \end{pmatrix}$ of determinant ± 1 can be obtained where a_{11} and a_{22} are odd and a_{12}, a_{21} even. In particular, we get NT^{-1}. The automorphisms NT^{-1}, $NT^{-1}S^{-1}T^{-1}$, $TST^{-1}T^{-1}S^{-1}$ generate Aut R. As the integer matrices (a_{ij}) with a_{11}, a_{22} odd, a_{12}, a_{21} even, form a normal subgroup of index 6 in Aut R, it follows that the kernel of ρ has order 4 and it is easy to determine its elements.

The geometric interpretation is simple: the elements change pairs of branch points of equal order in the cases (a + c + e), and rotate the triple in (b). These mappings are isotopic to the identity of S_1 but not fiber-isotopic.

The study of the way mapping class groups behave in the situation of a covering between surfaces was started by Birman-Hilden with the intention to answer questions about mapping classes with the help of braids, see [Birman-Hilden 1971, 1973]. They obtained the results of this section for unbranched and some special types of branched coverings. For the hyperbolic case the theorem was proved in [Maclachlan-Harvey 1975] using methods of complex analysis. The geometric approach via theorem 5.14.1 is from [Zieschang 1973], and we have copied that paper. The results can be extended to discontinuous groups that contain elements of finite order. This is easily done for the case where no reflections occur, by applying a quite general theorem about mapping classes and (unbranched) coverings [Birman-

Hilden 1972 , [Zieschang 1973'], an easy exercise in covering theory. If reflections exist some more geometric considerations have to be added, see [Zieschang 1973, § 10, 11].

Exercises: E 5.21,22

Additional literature to chapter 5:

[Alexander 1923], [Andrea 1967], [Bell 1976], [Bergau-Mennicke 1960], [Birman 1970], [Bourgin 1968], [Brödel 1935], [Bundgaard-Nielsen 1946], [Dehn 1938], [Dehn 1939], [Elvin-Short 1975], [Grossman 1974], [Jaco-Shalen 1977], [Kerékjartó 1943], [Kneser 1928], [Levine 1963], [Marden 1969], [Quine 1977], [Quintas 1965], [Ritter 1978], [Stillwell 1979'], [Stillwell 1979], [Turaev 1978], [Zarrow 1979], [Zieschang 1971], [Zieschang 1974], [Zimmermann 1977].

EXERCISES

E 5.1 Prove that in the free group $\langle t,u \mid \rangle$ the only solutions w,v of the equation $[w,v] = [t,u]$ are free generators (cf. E 1.21).

E 5.2 Solve the following equations in the free group $\langle x_1,x_2,\ldots \mid \rangle$

(a) $u^2v^2 = x_1^2$

(b) $u^2v^2w^2 = 1$

where u,v,w are words in $x_1,x_2,\ldots$. [Lyndon 1959].

E 5.3 Let $S = \langle x_1,x_2,\ldots \mid \rangle$ be a free group of rank $n \le \infty$. Let $w_1(x),\ldots,w_m(x) \in S$, $m \le n$ be such that

$$w_1^2(x) \ \ldots \ w_m^2(x) = x_1^2 \ldots x_m^2 \ .$$

Show that $w_1(x),\ldots,w_m(x)$ are free generators for the subgroup of S with basis $x_1,\ldots,x_m$. Prove the analogous result when

$$\Pi[w_{2i-1},w_{2i}] = \Pi[x_{2i-1},x_{2i}] \ .$$

E 5.4 Let S be the free group $\langle S_1,\ldots,S_n \mid \rangle$. Consider a binary product $\{X_1,\ldots,X_n;\Pi_X\}$ in S with $\Pi_X(X) = X_1^2\ldots X_n^2$ or $\Pi_X(X) = \prod_{i=1}^{n/2} [X_{2i-1},X_{2i}]$ and $\Pi_X(X) = 1$

Show that in the first case the binary product is related to $\{Y_1,\ldots,Y_{n-1}, Y_n; Y_1^2\ldots Y_n^2\}$ with

$$Y_{2i-1} = Y_{2i}^{-1} \quad i = 1,\ldots,n/2 \text{ for even } n,$$

$$Y_{2i-1} = Y_{2i}^{-1},\ Y_n = 1,\ i = 1,\ldots,(n-1)/2 \text{ for odd } n.$$

In the second case the binary product is related to $\{Y_1,\ldots,Y_n; \prod_{i=1}^{n/2} [Y_{2i-1},Y_{2i}]\}$ with $Y_{2i} = 1$, $i = 1,\ldots,n/2$.

E 5.5 Prove Theorem 5.3.7.

E 5.6 Use the Reidemeister-Schreier method to determine a basis for the commutator subgroup of $\langle t,u \mid \rangle$. Give a geometric interpretation and compare with 1.5.

E 5.7 Prove an analogue of Lemma 5.5.2 for groups with only orientation-preserving elements.

E 5.8 Let $\hat{F} = \langle T_1, U_1, \ldots, T_g, U_g \mid \rangle$, $F = \langle t_1, u_1, \ldots, t_g, u_g \mid \prod_{i=1}^{g} [t_i, u_i] \rangle$, $\hat{F} \to F$ the natural projection and $\Pi_* = \prod_{i=1}^{g} [T_i, U_i]$.

(a) Let $\hat{\alpha}: \hat{F} \to \hat{F}$ be an automorphism with $\hat{\alpha}(\Pi_*) = \Pi_*$ that induces the identity on F. Prove that $\hat{\alpha}$ is an inner automorphism with a power of Π_* as factor.

(b) Let A be the group of automorphisms of $\hat{F}$ that map Π_* to Π_* or Π_*^{-1}, and let T consist of the inner automorphisms with powers of Π_* as factor. Show that

$$\mathrm{Aut}\, F \cong A/T.$$

(Hint: use Theorem 5.6.1.)

Similar results are true for non-orientable surfaces and surfaces with boundaries ([Gramberg-Zieschang, unpublished]).

E 5.9 Give a geometrical proof of the equation $\omega(L_i)\varepsilon_i = \varepsilon = \pm 1$ from 5.7.2 for non-orientable surfaces.

E 5.10 Which of the following elements of $\langle t_1, u_1, t_2, u_2 \mid [t_1, u_1][t_2, u_2] \rangle$, the fundamental group of the closed orientable surface of genus 2, can be represented by simple closed curves?

$u_1^2 t_1 t_2^{-1} u_1 t_1 t_2^{-1}$,

$u_1^2 t_1 t_2^{-1} u_1 t_1 u_2 t_2^{-1}$,

$u_1^2 t_1^2 u_2^2 t_2^2$.

E 5.11 Prove Proposition 5.11.2. What is the situation for non-orientable surfaces?

E 5.12 Prove 5.11.7.

E 5.13 Prove that any non-nullhomotopic simple closed curve on the torus can be mapped onto any other by a homeomorphism of the torus
(a) using the classification of surfaces,
(b) using Dehn-Nielsen and Baer theorems.

E 5.14 Let F be a compact surface of genus $g \geq 1$, $h: F \to F$ a homeomorphism such that h^2 is homotopic to the identity. Prove that

(a) h is isotopic to a mapping h' for which there is a simple closed curve such that $h'(\gamma) \cap \gamma = \emptyset$ or $h'(\gamma) = \gamma^{\pm 1}$,

(b) h is isotopic to a homeomorphism h^* of order 2: $h^{*2} = \mathrm{id}$.

(Use elementary geometric-topological steps.)

E 5.15 Give an example of a mapping $f: F \to F$ of a surface F which is not homotopic to the identity but which induces the identity on $H_1(F)$ i.e. $\mathrm{id}_{\pi_1(F)} \neq f_\#: \pi_1(F) \to \pi_1(F)$ but $\mathrm{id}_{H_1(F)} = f_*: H_1(F) \to H_1(F)$.

E 5.16 Let F be a closed orientable surface of genus $g \geq 1$ and let $f: F \to F$ be a homeomorphism. Give several proofs for:

(a) $f_{*2}: H_2(F) \to H_2(F) \cong \mathbb{Z}$ maps 1 to 1, if f is orientation preserving, and to -1 if f reverses the orientation.

(b) $f_{*1}: H_1(F) \to H_1(F) \cong \mathbb{Z}^{2g}$ has determinant 1 if f preserves orientation. and -1 if f reverses the orientation.

(c) The intersection number $\nu(C_1,C_2)$, $C_i \in H_1(F)$, remains unchanged if f perserves the orientation and is multiplied with -1 if f reverses the orientation.

E 5.17 Prove Proposition 5.15.4.

E 5.18 Prove Proposition 5.15.5.

E 5.19 Let T be the torus.

(a) Determine the mapping classes of finite order of T and the finite subgroups of Homeot T.

(b) Prove that each mapping class of finite order of T contains a homeomorphism of this order. Prove that all finite subgroups of Homeot T can be realized by isomorphic groups of mappings.

E 5.20 Prove the statement in 5.11.6 that δ may be deformed by combinatorial isotopies into a curve that intersects Σ^* in the same points as γ.

E 5.21 Prove Theorem 5.16.2 using the results from the preceding sections.

E 5.22 Fill in the details for 5.16.10 and prove the results of table 5.16.13 for the other cases.

5.23 Give a direct proof of the assertion of Theorem 5.14.1, for the case when G is the automorphism group of the sequare net in the euclidean plane.

6. ON THE COMPLEX ANALYTIC THEORY OF RIEMANN SURFACES AND PLANAR DISCONTINUOUS GROUPS

6.1 INTRODUCTION

Our considerations until now have been of an essentially combinatorial kind, but dimension 2 is special in this respect. On the other hand, surfaces and planar discontinuous groups are much studied objects in complex analysis, as Riemann surfaces or discontinuous groups in the non-euclidean plane. Nevertheless certain theorems of these theories make no mention of the complex analytic structure, they are of a purely topological nature. In the previous sections we have established some of them by topological methods, but there remains a residue of purely topological theorems which require difficult theorems of analysis for their proofs.

It is now our intention to briefly describe the complex analytic theory of planar discontinuous groups and draw a few conclusions. At a first reading only the basic results of set theoretic topology and analysis are supposed, but the special literature is indispensible for a full understanding. This will be referred to at the appropriate places.

6.2 STRUCTURES ON SURFACES

We first recall the concept of a manifold.

6.2.1 Let M be a space with a countable basis and let $K = \{U,\phi_U\}$ be a system of subsets and one-to-one mappings $\phi_U\colon U \to Q^n$ of U on to the n-dimensional open unit cube of $\mathbb{R}^n$ (or $\mathbb{C}^n$, respectively). Let $\bigcup_{U\in K} U = M$. The intersection $U \cap U'$ for $U, U' \in K$ defines subsets $\phi_U(U \cap U')$ and $\phi_{U'}(U \cap U')$ of Q^n, and $\phi_U \circ \phi_{U'}^{-1}$ defines a mapping of these onto each other. We require the subsets to be open. Various assumptions are made about the mappings, of which the following are the most often used:

6.2.2 They are continuous.

6.2.3 They are m times continuously differentiable (or of class C^m).

6.2.4 They can be expanded locally in power series with real argument (class C^ω).

6.2.5 They can be expanded locally in complex power series.

Correspondingly, M is called a *topological, differentiable (of class* C^m*), real-analytic or (complex) analytic manifold of real or complex dimension* n. K is known as the *atlas with maps* (U, ϕ_U).

In what follows we confine ourselves to real dimension 2 or complex dimension 1 and speak of *topological, smooth* or *Riemann surfaces* according as 6.2.2, 6.2.3 or 6.2.5 holds. (We shall not be concerned with 6.2.4.)

In the preceding sections surfaces were defined as complexes. These may also be subsumed under the concept of manifold and indeed in various ways. It will suffice for us to identify the disk with the unit circular disk or a standard simplex, in such a way that common boundary segments define a continuous or differentiable transition. It is not so simple to give the complex an analytic structure.

6.2.7 Complexes, or triangulations as they are better called, are in close relation to another kind of mapping: a mapping is called *piecewise-linear* (or *semilinear*) when image and preimage may be divided into simplexes in such a way that the mapping is linear on each simplex. (See [Graeub 1950], [Schubert 1953].) *One can obviously give all 2-complexes a triangulation, and in fact one sufficiently refined that all the necessary mappings may be realized semi-linearly.* We shall not go into this further.

We confine ourselves to closed surfaces and the plane. (If not closed, we could suppose that the surface had a finite number of ends.) Questions naturally arise concerning the various structures and their relationships. Does a topological surface admit a triangulation, a differentiable or indeed an analytic structure? The answer is simple: "YES". The difficulties lie in the proof of

6.2.8 Theorem. *An arbitrary topological surface admits a triangulation.*

We will prove this important theorem of [Rado 1924] in the next chapter. In the preceding sections all topological surfaces have been regarded in this way. In addition, we shall see shortly in 6.4.7.

6.2.9 Theorem. *Each orientable closed surface may be provided with an analytic structure.*

□

Conversely, there is the question how many combinatorially inequivalent triangulations there are on a topological surface. Since orientability and genus characterize the complex and are determined by the homology group, we obtain:

6.2.10 Theorem. *A surface will only bear equivalent complexes, in other words the Hauptvermutung is correct in dimension 2.* □

Also, a surface possesses only one differentiable structure. The question of the possible analytic structures is very much more complicated: *this is the question of the moduli of Riemann surfaces.*

6.3 THE NON-EUCLIDEAN PLANE

We give a sketch of the non-euclidean geometry. Let $\mathbb{H}$ denote the upper half plane. It follows from Schwarz's lemma and the theorem of Mittag-Leffler that a one-to-one holomorphic mapping of $\mathbb{H}$ onto itself is linear fractional, and hence has the form $w = \frac{az+b}{cz+d}$. It also follows that such a linear fractional transformation can be written with real coefficients. We then have

6.3.1 Theorem. *An orientation-preserving and angle-preserving (holomorphic) mapping of the upper half plane on to itself may be written in the form*

$$w = \frac{az+b}{cz+d}, \ a,b,c,d \text{ real}, \ ad-bc = 1.$$

If $|a+d| < 2$ *then it has a fixed point in the upper half plane and a second which is its conjugate and in* $\mathbb{C}\setminus\mathbb{H}$, *when* $|a+d| = 2$ *there is exactly one fixed point, and it is on the real axis; for* $|a+d| > 2$ *there are exactly two fixed points on the real axis (to which is added* ∞*). Orientation reversing, angle-preserving mappings may be written in the form*

□

$$w = \frac{a\bar{z}+b}{c\bar{z}+d}, \ a,b,c,d \text{ real}, \ ad-bc = -1.$$

6.3.2 Theorem. *The cross-ratio* $\frac{z_1-z_3}{z_2-z_3} : \frac{z_1-z_4}{z_2-z_4}$ *of four distinct points is invariant under each linear fractional transformation. The four points lie on a circle – under which concept we also include straight lines closed by the point at* ∞ *– if and only if their cross-ratio is real. If the points lie on the circle in the order* z_1, z_2, z_3, z_4 *then the cross-ratio is greater than* 1. (*Proof* is exercise E 6.1 (b).)

□

6.3.3 Definition. The points of the open upper half plane $\mathbb{H}$ together with the parts of straight lines lying in $\mathbb{H}$ and the semicircles both perpendicular to the real axis are the *points* and *lines* of an image of the *non-euclidean plane*, the *Poincaré model*. Each noneuclidean line then has two *ends* on the real axis closed by ∞. The non-euclidean distance $d(z_1,z_2)$ between two points $z_1, z_2 \in \mathbb{H}$ is defined as follows: let z_3,z_4 be the ends of the non-euclidean line through z_1 and z_2 and suppose these points occur in the order z_4, z_1, z_2, z_3.

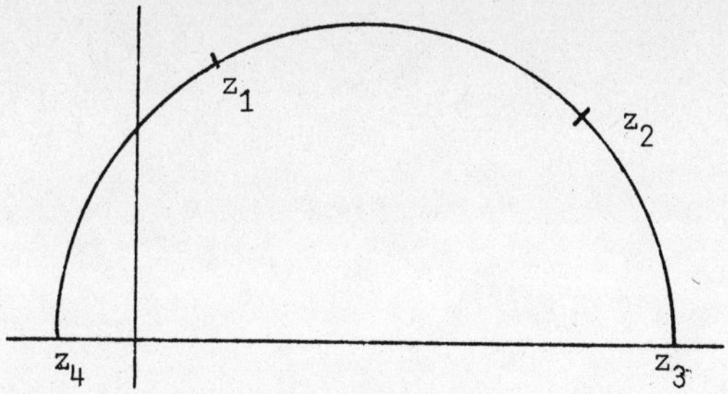

The *distance* between z_1 and z_2 ist then

$$\log \frac{z_1-z_3}{z_2-z_3} : \frac{z_1-z_4}{z_2-z_4} .$$

If one lets z_2 approach z_1, then $d(z_1,z_2)$ coverges to 0. Therefore it is meaningful to set $d(z,z) = 0$. For $z_1 \neq z_2$, $d(z_1,z_2) > 0$. Moreover, we have the triangle inequality $d(z_1,z_0) \leq d(z_1,z_2) + d(z_2,z_0)$. Equality only holds when z_2 lies between z_1 and z_0 on a non-euclidean line.

The *non-euclidean angle measure* is simply the measure of the euclidean angle between the appropriate euclidean circles or straight lines. *Angle-preserving one-to-one mappings are then just the linear fractional substitutions of Theorem 6.3.1, which may also be characterized as the length-preserving mappings.* We therefore have

6.3.4 Corollary. *The orientation-preserving motions of the non-euclidean plane are given by the transformations*

$$w = \frac{az+b}{cz+d}, \quad a,b,c,d \text{ real}, \quad ad-bc = 1,$$

those which reverse orientation by

$$w = \frac{a\bar{z}+c}{c\bar{z}+d}, \quad a,b,c,d \text{ real}, \quad ad-bc = -1.$$

□

6.3.5 Definition. Orientation-preserving motions with $|a+d| < 2$ are called *elliptic* or rotations, those with $|a+d| = 2$ *parabolic,* and those with $|a+d| > 2$ *hyperbolic,* or displacements. The orientation-reversing motions of order 2 $(a+d = 0)$ are called *reflections,* the others *glide reflections.*

6.3.6 Lemma. *If* x *is a parabolic transformation then there is a sequence* P_i *of points* $\mathbb{H}$ *with*

$$\lim_{i\to\infty} d(P_i,P_ix) = 0.$$

Proof: By suitable conjugation we can arrange that the parabolic transformation has the form

$$x(z) = z + 2b, \quad b \text{ real}$$

(cf. say [Siegel 1964, Ch. 3, § 1]). The distance of the point $-b+i\sqrt{t^2-b^2}$ ($t > |b|$) from its image $b+i\sqrt{t^2-b^2}$ is

$$\log\left(\frac{-b+i\sqrt{t^2-b^2}-t}{b+i\sqrt{t^2-b^2}-t} : \frac{-b+i\sqrt{t^2-b^2}+t}{b+i\sqrt{t^2-b^2}+t}\right) .$$

As $t \to \infty$ this expression converges to 0.

□

6.3.7 Definition. The *non-euclidean arc element* is given by

$$ds = \frac{|dz|}{y}$$

where $z = x + iy$. The continuously differentiable functions which leave this element invariant are exactly the motions of non-euclidean geometry. The *non-euclidean surface element*

$$dw = \frac{dxdy}{y^2}$$

is likewise invariant under non-euclidean motions. By a *non-euclidean polygon* we mean a region in the non-euclidean plane bounded by non-euclidean line segments. Using the Green formula or the Gauss-Bonnet formula one obtains

6.3.8 Theorem. *The non-euclidean area of a non-euclidean polygon with* n *vertices and internal angles* $\alpha_1,\dots,\alpha_n$ *is*

$$(n-2)\pi - (\alpha_1 + \dots + \alpha_n).$$

Proof is exercise E 6.6. □

Furthermore, vertices may also lie on the real axis, in which case the corresponding sides will be half lines with the same end, Then the angle is reckoned to be 0.

For a detailed exposition of the material in this paragraph see textbooks on function theory, especially [Lehner 1964], [Siegel, 1964'], [Zieschang, 1980].

Exercises: E 6.1-6, E 6.13.

6.4 PLANAR DISCONTINUOUS GROUPS

6.4.1 If a group $\mathcal{B}$ operates on a topological space R then $\mathcal{B}$ can be given a topology which reflects the fact that "small" changes in the mappings displace the image points only slightly. A topology which is frequently considered has a subbasis consisting of the sets $\{x \in \mathcal{B}: xU \subset V\}$ where the U are the compact, and the V the open, subsets of R. Obviously subgroups receive the induced topology. If $\mathcal{B}$ consists of motions of the euclidean or non-euclidean plane then one obtains the topology naturally associated with the parameters a, b, c, d.

6.4.2 Definition. A group $\mathcal{B}$ of homeomorphisms is called *discontinuous* when there is no point $P \in R$ for which the set $\{xP: x \in \mathcal{B}\}$ has an accumulation point. In particular, a point can only be a fixed point finitely often.

6.4.3 Theorem. *A subgroup of the motions of the euclidean or non-euclidean plane is discrete if and only if it operates discontinuously on the plane.*

The proof is an exercise.

□

The following theorem of [Nielsen 1940] may also be mentioned, see [Siegel 1950, 1964' pg. 39].

6.4.4 Theorem. *If a non-commutative subgroup of the motions of the non-euclidean plane contains only hyperbolic transformations, then it is discrete.* □

We say that a discontinuous group has *compact fundamental domain* when the images of some compact subset cover the whole space. This is equivalent to saying that the factor space modulo the group is compact.

6.4.6 Theorem. *If a discontinuous group G of motions of the euclidean or non-euclidean plane has compact fundamental domain then there is a net in the plane, the edges and faces of which are line segments and convex polygons of the corresponding geometry.*

The requirement that the group has compact fundamental domain is unnecessarily strong. The idea of the proof may be carried over to the general case, and one likewise obtains a net, the faces of which may however have half lines and possibly pieces of the real axis in their boundaries, indeed this must happen if the fundamental domain is not compact, cf. [Siegel 1964', p. 42].

Proof. Since the group is discontinuous there is a point $P \in \mathbb{H}$ or $\mathbb{C}$ respectively which is not fixed under any transformation in G except the identity. Px again denotes the image of P under application of $x \in G$. For each $x \in G$, $x \neq 1$ we take the closed half-plane consisting of the points of the plane which are no further from P than from Px. (Geometrical terms refer to the appropriate geometry.) This is the half-plane defined by the perpendicular bisector of the line from P to Px and containing P. Let the intersection of all these half-planes be F; it is the set of points which are at least as far from each of the other points Px $(x \neq 1)$ as they are from P. If $Q \in F$ then Q is nearer to P than any other Qx $(x \neq 1)$. If two points from the same equivalence class are both in F then they must lie on the boundary of F. Therefore F is a fundamental domain and contains the points from each equivalence class which are of minimal distance from P.

Now there is a circle around P which contains a point from each class, hence, it contains F;(this is where the compactness of the fundamental domain comes in). If we then take the equivalents of the point P which are at a distance of more than two radii from P, then the corresponding circles do not meet the one centered on P. But since G is discontinuous, we have omitted only finitely many transformations, i.e. F is the intersection of finitely many half-planes and compact, thus a convex polygon. If one constructs the corresponding polygon for each point equivalent to P, then one obtains a planar net on which the group acts. This is a consequence of the fact that F contains exactly the points of minimal distance from P from each equivalence class.

□

We have shown in this way that each discontinuous group of motions of the plane with compact fundamental domain carries a planar net into itself, and thus is of the type treated algebraically in Chapter 4. Conversely, it also holds that each of these groups may be realized by motions:

6.4.7 Theorem. *A group which is given by generators and defining relations of the type* A *or* B *in Theorem 4.8 is realizable as a discontinuous group of motions on the sphere* (>), *the euclidean* (=) *or non-euclidean plane* (<) *according as, in case* A

$$0 \gtreqless 2\sum_{i=1}^{m}\left(1-\frac{1}{h_i}\right) + \sum_{i=1}^{q}\sum_{j=1}^{m_i}\left(1-\frac{1}{h_{ij}}\right) + 4g - 4 + 2q$$

or in case B

$$0 \gtreqless 2\sum_{i=1}^{m}\left(1-\frac{1}{h_i}\right) + \sum_{i=1}^{q}\sum_{j=1}^{m_i}\left(1-\frac{1}{h_{ij}}\right) + 2g - 4 + 2q$$

Proof. First we choose a polygon, on the sphere, euclidean plane or non-euclidean plane respectively, the sides of which are denoted by symbols as in Theorem 4.3.6, so that it takes the form 4.3.7 or 4.3.8. Segments denoted by the same Greek letters and indices will have the same length, the segments denoted by σ_i and σ_i' will meet in the angle $\frac{2\pi}{h_i}$ those denoted by $\gamma_{k,i}$ and $\gamma_{k,i+1}$ in the angle $\frac{\pi}{h_{ki}}$, the sum of the angles between η_k' and γ_k and between γ_{k,m_k+1} and η_k will be π, and the sum of the remaining angles will be 2π. It is known that such polygons exist on the sphere and in the euclidean plane; moreover they correspond to well-known groups. In the non-euclidean plane one proves the existence of such a polygon by a continuity argument. (Small polygons are "nearly euclidean", hence can have angle sum arbitrarily close to the euclidean value, while large polygons have angle sum arbitrarily close to 0. Then by continuity we can get any value in between.)

In Theorem 4.7.1 we began with the groups, and constructed planar nets on which they act. We take the dual net to that described in the theorem. Here the group acts simply transitively on the faces, the boundaries of which may be naturally written in the form 4.3.7 or 4.3.8. If we replace the faces of the net by the polygon and its labelling then we obtain a Riemann surface, since for each point we can find a neighbourhood holomorphic to the unit circle. For interior points of the polygon this is trivial, for boundary points which are not vertices it follows from the fact that segments with the same label have equal length; and the angle conditions take care of the vertices. Moreover, the Riemann surface is open and simply connected, and G operates conformally on it. Then on the basis of the Riemann mapping theorem G may be realized by a discontinuous group of motions of euclidean or non-euclidean geometry.

□

<u>6.4.8 Remarks</u>. a) It also follows from the inequalities for the angle sum of a convex polygon that one can determine the plane on which the group operates analytically from the inequalities in the theorem.

b) It is natural, possible and frequently done in the literature (though often carried out incorrectly), to construct the net in the plane by successively laying down polygons; for locally congruent polygons can be laid next to each other as required, and generating motions defined as a result, which satisfy the defining relations. Beginning with a polygon, this step-by-step process gives us a holomorphic mapping of the net constructed in the euclidean or non-euclidean plane, and in fact a covering without branching. Since the plane is simply connected and since one can reach every point in the plane by a suitable chain of congruent polygons, the mapping turns out to be "onto" and our Riemann surface is holomorphic to the plane. The discontinuous group therefore consists of motions of the geometry.

c) The Riemann-Hurwitz formula 4.14.23 is immediate in this context. It simply compares the areas of fundamental domains for group and subgroup.

Exercises: E 6.7-9

6.5 ON THE MODULAR PROBLEM

In this section questions concerning the complex-analytic structure of Riemann surfaces and planar discontinuous groups are discussed. From the Riemann mapping theorem [Behnke-Sommer 1955, pg. 442], [Ahlfors-Sario 1960, III.11] it follows that:

6.5.1 Theorem. *Any two Riemann surfaces of genus 0 are conformally equivalent.*

□

6.5.2 The situation is already different for genus 1. A group of motions isomorphic to $\mathbf{Z}^2$ operates as a covering transformation group on the universal covering. It easily follows that hyperbolic transformations commute only when they have equal displacement axes, parabolic ones when they have the same fixed point, and hyperbolic and parabolic do not commute with each other. Consequently $\mathbf{Z}$ is the only possible commutative discontinuous group of the hyperbolic plane without elements of finite order. The universal covering of a torus must therefore be holomorphic to the euclidean plane $\mathbb{C}$. As a result the *covering transformation group is mapped onto a subgroup of the group of translations.*

6.5.3 By looking for the "next" linearly independent points equivalent to a given point one finds two numbers ω_1,ω_2 such that the group consists of the translations $w = z + m\omega_1 + n\omega_2$, $n,m \in \mathbb{Z}$. The conformal classification of surfaces now can be restated as: two groups G and G' are conformally equivalent when there is a conformal mapping ζ of the plane onto itself with

$$G' = \zeta G \zeta^{-1}.$$

The equivalence of the groups for $(k\omega_1, k\omega_2)$, $k \neq 0$ and (ω_1,ω_2) is shown by $z \mapsto kz$. In addition we can distribute the indices between ω_1 and ω_2 in such a way that $\operatorname{Im} \frac{\omega_2}{\omega_1} > 0$; $z \mapsto z' = z + \omega_1$, $z \mapsto z' = z + \omega_2$ constitute a "positive" pair of generators. We now associate with the pair (ω_1,ω_2) the pair $(1,\omega)$ where $\omega = \frac{\omega_2}{\omega_1}$, since these are conformally equivalent. Accordingly *each positive generator pair of an arbitrary torus group corresponds to a point of the upper half plane, and to only one,* since a holomorphic mapping of the plane onto itself with fixed points 0 and 1 is constant.

6.5.4 Thus the translations $z \mapsto z' = z + (a\omega_1 + b\omega_2)$, $z \mapsto z' = z + (c\omega_1 + d\omega_2)$, a,b,c,d integers and $ad - bc = 1$ constitute a positive generator system for the group generated by $z \mapsto z' = z + \omega_1$, $z \mapsto z' = z + \omega_2$.

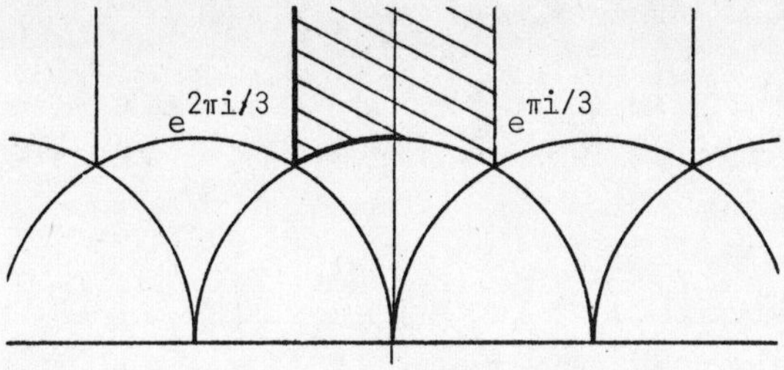

Therefore ω *may be replaced by* $\frac{a\omega+b}{c\omega+d}$, a,b,c,d *integers and* $ad - bc = 1$ *and one remains on the same Riemann surface. There are no more conformal equivalences for the group. Thus the fundamental domain for the* modular group SLF(2,$\mathbb{Z}$), *the group of linear fractional transformation with integer coeffizients, gives the different complex-analytic structures on the torus.* For details see textbooks of function theory cf [Siegel 1964 ch. I, § 3] or E 6.10.

The modular problem becomes much more complicated for higher genus. However, the analogue of the upper half plane, namely the space of canonical systems of generators, has been found for the various cases. First, [Fricke-Klein 1897, 1912] gave a system of modules which are the traces of the generators and some other transformations. These modules are again treated in [Keen 1966, 1971] and in the continuation of this book [Zieschang 1980], see also [Kroll 1974], [Vollmer-Bekemeier 1978]. Another approach was started in [Teichmüller 1940, 1943]: he uses quasiconformal mappings. In later work, [Ahlfors 1953, 1960], [Bers 1958, 1960, 1973], this theory was completed and extended and has grown to a major discipline of complex analysis. A good collection of literature is in [Kra 1972].

In the next section we describe the schema of Teichmüller's theory, following [Ahlfors 1953], and later give a geometric foundation for it. This is closely related to the topological theory of canonical dissection on a surface which we have considered before. It has been generalized to arbitrary Fuchsian groups with compact fundamental region in [Coldewey 1971].

Exercises: E 6.9-11

6.6 TEICHMÜLLER THEORY AND ITS CONSEQUENCES

A similar solution to the modular problem for higher genus will be discussed now.

6.6.1 After assigning an orientation to the plane, each Riemann surface also recieves an orientation in a natural way; for holomorphic mappings preserve orientation. As a result canonical curve systems or dissections may be divided into two classes, where two belong to the same class if they may be carried into each other by orientation-preserving mappings. We designate one of these classes as "positive" and in what follows we shall consider only dissections from the latter class.

6.6.2 Definition. A pair (R,Σ) consisting of a Riemann surface R and an isotopy class Σ of a (positive) canonical dissection is called a *marked Riemann surface*.

On the basis of the theorems of J. Nielsen and R. Baer one may just as well regard Σ as a set of mutually conjugate canonical generator systems, and we shall freely do this. However, these theorems are not needed in what follows.

6.6.3 Definition. Two marked Riemann surfaces (R,Σ) and (R',Σ') are called *holomorphically equivalent* when there is a holomorphic mapping $\varphi\colon R \to R'$ with $\varphi\Sigma = \Sigma'$, where Σ is defined in the obvious way. A homeomorphism $\varphi\colon R \to R'$ with $\varphi\Sigma = \Sigma'$ will be denoted simply by $\varphi\colon (R,\Sigma) \to (R',\Sigma')$. The set of marked Riemann surfaces holomorphically equivalent to (R,Σ) will be denoted by $[R,\Sigma]$ and the space of these classes will be called *Teichmüller space*. Later we shall give it a topology under which it is homeomorphic to $\mathbb{R}^{6g-6}$, if g is the genus of R.

6.6.4 Definition. A set T of homeomorphisms between two Riemann surfaces of genus g is the *basis for a Teichmüller theory* of this genus when (a-c) are satisfied.

(a) For each two marked Riemann surfaces (R,Σ) and (R',Σ') of genus g there is exactly one mapping $\varphi\colon (R,\Sigma) \to (R',\Sigma')$ in T.

(b) If there is a holomorphic mapping $\varphi\colon (R,\Sigma) \to (R',\Sigma')$ this is the mapping of T.

(c) If $\varphi\colon (R,\Sigma) \to (R',\Sigma')$ is in T and if $\psi_1\colon (\bar{R},\bar{\Sigma}) \to (R,\Sigma)$ and $\psi_2\colon (R',\Sigma') \to (\bar{R}',\bar{\Sigma}')$ are conformal and either both orientation-preserving (and therefore holomorphic) or both orientation reversing, then $\psi_2\varphi\psi_1$ is in T.

The homeomorphisms in T will be called *T-mappings*.

Under the conditions (b) and (c), (a) splits into assertions of existence and uniqueness. If we regard Σ as a set of isotopic canonical curve systems, then the existence statement is the simpler part in the literature; if we regard Σ as conjugation classes of canonical generator systems then it already follows from the theorem of Nielsen. The burden of the foundation of Teichmüller theory in literature rests on the uniqueness statement. In our solution, close to the approach of Fricke, both assertions will be handled somewhat similarly.

In addition 6.6.4 (a,b) together say that an isotopy class contains at most one holomorphic mapping. This is correct for genus $g \geq 2$. We give two proofs of this, one on the basis of a Teichmüller theory, the other by application of the fixed point formula.

We now make the

6.6.5 Hypothesis: *we can find a basis for a Teichmüller theory for arbitrary genus* $g \geq 2$, and we shall show that we can then also find a basis for a Teichmüller theory of discontinuous groups with compact fundamental domain. The sets T for different genus need have nothing to do with each other. Our hypothesis corresponds to the solution of problem A in [Ahlfors 1953], and our argument follows the steps in the solution of his problem C.

6.6.6 The universal covering of R is the non-euclidean plane $\mathbb{H}$, and the fundamental group F acts on $\mathbb{H}$ as the covering transformation group. The pair $(\mathbb{H},F)$ is therefore associated with R, and indeed up to holomorphic equivalence: $(\mathbb{H},F)$ *is equivalent* to $(\mathbb{H},F')$ *when there is a holomorphic mapping* $\zeta\colon \mathbb{H} \to \mathbb{H}$ *with* $\zeta F \zeta^{-1} = F'$. Corresponding to the Σ of a marked Riemann surface (R,Σ) there is a conjugation class of the canonical generator system of F (by the theorem of Baer) which we shall likewise denote by Σ, and consequently, there is a triple $(\mathbb{H},F,\Sigma)$ corresponding to the pair (R,Σ). Two such triples $(\mathbb{H},F,\Sigma)$ and $(\mathbb{H},F',\Sigma')$ are holomorphically equivalent when there is a holomorphic mapping $\zeta\colon \mathbb{H} \to \mathbb{H}$ such that $\zeta F \zeta^{-1} = F'$ and $\zeta\Sigma\zeta^{-1} = \Sigma'$. $[\mathbb{H},F,\Sigma]$ denotes the equivalence class. Obviously $[R,\Sigma]$ may be identified with $[\mathbb{H},F,\Sigma]$. If we regard R as $\mathbb{H}/F$, then conversely we obtain marked Riemann surface from triples.

By a *homeomorphism* of a triple $(\mathbb{H},F,\Sigma)$ onto $(\mathbb{H},F',\Sigma')$ we mean a homeomorphism $\phi\colon \mathbb{H} \to \mathbb{H}$ with

$$f\phi = \phi f^{\alpha} \tag{6.6.7}$$

where $\alpha\colon F \to F'$ is an isomorphism with $\alpha\Sigma = \Sigma'$. We have denoted the image of f under α by f^{α}. The different isomorphisms α with $\alpha\Sigma = \Sigma'$ are distinguished only by inner automorphisms of F', since the image under α of a fixed generator system in Σ can only be altered by conjugation. If we replace α by α_* with $f^{\alpha_*} = f_o'^{-1} f^{\alpha} f_o'$, where $f_o' \in F'$ is a fixed element, then we have to replace ϕ by $\phi_* = \phi f_o'$, which means altering it by a (real) displacement. Conversely, if we change ϕ by a displacement to $\phi_* = \phi f_o'$ then we must also replace α by α_* with $f^{\alpha_*} = f_o'^{-1} f^{\alpha} f_o'$.

If a homeomorphism ϕ satisfies equations 6.6.7 then it induces a homeomorphism $\varphi\colon \mathbb{H}/F \to \mathbb{H}/F'$. As a result the φ does not change when ϕ is replaced by $\phi f_o'$ and α is replaced by α_*. Conversely, each homeomorphism $\varphi\colon (R,\Sigma) \to (R',\Sigma')$ may be lifted to a homeomorphism $\phi\colon (\mathbb{H},F,\Sigma) \to (\mathbb{H},F',\Sigma')$, and this ϕ is determined up to displacements. A homeomorphism ϕ is called a T-mapping when the φ induced on the Riemann surface is a T-mapping. Our discussion shows

6.6.8 Theorem. *Assuming the Hypothesis 6.6.5 holds, then:*
(a) For any two triples $(\mathbb{H},F,\Sigma)$ *and* $(\mathbb{H},F',\Sigma')$ *there is exactly one* T*-mapping*

ϕ: $\mathbb{H} \to \mathbb{H}$ such that

6.6.7 $f\phi = \phi f^{\alpha}$, $f \in F$, *where* α: $F \to F'$ *is an isomorphism with* $\alpha\Sigma = \Sigma'$.

(b) If there is a holomorphic mapping ϕ which satisfies 6.6.7, then it is the T-mapping.

(c) If ψ_1 and ψ_2 are arbitrary conformal mappings, either both holomorphic or both non-holomorphic, from $\mathbb{H}$ to $\mathbb{H}$, then along with ϕ, $\psi_1\phi\psi_2$ is also a T-mapping.

□

Remark: In order for (c) to be meaningful, ψ_1 must be a mapping from $(\mathbb{H}, \psi_1^{-1} F \psi_1, \psi_1^{-1} \Sigma \psi_1)$ to $(\mathbb{H}, F, \Sigma)$ and ψ_2 must be a mapping from $(\mathbb{H}, F, \Sigma)$ to $(\mathbb{H}, \psi_2 F \psi_2^{-1}, \psi_2 \Sigma \psi_2^{-1})$.

The generator systems Σ and Σ' are strictly no longer necessary in Theorem 6.6.8, but only the isomorphism α. Then we express the theorem directly for all discontinuous groups with compact fundamental domain and obtain:

6.6.9 Theorem. *If the Hypothesis 6.6.5 is satisfied then there is a Teichmüller theory for all discontinuous groups of motions with compact fundamental domain in the hyperbolic plane: for each algebraic type of such a discontinuous group of motions in the non-euclidean plane there is a set T of homeomorphisms $\mathbb{H} \to \mathbb{H}$ with the following properties:*

(a) If α: $G \to G'$ is an isomorphism then there is exactly one mapping $\phi \in T$ with $g\phi = \phi g^{\alpha}$, $g \in G$.

(b) If a holomorphic mapping satisfies these equations, then it is in T.

(c) If ψ_1 and ψ_2 are conformal mappings $\mathbb{H} \to \mathbb{H}$, either both holomorphic or both non-holomorphic, then along with ϕ, $\psi_1\phi\psi_2$ is also a T-mapping.

□

Proof. Let G, G' be groups as required and α: $G \to G'$ an isomorphism. By Theorem 4.10.9 G contains a normal subgroup F of finite index which is isomorphic to the fundamental group of an orientable surface of genus $g \geq 2$. Let $F' = \alpha F$. We now consider the T-mappings for genus g. By Theorem 6.6.8 there is a T-homeomorphism ϕ: $\mathbb{H} \to \mathbb{H}$ such that

$$f\phi = \phi f^{\alpha} \quad \text{for } f \in F.$$

This ϕ is the single T-mapping which satisfies these equations, and it will be taken as the T-mapping for α: $G \to G'$. For $g \in G$, set $\phi_* = g^{-1}\phi g^{\alpha}$. Now if $f \in F$ then

$$f\phi_* = fg^{-1}\phi g^{\alpha} = g^{-1}gfg^{-1}\phi g^{\alpha} = g^{-1}\phi(gfg^{-1})^{\alpha}g^{\alpha} = g^{-1}\phi g^{\alpha}f^{\alpha} = \phi_* f^{\alpha}$$

since $gfg^{-1} \in F$. Thus ϕ_* likewise satisfies the functional equations. Since g^{-1} and

g^α are conformal and either both holomorphic or both non-holomorphic, ϕ_* is likewise a T-mapping for genus g and hence identical with ϕ:

$$g\phi = \phi g^\alpha, \ g \in G.$$

There ϕ satisfies the equations; uniqueness already results from the action on the subgroup.

□

6.6.10 Corollary. *The algebraic isomorphism of planar discontinuous groups with compact fundamental domain implies the geometrical isomorphism.*

Other proofs of this corollary are in [Macbeath 1967], [Tukia 1972], [Keller 1973].

Using the same trick as in the proof of 6.6.9, we can easily prove part (a) of theorem 5.16.2. Let us reformulate and slightly generalize the statement:

6.6.11 Theorem. *Let G be a finitely generated planar discontinuous group with compact fundamental domain and let F be a subgroup of finite index which is of hyperbolic type, i.e. the centralizer of any non-trivial element of F is cyclic. Then a geometrical automorphism α of F is induced by a G-fiber preserving homeomorphism ϕ of the (hyperbolic) plane $\mathbb{H}$ if and only if α is induced by an automorphism $\tilde{\alpha}$ of G.*

□

Exercises: 6.12-13.

6.7 A BASIS FOR A TEICHMÜLLER THEORY

The universal covering of a closed Riemann surface of genus $g \geq 2$ is conformally equivalent to the upper half plane. Noneuclidean geometry defines a Riemannian metric on the surface so that the geodesics and non-euclidean lines correspond. It follows that

6.7.1 Lemma. *In a homotopy class of paths (with fixed initial point) there is exactly one geodesic.*

□

If one has a dissection of the surface, then it defines a polygon in the plane (in general curvilinear) which serves as fundamental domain of the covering motion group. We now carry over length and angle measure from non-euclidean geometry:

6.7.2 Definition. $d(\omega)$ denotes the *length of a curve* ω. If the curves of a dissection are geodetic (beginning and end need not be geodetically connected to each other, however), then we speak of a *geodetic dissection*. If the sum of two consecutive angles in the star around the basepoint does not exceed π then the *dissection* is called *doubly-convex*.

Assuming a geodetic dissection, double convexity in the universal covering surface means that a union of two adjacent polygons of the net is still convex. Our aim is now to prove the following Theorem 6.7.3, which will have to be preceded by a few lemmas.

6.7.3 Theorem. *On any closed Riemann surface* R *there is a doubly-convex geodetic dissection.*

6.7.4 Lemma. *There is a closed, non-null-homologous geodesic on* R *which has minimal length among all non-null-homologous curves.*

Proof. Let F denote the fundamental group of R, considered as group of covering transformations of the universal cover $\mathbb{H}$. Given $Q \in \mathbb{H}$ we construct

$$\text{infimum } \{d(Q,Qx): x \in F,\ x \neq 1\}.$$

Because of the discontinuity, the infimum distance for each Q is realized for only finitely many x, and we therefore obtain a continuous function on $\mathbb{H}$ which is also periodic under F. Since the fundamental domain is compact, this function takes a minimum. Since there are no rotations in F this minimum is positive, and by Lemma 6.3.6 there can be no parabolic transformations in F.

One now considers

$$\text{infimum } \{d(Q,Qx): x \in F,\ x \notin [F,F]\}$$

Again one obtains a continuous function on $\mathbb{H}$, periodic relative to F, which takes a minimum at P, say. Because of the discontinuity of F there is then a hyperbolic $x_1 \in F$, $x_1 \notin [F,F]$ such that

$$d(P,Px_1) \leq d(Q,Qx)$$

for $x \in F$, $x \notin [F,F]$, $Q \in \mathbb{H}$. Since the points of the displacement axis of x_1 are transported the smallest distance by x_1, the points Px_1^n lie on a line, the displacement axis of x_1. The image of the latter on R traverses a closed geodesic η_1, which because of the extremal conditions is simple, and of minimal length among all non-null-homologous curves.

□

6.7.5 *Remark.* Naturally P and x_1 are not uniquely determined by the minimality condition; at the very least all elements conjugate to x_1 are permissible, and P can be chosen arbitrarily on its displacement axis. However, these displacement axes yield the same geodesic on R.

6.7.6 Lemma. *Among the curves which have algebraic intersection number* ± 1 *with* η_1 *there is one of minimal length, and this is then a simple closed geodesic which meets* η_1 *exactly once.*

Proof. In each homotopy class (with movable initial point) there is a closed geodesic, namely that derived from the displacement axis, and only one, and this has minimal length among all curves of the homotopy class. Among the closed geodesics which have algebraic intersection number ± 1 with η_1 we now choose the one with minimal length. We call it κ and we shall show that it meets η_1 exactly once and is simple. If it met η_1 several times then we would have a situation as in the figure, where the upper arc β on κ does not meet η_1 at other points. Then β would be longer than α, otherwise we could shorten η_1 by replacing α by β, obtaining a curve which had intersection number ± 1 with one of the others, and which would in particular not be null-homologous. If then β were replaced by α in κ we would obtain a curve shorter than κ which had algebraic intersection number ± 1 with η_1. Thus κ must meet η_1 only once. If κ were not simple and closed, then it would contain a shorter part which cuts η_1 only once.

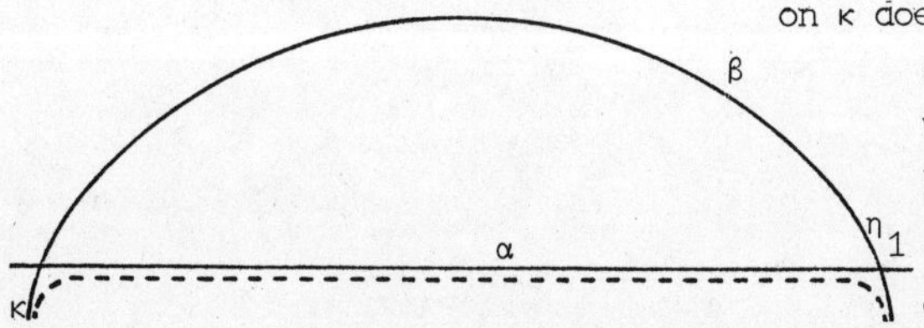

□

6.7.7 Now let P be the point of intersection of η_1 and κ. We shall now show inductively that the following statements (E_i), $1 \le i \le 2g$, hold.

(E_i) $\eta_1,\ldots,\eta_i$ *are simple closed geodesics on* R *with the common initial point* P, *which do not meet each other except at* P. *The curves* $\eta_1,\ldots,\eta_i$ *are independent with respect to homology and* $\eta_1,\ldots,\eta_{i-1}$ *satisfy* (E_{i-1}). *If* ζ *is a closed curve beginning at* P *and independent of* $\eta_1,\ldots,\eta_{i-1}$ *with respect to homology then* $d(\zeta) \ge d(\eta_i)$. *If an* η_j *has a non-zero algebraic intersection number with* η_1 *then* κ *occurs among* $\eta_2,\ldots,\eta_j$.

6.7.8 Lemma. (E_i) *holds for* $i = 1,\ldots, 2g$.

Proof. (E_1) holds because of Lemma 6.7.4. Now assume (E_i), and we shall deduce the validity of (E_{i+1}) for $i < 2g$. In any case there are curves independent of

$\eta_1,\dots,\eta_i$ with respect to homology. Among those which emanate from P there will be one, ω, of minimal length. If no curve $\eta_2,\dots,\eta_i$ has algebraic intersection number ± 1 with η_1, and if however $d(\omega) = d(\kappa)$, then we take $\omega = \kappa$. We shall show that we can take ω as η_{i+1}. Obviously ω is geodetic and satisfies the minimality requirement. If ω were not simple and closed then we would obtain a contradiction to the minimality of ω as follows: we have $\omega = \omega_1\omega_2\omega_3$ where ω_2 is closed and $d(\omega_2) > 0$. Since $d(\omega_1\omega_3) < d(\omega)$, the closed path $\omega_1\omega_3$ must be homologous to a sum in $\eta_1,\dots,\eta_i$. If say $d(\omega_1) \le d(\omega_3)$ then $\omega_1\omega_2\omega_1^{-1}$ is at most as long as ω and homologous to no sum in $\eta_1,\dots,\eta_i$. Therefore $\omega_1\omega_2\omega_1^{-1}$ must likewise be geodetic, and ω_3 must $= \omega_1^{-1}$. However, this is an inconsistency: ω must run to its midpoint and then reverse itself.

Now we have ω meeting η_k again outside P, in Q. Thus ω and η_k are divided by P and Q into pairs of arcs ω_1,ω_2 and η_{k1},η_{k2} respectively, so that $\eta_k = \eta_{k1}\eta_{k2}$ and $\omega = \omega_1\omega_2$ and Q is the endpoint of ω_1 and η_{k1}. Then the following four paths begin and end in P: $\eta_{k1}\omega_2, \eta_{k1}\omega_1^{-1}, \eta_{k2}^{-1}\omega_2, \eta_{k2}^{-1}\omega_1^{-1}$. All these paths are non-geodetic, since a proper intersection must occur at Q. The corresponding geodetic paths therefore have strictly smaller length.

Because $d(\omega) \ge d(\eta_k)$, an inequality $d(\omega_j) \ge d(\eta_{k\ell})$ must hold, say $d(\omega_1) \ge d(\eta_{k1})$. Consequently $d(\eta_{k1}\omega_2) \le d(\omega)$, and by making it geodetic, properly smaller. Therefore

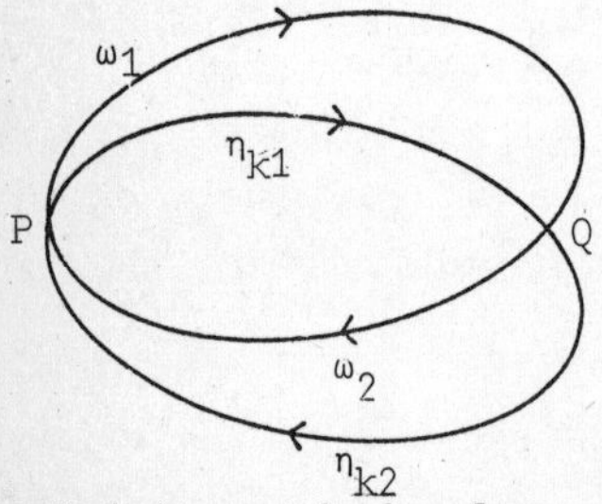

$\eta_{k1}\omega_2$ is homologous to a sum in $\eta_1,\dots,\eta_i$ and we may replace ω_2 by η_{k1}^{-1} without violating the condition on ω. Because of the minimality of ω, $d(\eta_{k1}) > d(\omega_2)$. But then the curve $\omega_2^{-1}\eta_{k2}$ is shorter than η_k, and hence by (E_k) it must be homologous to a sum in $\eta_1,\dots,\eta_{k-1}$. Computing homologically, $\omega_1\eta_{k2}$ and ω differ only by a sum of $\eta_1,\dots,\eta_{k-1}$ and thus $\omega_1\eta_{k2}$, like ω, is homologically independent of $\eta_1,\dots,\eta_{k-1}$ for $i \ge k$. Therefore we must have $d(\eta_{k2}) > d(\omega_2)$ and η_k may be replaced by the shorter $\eta_{k1}\omega_2$ and (E_k) cannot hold. Thus if we take ω as η_{i+1}, (E_{i+1}) will hold.

□

Proof of Theorem 6.7.3. By means of Lemma 6.7.8 we finally find a geodetic dissection which contains two closed geodesics - η_1 and κ here. As a result, the sum of two adjacent angles cannot exceed π, and the dissection is doubly convex.

□

6.7.9 Lemma. *Let* Z *be a doubly-convex geodetic dissection of* R *and* F *a fundamental domain of* Γ *in* $\mathbb{H}$ *the boundary of which is mapped onto* Z. *Assume that* $F \cap t(F)$, $t \in \Gamma$, *is the segment* $\overline{PQ}$.
(a) Then there is a point $A \in \overline{PQ}$ *such that* $t^{-1}(A), A, t(A)$ *are on one line and this is*

the axis of t, *see figure.*

(b) *If* t_1, t_2 *are two generators defined by* F *and if they correspond to curves on* R *with algebraic intersection number* ± 1 *then the axes of* t_1 *and* t_2 *cross each other in* F.

Proof. For the symbols used see figure . The doubly-convexity of Z implies that the angles α and β do not exceed π. Now we take a point $X \in \overline{PQ}$ and consider the angle between $t^{-1}(X)X$ and $Xt(X)$. It varies continuously from $\alpha \leq \pi$ to $2\pi - \beta \geq \pi$. Hence there is an A as required. If, for a linear fractional transformation t, the points A, $t^{-1}(A) \neq t(A)$ are on one line then the transformation is hyperbolic and the line equals the axis of t; for a proof see [Zieschang 1980, 14.17] or E 6.13. Now (b) is a direct consequence of (a). □

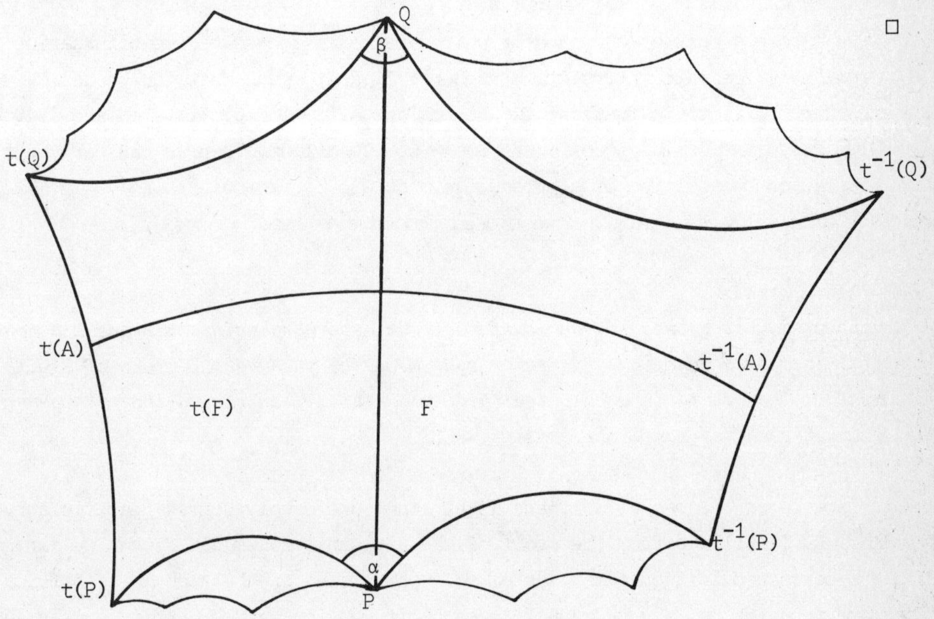

6.7.10 Lemma. *Let* Z *be a doubly-convex geodetic dissection. Then each point of the surface can serve as the basepoint of a geodetic dissection dual to* Z *or isotopic to* Z *respectively.*

Proof. We again consider the universal covering and divide a fundamental domain F defined by Z into pieces by the following rule: F contains the point Q over the desired basepoint, and let Fx be a neighbour of F with a common edge. Because F is doubly-convex, F and Fx together constitute a convex polygon, in the interior of which Q and Qx may be connected by a line segment which cuts the common edge once. Since the single polygon is likewise convex, the paths to the various boundary segments of F meet only at Q. If we carry out this construction for all cells then we obtain a dual net of geodetic edges, the vertices of which lie over the prescribed point, and upon which the group likewise acts.

There are two generators t_1, t_2 of the dissection Z which have the algebraic intersection number ± 1 with each other, i.e. F and Ft_i have a common edge τ_i^{ε} and the pairs τ_1, τ_1^{-1} and τ_2, τ_2^{-1} meet in the boundary of F. We choose the point of intersection Q of the displacement axes of t_1 and t_2 as the basepoint for the construction of the dual net Z_*. By 6.7.9 it contains two intersecting lines, so it is doubly-convex. Starting with an arbitrary point we can construct the net dual to Z_* by geodetic nets. These are isotopic to each other and Z is among them.

□

6.7.11 Lemma. *Let* Z *be a doubly-convex geodetic dissection and let* Z' *result from* Z *by bifurcation. Then one can find a point* Q *and dissections* Z_1 *and* Z_1' *originating from it, isotopic to* Z *and* Z' *respectively and both geodetic and doubly-convex.*

Proof. We choose a pair of curves of Z which are not affected by the bifurcation and which have algebraic intersection number ± 1, and Q such that the geodesics emanating from Q isotopic to these two curves are also closed. Z_1 is chosen isotopic to Z with basepoint Q and geodetic, as Lemma 6.7.10 allows. Then we can carry out the bifurcation by addition of a geodesic and obtain Z_1' isotopic to Z'. Again, Z_1' is geodetic and, since it contains two closed geodesics, doubly-convex also.

□

6.7.12 Theorem. *Let* R *be a Riemann surface,* P *an arbitrary point on* R *and* Σ *a conjugation class of a canonical generator system of the fundamental group of* R *with basepoint* P. *Then there is a canonical geodesic system* Z *on* R *such that the generator system which results from* Z *corresponds to* Σ.

Proof. Because of Theorem 6.7.3 there is at least one doubly-convex geodetic curve system on R. It follows from the proof of the Nielsen Theorem 5.6.3 that we can go from the corresponding generator system to a prescribed one by a chain of bifurcations and renumberings. By Lemma 6.7.11 each bifurcation may be carried out in such a way that we again obtain a doubly-convex geodesic system. In doing so we must necessarily change the basepoint. By Lemma 6.7.10 we can make P the basepoint at the end. Because of the loose handling of the basepoint our statement holds only up to inner automorphism.

□

The application of Nielsen's Theorem would not be necessary if Z was taken to be a class of canonical dissections, and if it was shown that one can go from one canonical dissection to another by a chain of bifurcations.

6.7.13 Let R be a Riemann surface and let $\phi: \mathbb{H} \to R$ be the universal covering mapping.

Since an element $x \in F$ is a hyperbolic transformation it has a displacement axis and its image is a closed geodesic, possibly with double points. The displacement axes of elements conjugate to x are the images of the displacement axis of x under application of the group F and consequently have the same image on the surface. Conversely, these curves or a suitable power uniquely represent the conjugation class of x. If the element x of the fundamental group contains a simple closed curve and if there is no point equivalent to the endpoints on an interval between equivalent points on the axis of x then its image is a simple closed geodesic. For non-separating simple closed curves this is a consequence of Theorem 6.7.12 since they may be built into dissections; separating but noncontractible curves can be regarded as "diagonals" of the dissection, and it follows again from the same theorem.

6.7.14 Now let F be given by canonical generators $t_1,\dots,u_g$ with $\prod_{i=1}^{g}[t_i,u_i] = 1$. Since the algebraic intersection number of t_1 and u_1 equals 1 the displacement axes of t_1 and u_1 intersect in a point, so corresponding images intersect in a point of F; call it P. If we replace the generator system by a conjugate one the displacement axes are moved by a transformation from the group; so the intersection point always lies over P. This point is therefore uniquely determined by the conjugation class Σ of the generator system. We now take P as basepoint and represent the generators by a geodetic curve system. As a result t_1 and u_1 appear as closed geodesics, though the others do not, since they only touch t_1 and u_1 at P. At any rate, this curve system is doubly-convex. By cutting, one obtains a convex polygon of the non-euclidean plane which is naturally associated with the generator system as a fundamental domain in the universal covering. Conversely, the curve system determines the generator system up to conjugation.

If Σ' is a conjugation class of a canonical generator system for a surface R' and if there is a holomorphic mapping $R \to R'$ which carries Σ into Σ' then the fundamental polygons associated with Σ, Σ' are congruent with retention of the labelling; for the curve system distinguished above is transported with the mapping and therefore represents a system like the one which now represents the conjugation class Σ'. All angles and non-euclidean lengths are preserved.

<u>6.7.15 Definition</u>. To formulate the outcome of the considerations above we define a *canonical polygon* to be a 4g-gon with inner angle sum 2π (i.e. with area $(4g-2)\pi - 2\pi$), the boundary segments of which are labelled with the symbols $\tau_1,\mu_1,\tau_1^{-1},\mu_1^{-1},\dots,\tau_g,\mu_g,\tau_g^{-1},\mu_g^{-1}$, so that the boundary path is $\prod_{i=1}^{g}\tau_i\mu_i\tau_i^{-1}\mu_i^{-1}$ and the segments denoted τ_i and τ_i^{-1}, and likewise μ_i and μ_i^{-1}, have equal length.

We understand *congruent canonical polygons* to be ones which may be conformally mapped onto each other so that identically labelled segments correspond. With these definitions we can formulate our result as

6.7.16 Theorem. *Each marked Riemann surface corresponds to a unique doubly-convex geodetic canonical curve system and hence to a doubly-convex non-euclidean canonical polygon.*

□

6.7.17 Canonical curve systems associated with holomorphically equivalent marked surfaces are holomorphically equivalent, and their canonical polygons are congruent. Obviously, congruent canonical polygons define holomorphically equivalent marked Riemann surfaces.

6.7.18 Corollary. *A holomorphic mapping of a closed Riemann surface* R *of genus* $g > 1$ *onto itself which is isotopic to the identity is the identity.*

Proof. Namely, if we take a Σ on R, then the mapping preserves the Σ, and hence also the images of the displacement axes of t_1 and u_1 and their point of intersection P. In addition, the mapping is conformal so it is the identity on the images of the displacement axes in R and hence everywhere.

□

See also [Kerékjartó 1934].

6.7.19 A construction. We shall now construct a certain mapping between any two canonical polygons of the same genus, and they will then yield the T-mappings for Riemann surfaces.

First we choose a point Q in the interior of a canonical polygon invariant under congruences, say as follows: we connect the midpoints of the sides denoted τ_1 and τ_1^{-1}, μ_1 and μ_1^{-1} respectively, and let Q be the intersection of these connecting segments. If one displaces the canonical polygon by a conformal mapping then the midpoints go into the midpoints of the segments with the same labels, connecting segments go into connecting segments, and hence their intersection goes into the intersection.

Now we introduce polar coordinates relative to Q, say as follows: we parametrize the boundary so that the initial points of the segments take parameter values according to the following table:

Initial point of	τ_i	μ_i	τ_i^{-1}	μ_i^{-1}
Parameter value	$\frac{4(i-1)}{4g}$	$\frac{4(i-1)+1}{4g}$	$\frac{4(i-1)+2}{4g}$	$\frac{4(i-1)+3}{4g}$

and the boundary segments are linearly (in the sense of noneuclidean length) parametrized (so e.g. the midpoint of τ_g has the parameter value $\frac{4(g-1)+1/2}{4g}$). Similarly we parametrize the segments from Q to the boundary of the polygon linearly, so that Q has the value 0 and boundary points take the value 1. Now if the point X lies on the segment which connects Q to the boundary point with parameter value t, and if X has the value s on its segment, then we map X to the point $s \cdot e^{2\pi it}$.

For any canonical polygon P this defines for us a fixed topological mapping ϕ_P: $P \to \{z: |z| \leq 1\}$. We take $\phi_{P_2}^{-1} \phi_{P_1}$ as the T-mapping from the polygon P_1 to P_2. If we paste the corresponding sides of P_1 and P_2 together, then the images of identified points are also identified, and we obtain a topological mapping of the corresponding Riemann surfaces onto each other. Let T be the set of mappings defined in this way between Riemann surfaces. Now since holomorphic mappings of the unit circle onto itself are the same as motions of the non-euclidean plane, not only are the distinguished points Q carried into each other by holomorphic mappings, but the linearization is also preserved. Thus we have

6.7.20 Theorem. *T is the basis of a Teichmüller theory.*

□

Exercise: E 6.13

6.8 THE CLASSICAL FOUNDATION

This is based on a new concept, that of the *quasiconformal mapping*. Intuitively speaking, a diffeomorphism carries a "small circle" in the neighbourhood of a point into a "small ellipse" in the neighbourhood of the image point; a mapping is conformal when it remains a circle. If the quotient of the major and minor axes of the ellipse for each point remains bounded (by K say) then the mapping is called (K)-quasiconformal. This can be made more precise by giving the tangent space the metric determined by the complex structure. Then a diffeomorphism defines a linear mapping at each point of the tangent space and hence a dilatation.

Since we want to extend the theory to non-differentiable mappings, the following definition is cleverer: we understand a quadrilateral Q to be a region of the complex plane, the boundary of which is a simple closed curve divided into four arcs $\alpha_1, \beta_1, \alpha_2, \beta_2$. There is then a topological mapping onto a rectangle with sides of length a and b, which is holomorphic in the interior and which maps the arcs α_i on to the sides of length a, the β_i on to the sides of length b. Two rectangles may then be mapped onto each other by a homeomorphism which is holomorphic in the interior and which maps boundary sides to those with the same labels if and only if the quotients of the side lengths are equal; hence the quotient a/b is a conformal invariant of the quadrilateral Q: its *modulus*. A topological mapping is called K-*quasiconformal* when an arbitrary quadrilateral of modulus m has image with modulus $m' \leq Km$. If one exchanges a and b, then one obtains the moduli $\frac{1}{m}$ and $\frac{1}{m'}$, so that K must be ≥ 1. A topological mapping is then conformal just in case it is 1-quasiconformal. Among the theorems on quasi-conformal mappings we mention only one: *a mapping is* K-*quasiconformal when it is for all "small" quadrilaterials* [Ahlfors 1953]. Therefore one can extend the concept of quasiconformality to mappings between Riemann surfaces.

By the *dilatation of a quadrilateral under a mapping* we mean the quotient of the moduli of image and preimage. A mapping has an associated maximal dilatation. The most important result for us is the following theorem of O. Teichmüller:

6.8.1 Theorem. *Let* R *and* R' *be two closed Riemann surfaces of genus* $g \geq 2$ *and let* $\varphi: R \to R'$ *be a homeomorphism. Then there is exactly one homeomorphism isotopic to* φ, *the maximal dilatation of which is minimal among those of mappings isotopic to* φ.

Proofs of this difficult theorem are found in [Teichmüller 1940, 1943], [Ahlfors 1953], [Bers 1960]. The obtained mappings are called *extremal quasiconformal*. □

It follows directly from the definition of quasiconformality that the product of K_1- and K_2-quasiconformal mappings is K_1K_2-quasiconformal, hence, conformal mappings do not change the maximal dilatation. And Theorem 6.8.1 says:

6.8.2 Theorem. *The set* $\mathcal{T}$ *of extremal quasiconformal mappings is the basis of a Teichmüller theory.*

This basis has an advantage over the one given in 6.7. It is derived from a local property which is not as arbitrary as the labelling of a polygon, the choice of a centrum Q, and the introduction of polar coordinates depending on it. In particular, the $\mathcal{T}$-mappings for surfaces of different genus are of the same kind.

Naturally, the question of the meaning of our arbitrary labelling presents itself. It should be emphasized however, that the product of two of our $\mathcal{T}$-mappings, if defined at all, is again a $\mathcal{T}$-mapping.

6.9 TEICHMÜLLER SPACES

By Theorem 6.7.16 each marked Riemann surface of genus g is associated with a 4g-gon, exact up to congruence, in the non-euclidean plane. This permits a topology to be given to the space of these marked surfaces, for example as follows:

6.9.1 Construction. We fix the initial point and the direction of the first segment τ_1, then the polygon corresponds to a unique point in $\mathbb{H}^{4g}$ and we obtain a closed subset. The induced topology is taken for the space of marked Riemann surfaces. This topology is also obtained as follows: if one introduces the angle and arc measures as parameters then for the g-th commutator one obtains 5 free parameters, and 6 for the next (g-2) commutators. A pentagon for the first commutator is then determined as shown. The length of K is determined by the last (g-1) commutators. In addition, the angle sum of the pentagon is determined by the total content. One sees that if α and

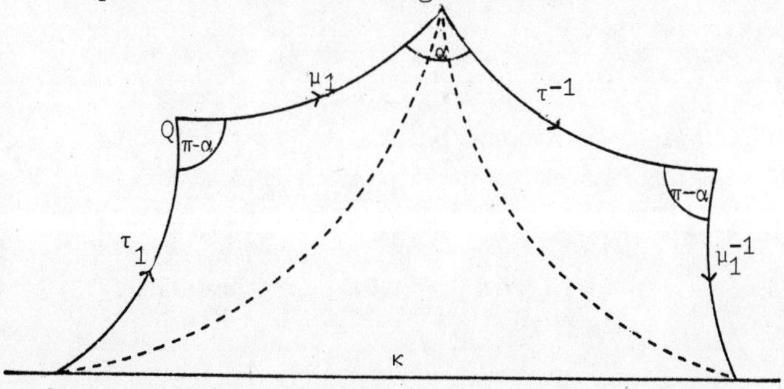

the lengths of τ_1 and μ_1 are given then the dotted lines are uniquely determined and hence so is the pentagon itself. Because of the condition that the angle at Q is exactly $\pi-\alpha$

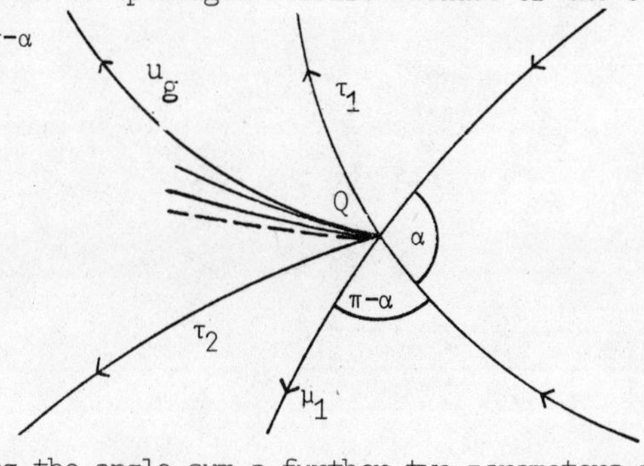

(on the Riemann surface)

and by fixing the angle sum a further two parameters are lost, there are altogether 6(g-1) free parameters left. By being more careful one obtains $\mathbb{R}^{6g-6}$, see 6.9.2.

One obtains a natural definition of the topology when one takes the ends and lengths of the displacement axes as parameters [Bers 1960],[Zieschang 1980]. The space can also be given a metric, and in fact the distance between two marked Riemann surfaces is the logarithm of the maximal dilatation of the extremal quasi-conformal mapping between them. As a result, the Teichmüller space receives a finer structure [Ahlfors 1960].

6.9.2 Theorem. *The space of marked Riemann surfaces of genus* g *with the topology described above is homeomorphic to* $\mathbb{R}^{6g-6}$.

We will prove this theorem in 6.11.11.*

It obviously suffices to do this for the space of the canonical polygons. In the proof which follows a canonical polygon will be dissected in such a way that one obtains an overview of the possible subpieces, and a means of reconstructing a canonical polygon step-by-step from its dissection. The reconstruction process leads in a natural way to a fibering of the space of the canonical polygons, from which the homeomorphism with $\mathbb{R}^{6g-6}$ follows easily.

It is in the nature of the proof that it makes essential use of methods from non-euclidean geometry (in particular the Poincaré model of hyperbolic geometry) including a few special formulae (distance between two points as a function of the coordinates, area of a triangle as a function of the lengths of the sides) which are not proved here.

In this connection the reader is referred to [Siegel 1964'] and [Liebmann 1912].

6.10 SOME LEMMAS FROM HYPERBOLIC GEOMETRY

6.10.1 Lemma. *The area of a hyperbolic triangle* ABC *with side* a = BC *may be given as a function of* $|a|$, $s = \frac{1}{2}(|a| + |b| + |c|)$ *and* $\Delta = |b| - |c|$. *The area grows with increasing* s *for fixed* Δ, *and with decreasing* Δ *for fixed* s.

Proof. By the "Heron" formula the area F of the triangle is given by

$$\sin \frac{F}{2} = \frac{\sqrt{\sinh s \sinh(s-|a|) \sinh(s-|b|) \sinh(s-|c|)}}{2 \cosh \frac{|a|}{2} \cosh \frac{|b|}{2} \cosh \frac{|c|}{2}} ,$$

see [Liebmann 1912, pg. 129]. Now

---

* We thank Herr C.L. Siegel for his friendly critique and especially for his help in preparing sections 6.10 and 6.11.

$$s - |b| = \frac{1}{2}(|a| + |c| - |b|) = \frac{1}{2}(|a| - \Delta)$$

$$s - |c| = \frac{1}{2}(|a| + |b| - |c|) = \frac{1}{2}(|a| + \Delta) \text{ and hence}$$

$$\sinh(s-|b|)\sinh(s-|c|) = \sinh\frac{1}{2}(|a|-\Delta)\sinh\frac{1}{2}(|a|+\Delta)$$

$$= \frac{1}{2}(\cosh|a| - \cosh\Delta) \ ;$$

and the same way:

$$\sinh s \sinh(s-|a|) = \sinh\frac{1}{2}(2s-|a|+|a|)\sinh\frac{1}{2}(2s-|a|-|a|)$$

$$= \frac{1}{2}(\cosh(2s-|a|) - \cosh|a|),$$

$$\cosh\frac{|b|}{2}\cosh\frac{|c|}{2} = \cosh\frac{1}{2}(s-\frac{|a|}{2}+\frac{\Delta}{2})\cosh\frac{1}{2}(s-\frac{|a|}{2}-\frac{\Delta}{2})$$

$$= \frac{1}{2}(\cosh(s-\frac{|a|}{2}) + \cosh\frac{\Delta}{2}).$$

This gives

$$\sin\frac{F}{2} = \frac{1}{2\cosh\frac{|a|}{2}} \cdot \frac{\sqrt{(\cosh(2s-|a|)-\cosh|a|)(\cosh|a|-\cosh\Delta)}}{(\cosh(s-\frac{|a|}{2})+\cosh\frac{\Delta}{2})}$$

or

$$\tilde{F} := 4\cosh^2\frac{|a|}{2}\sin^2\frac{F}{2} = \frac{(\cosh(2s-|a|)-\cosh|a|)(\cosh|a|-\cosh\Delta)}{(\cosh(s-\frac{|a|}{2})+\cosh\frac{\Delta}{2})^2}$$

But since $0 < \frac{F}{2} < \frac{\pi}{2}$, $\tilde{F}$ increases or decreases for fixed a as F increases or decreases, respectively.

$$\frac{\partial\tilde{F}}{\partial\Delta} = -\frac{\cosh(2s-|a|)-\cosh|a|}{(\cosh(s-\frac{|a|}{2})+\cosh\frac{\Delta}{2})^3}\left[\sinh\frac{\Delta}{2}(\cosh|a| - \cosh\Delta) + \sinh\Delta(\cosh(s-\frac{|a|}{2})+\cosh\frac{\Delta}{2})\right]$$

$$\Rightarrow \operatorname{sgn}\left(\frac{\partial\tilde{F}}{\partial\Delta}\right) = -\operatorname{sign}\Delta \qquad (\text{because } 2s > |a| > |\Delta|).$$

This simply means that $\tilde{F}$ (and hence also F) increases monotonically when $|\Delta|$ decreases.

$$\frac{\partial\tilde{F}}{\partial s} = \frac{2(\cosh|a|-\cosh\Delta)}{(\cosh(s-\frac{|a|}{2})+\cosh\frac{\Delta}{2})^3}\ [\sinh(2s-|a|)\cosh(s-\frac{|a|}{2})$$

$$-\cosh(2s-|a|)\sinh(s-\frac{|a|}{2}) + \sinh(2s-|a|)\cosh\frac{\Delta}{2}$$

$$+\sinh(s-\frac{|a|}{2})\cosh|a|]$$

$$= \frac{2(\cosh|a|-\cosh\Delta)}{(\cosh(s-\frac{|a|}{2})+\cosh\frac{\Delta}{2})^3}\ [(1+\cosh|a|)\sinh(s-\frac{|a|}{2}) + \cosh\frac{\Delta}{2}\sinh(2s-|a|)]$$

$\Rightarrow \frac{\partial\tilde{F}}{\partial s} > 0$, and therefore, since $\operatorname{sgn}\frac{\partial F}{\partial s} = \operatorname{sgn}\frac{\partial\tilde{F}}{\partial s}$, the area F of the triangle grows with increasing s.

□

This lemma will be applied when the third vertex A moves on an altitude h_a perpendicular to a. The altitude h_a is defined as follows: Let g_a be the line containing the segment a. A line h_a which intersects g_a perpendicularly at a point D is called an *altitude on* a *erected at* D.

Then we have

6.10.2 Lemma. *Let* a = BC *be a line segment,* h_a *an altitude on* a *erected at* D*, and* A *a point on* h_a*. If one moves* A *on* h_a *away from* D*, then* $|b| = |CA|$ *and* $|c| = |BA|$ *increase strictly monotonically, while* $||b|-|c||$ *decreases monotonically (in fact strictly monotonically if* $|b| \neq |c|$*).*

Proof. To prove this we map the interior of the unit circle conformally onto the upper half plane so that the points B, C are mapped on to the real axis and D onto the point (0,0). The altitude on which the third point varies is then the imaginary axis.

The distance formula, see [Siegel 1964', § 2]

$$d(z_1,z_2) = \log\frac{1+\left|\frac{z_2-z_1}{1-\bar{z}_1 z_2}\right|}{1-\left|\frac{z_2-z_1}{1-\bar{z}_1 z_2}\right|}$$

with the particular values

$$z_1 = x,\ z_2 = ti;\quad x,\ t \in \mathbb{R}$$

gives

$$d(x,ti) = \log \frac{\sqrt{1+x^2t^2} + \sqrt{x^2+t^2}}{\sqrt{1+x^2t^2} - \sqrt{x^2+t^2}} \quad .$$

Here we are interested in the way $d(x,ti)$ varies with t. A short computation gives

$$\frac{\partial d(x,ti)}{\partial t} = \frac{2t}{1-t^2} \frac{(1+x^2)}{\sqrt{(x^2+t^2)(1+x^2t^2)}} \quad \text{with } |x| < 1,\ 0 < t < 1.$$

Thus $d(x,ti)$ increases strictly monotonically with increasing $|t|$, and, on symmetry grounds, likewise with increasing $|x|$.

But $\frac{\partial d(x,ti)}{\partial t}$ also depends on x and we have

$$\frac{\partial^2 d(x,ti)}{\partial x \partial t} = -2 \frac{xt(1-x^2)(1-t^2)}{(x^2+t^2)^{3/2}(1+x^2t^2)^{3/2}}$$

which says: $\frac{\partial d(x,ti)}{\partial t}$ is a monotone strictly decreasing function of $|x|$ for fixed t.

It follows in particular that when $|t|$ increases the shorter of the two sides AC and AB in the triangle ABC with

$$A = (0,t),\ B = (x,0),\ C = (y,0),\ |x| \neq |y|$$

increases in length faster than the longer side. Consequently $|\Delta| = ||AC|-|AB|| = ||b|-|c||$ is a monotone strictly decreasing function of t.

For $|x| = |y|$ we have $\Delta = 0$.

□

Thus Lemma 6.10.2 is proved and one sees, by reference to Lemma 6.10.1 that the area of a triangle ABC properly increases when A moves along h_a away from the lines through B and C. However, there is still the question of the upper limit for the area of the triangle ABC when one fixes the points B and C and moves A on a fixed altitude.

6.10.3 Lemma. *The upper limit for the area with a fixed* Δ *is a function of* Δ *and* $|a|$ *which increases strictly monotonically with* $|a|$*, and strictly decreases when* $|\Delta|$ *increases. If* A *lies on the altitude* h_a *erected at* D*, then the upper limit grows strictly monotonically as* D *approaches the midpoint of the segment* a.

Proof. First let Δ be fixed. Then

$$\sin^2 \frac{F}{2} = \frac{(\cosh|a|-\cosh \Delta)}{4\cosh^2 \frac{|a|}{2}} \quad \frac{\cosh(2s-|a|)-\cosh|a|}{(\cosh(s-\frac{|a|}{2})+\cosh \frac{\Delta}{2})^2}$$

$$\lim_{s\to\infty} \frac{\cosh(2s-|a|)-\cosh|a|}{(\cosh(s-\frac{|a|}{2})+\cosh \frac{\Delta}{2})^2} = \lim_{s\to\infty} \frac{\sinh(2s-|a|)}{\sinh(s-\frac{|a|}{2})(\cosh(s-\frac{|a|}{2})+\cosh \frac{\Delta}{2})}$$

$$= \lim_{s\to\infty} 2 \frac{\cosh(s-\frac{|a|}{2})}{\cosh(s-\frac{|a|}{2})+\cosh \frac{\Delta}{2}}$$

$$= 2$$

$$\Rightarrow \lim_{s\to\infty} \sin^2 \frac{F}{2} = \frac{1}{2} \frac{\cosh|a|-\cosh \Delta}{\cosh^2 \frac{|a|}{2}}$$

$$= \frac{\cosh|a|-\cosh \Delta}{\cosh|a|+1}$$

$$= 1 - \frac{\cosh \Delta +1}{\cosh|a|+1}$$

$$= 1 - \frac{\cosh^2 \frac{\Delta}{2}}{\cosh^2 \frac{|a|}{2}}$$

$$\sin^2 \frac{F_{max}}{2} = 1 - \frac{\cosh^2 \frac{\Delta}{2}}{\cosh^2 \frac{|a|}{2}} \Rightarrow \left|\cos \frac{F_{max}}{2}\right| = \frac{\cosh \frac{\Delta}{2}}{\cosh \frac{|a|}{2}}$$

and because $0 \le \frac{F_{max}}{2} \le \frac{\pi}{2}$ it follows that

$$F_{max} = 2 \operatorname{arc\,cos} \frac{\cosh \frac{\Delta}{2}}{\cosh \frac{|a|}{2}}.$$

This proves the first part of the assertion and one sees in addition that the upper limit of all triangles with base a is exactly $2 \operatorname{arc\,cos} \frac{1}{\cosh \frac{|a|}{2}}$. One obtains the improper triangle with this area when one places A at the intersection of the real axis with the perpendicular bisector of a.

In order to prove the second part of the assertion, the side a is mapped onto the imaginary axis by a linear fractional transformation in such a way that C is mapped onto the point (0,1). Let the coordinates of B be (0,t). The altitudes on a are then the euclidean semicircles with midpoint (0,0). Let Δ^r be the limit approached by Δ when A goes to the real axis on the circle

$x^2 + y^2 = r^2$ (or $z\bar{z} = r^2$). It then suffices to show that, for $r \geq \sqrt{t}$, Δ^r grows monotonically with r.

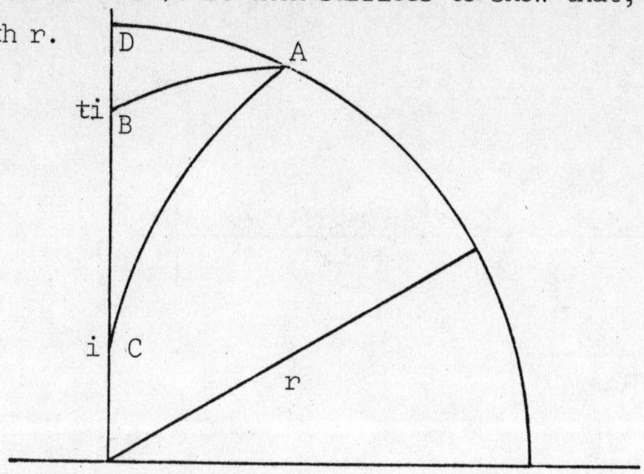

The distance between the points $z_1 = x_1 + iy_2 = r_1 e^{i\phi_1}$

$$z_2 = x_2 + iy_2 = r_2 e^{i\phi_2}$$

is $d(z_1,z_2) = \text{arc cosh}\,\frac{r_1^2+r_2^2-2x_1x_2}{2y_1y_2}$

and consequently, with $A = (x,y)$, $x^2 + y^2 = r^2$

$$\Delta = |b|-|c| = \text{arc cosh}\,\frac{1+r^2}{2y} - \text{arc cosh}\,\frac{t^2+r^2}{2ty}$$

$$= \text{arc cosh}\left[\frac{(1+r^2)(t^2+r^2)}{4ty^2} - \sqrt{\frac{((1+r^2)^2-4y^2)((t^2+r^2)^2-4t^2y^2)}{(4y^2t)^2}}\right]$$

$$= \text{arc cosh}\,\frac{(1+r^2)(t^2+r^2)-\sqrt{((1+r^2)^2-4y^2)((t^2+r^2)^2-4t^2y^2)}}{4y^2t}$$

$$\Rightarrow \cosh \Delta^r = \lim_{y\to 0} \frac{8y((t^2+r^2)^2-4t^2y^2)+8yt^2((1+r^2)^2-4y^2)}{16yt\sqrt{((1+r^2)^2-4y^2)((t^2+r^2)^2-4t^2y^2)}}$$

$$= \lim_{y\to 0}\frac{1}{2}\,\frac{t^2(1+r^2)^2+(t^2+r^2)^2-8t^2y^2}{t\sqrt{((1+r^2)^2-4y^2)((t^2+r^2)^2-4t^2y^2)}}$$

$$= \frac{1}{2}\,\frac{t^2(1+r^2)^2+(t^2+r^2)^2}{t(1+r^2)(t^2+r^2)}$$

$$\Rightarrow \Delta^r = \text{arc cosh}\,\frac{1}{2}\left[\frac{t(1+r^2)}{t^2+r^2} + \frac{t^2+r^2}{t(1+r^2)}\right].$$

It follows that $\Delta^r = 0 \Leftrightarrow (r^2-t)^2(t-1)^2 = 0$ or, since $t \neq 1$, $\Delta^r = 0 \Leftrightarrow r = \sqrt{t}$.

Hence when $r > \sqrt{t}$, $\frac{\partial \Delta^r}{\partial r} > 0 \Leftrightarrow \frac{\partial(\cosh \Delta^r)}{\partial r} > 0$. Now

$$\frac{\partial \cosh \Delta^r}{\partial r} = \frac{1}{2}\left[\frac{2tr(t^2+r^2)-2tr(1+r^2)}{(t^2+r^2)^2} + \frac{2tr(1+r^2)-2tr(t^2+r^2)}{t^2(1+r^2)^2}\right]$$

$$= rt(t^2-1)\left[\frac{1}{(t^2+r^2)^2} - \frac{1}{t^2(1+r^2)^2}\right]$$

$$= \frac{rt(t^2-1)}{t^2(1+r^2)^2(t^2+r^2)^2}\,(t^2+2t^2r^2+t^2r^4-t^4-2t^2r^2-r^4)$$

$$= \frac{r(t^2-1)^2(r^4-t^2)}{t(1+r^2)^2(t^2+r^2)^2}\,.$$

$$r > \sqrt{t} \Rightarrow r^4 > t^2 \Rightarrow \frac{\partial \cosh \Delta^r}{\partial r} > 0.$$

□

We have also proved:

6.10.4 Lemma. *Let* a *be a given segment. The upper limit for the area of all triangles with base* a *is*

$$F_{max} = 2\arccos \frac{1}{\cosh \frac{|a|}{2}}\,.$$

The upper limit for the triangles with fixed Δ *is*

$$F^{\Delta}_{max} = 2\arccos \frac{\cosh \frac{\Delta}{2}}{\cosh \frac{|a|}{2}}\,.$$

The upper limit for the area of all triangles with base a *whose third vertex* A *lies on the altitude* h_a *erected at* D *grows monotonically when* D *moves to the midpoint of* a.

□

Lemma 6.10.4 concludes the preliminaries on the geometry of triangles. But before considering the special properties of canonical polygons we point out the following property of polygons, which is immediate from Lemma 6.10.4.

6.10.5 Lemma. *The upper limit of the area of all polygons with* n *vertices with a given base side grows continuously and strictly monotonically with the length of this side.*

□

A similarly proved assertion about special polygons will be used later.

Exercises: E 6.14-15

6.11 CANONICAL POLYGONS

In what follows we consider canonical polygons K_g of closed Riemann surfaces of genus g.

6.11.1 K_g consists of a g-gon to each side of which is attached a certain 5-gon, a "commutator surfaces". The commutator surfaces have the property that when the outer sides are denoted successively by s_1, s_2, s_3, s_4 then $|s_1| = |s_3|$ and $|s_2| = |s_4|$. One of these commutator surfaces, which will be called *unfree,* also satisfies the angle conditions

$$\measuredangle(s_1,s_2) = \measuredangle(s_3,s_4), \quad \measuredangle(s_1,s_2) + \measuredangle(s_2,s_3) = \pi.$$

The sum of the interior angles of K_g is 2π, and its area therefore $(4g-2)\pi - 2\pi = 4(g-1)\pi$.

6.11.2 Definition. The space of g-fons is topologized by the embedding into $\mathbb{H}^g \cong \mathbb{R}^{2g}$ with the vertices as coordinates. This topology corresponds to that of the space of marked Riemann surfaces from 6.9.1.

6.11.3 A construction. Since the unfree commutator surface already contains two angles whose sum is π, all the angles in it are smaller than π. A dissection of the polygon is now carried out as follows.

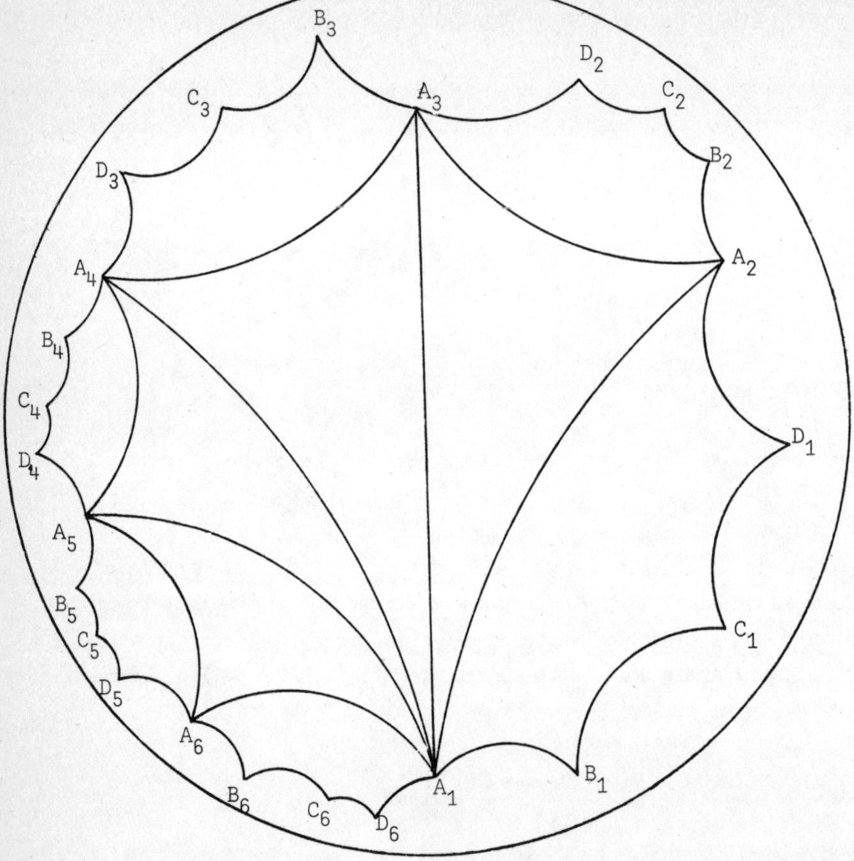

Starting at the vertex A_1 of the unfree commutator surface $A_1B_1C_1D_1A_2$ we first divide the polygon into g-2 hexagons $A_1A_iB_iC_iD_iA_{i+1}$ $(1 < i < g)$ (commutator hexagons) and two pentagons (the unfree commutator surface $A_1B_1C_1D_1A_2$ and the free commutator surface $A_1A_gB_gC_gD_g$) as shown in the sketch. The commutator hexagon then divides into a triangle $A_1A_iA_{i+1}$ and an attached commutator surface $A_iB_iC_iD_iA_{i+1}$ $(1 < i < g)$. Since the angle sum in the polygon is 2π and the unfree commutator surface already contains two angles whose sum is π, one obtains the following bounds on the areas J_ℓ of the surfaces:

6.11.4 (a) unfree commutator surface $\pi < J_1 < 2\pi$

(b) free commutator surface $2\pi < J_2$ $(< 3\pi)$

(c) commutator hexagon $3\pi < J_3$ $(< 4\pi)$.

Here the upper bounds for the free commutator surface and the commutator hexagon are general bounds for hyperbolic pentagons and hexagons respectively, so they actually impose no restrictions on the surfaces.

These simple bounds were the main reason for choosing this dissection. Namely, if one constructs the canonical polygon by beginning with the unfree commutator surface, then in the course of the construction one obtains exclusive conditions, which bound the areas of the surfaces to be constructed below, but not above. Therefore at each step one only has to take care that the side along which the construction is to be continued is sufficiently long in order to be able to achieve the construction of a correspondingly large surface; cf. Lemma 6.10.5. In what follows we first investigate these individual surface elements more precisely.

6.11.5 *The unfree commutator surface*

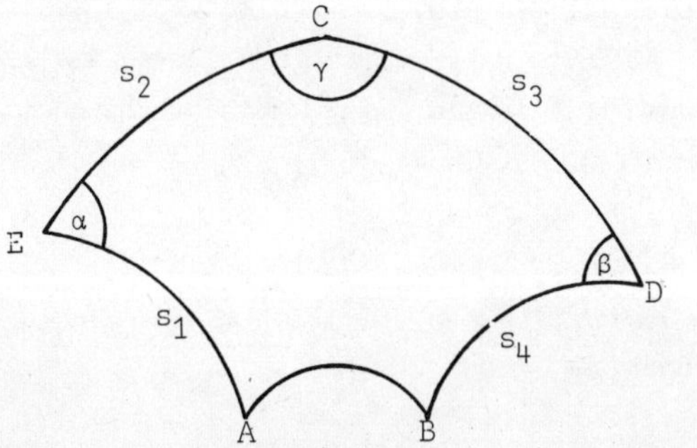

$|s_1| = |s_3|$

$|s_2| = |s_4|$

$\alpha = \beta$, $\alpha + \gamma = \pi$

The triangles ACE and CBD are congruent. Thus $|AC| = |BC|$ and C lies on the perpendicular bisector of AB. (One also sees that the points A, B, C and E determine the point D.)

Because of the special position of C one can use a motion of the upper half plane to bring the perpendicular bisector of AF onto the imaginary axis, so that C lies symmetrically over AB

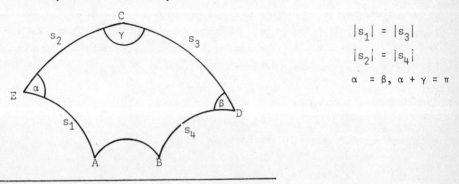

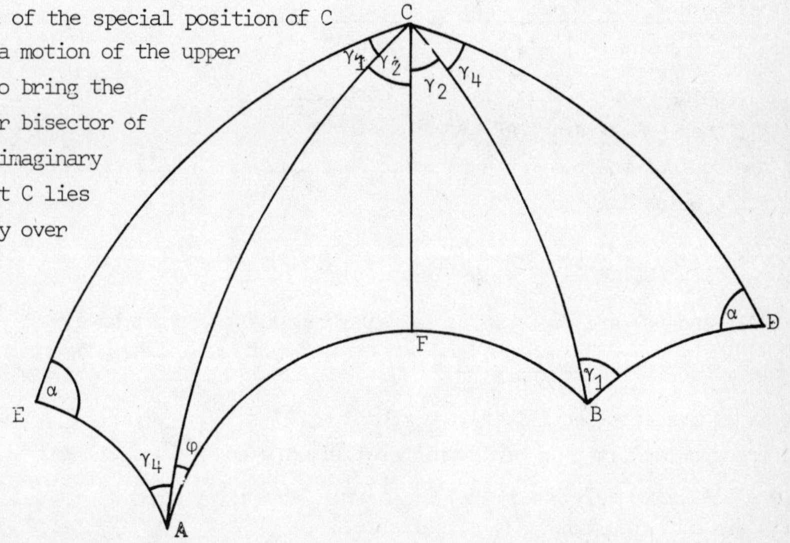

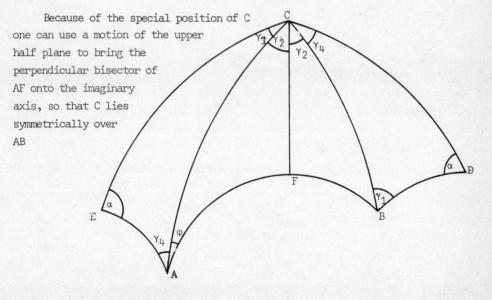

Once C is fixed, so is γ_2. Because $\alpha + \gamma_1 + 2\gamma_2 + \gamma_4 = \pi$ or $2\gamma_2 = \pi - (\alpha + \gamma_1 + \gamma_4)$, the area of the triangle ACE is thereby determined: $|ACE| = 2\gamma_2$.

The area of the commutator surface is $J = \pi - 2\gamma_2 - 2\varphi + 4\gamma_2 = \pi + 2(\gamma_2 - \varphi)$. What conditions on C and E may be derived from this when AB is given?

Since the area is to be greater than π, we must first of all have $\gamma_2 > \varphi$, or equivalently

(1) $$|CF| < \frac{1}{2}|AB| = |AF|,$$

so $|AF|$ is a bound for $|CF|$. Furthermore, $|CF|$ is also bounded below. The area of the triangle to be constructed on base AC is to be $2\gamma_2$, but must also be smaller than $2 \operatorname{arc\,cos} (\cosh \frac{|AC|}{2})^{-1}$, because this is the upper limit for the area of all possible triangles on base AC (Lemma 6.10.4). Thus

(2) $$\cos \gamma_2 > \frac{1}{\cosh \frac{|AC|}{2}} .$$

By the Sinus-theorem for hyperbolic functions (see E 6.16),

$$\sinh |AF| = \sinh |AC| \cdot \sin \gamma_2;$$

from this and $\cosh^2 x - \sinh^2 x = 1$ we get

$$\cos^2 \gamma_2 = 1 - \sin^2 \gamma_2 = 1 - \frac{\sinh^2 |AF|}{\sinh^2 |AC|} = \frac{\cosh^2 |AC| - \cosh^2 |AF|}{\cosh^2 |AC| - 1} .$$

Moreover, $\cosh^2 \frac{|AC|}{2} = \frac{\cosh |AC| + 1}{2}$ by E 6.14. Hence (2) is equivalent to

$$\frac{\cosh^2 |AC| - \cosh^2 |AF|}{\cosh^2 |AC| - 1} > \frac{2}{\cosh |AC| + 1} \iff$$

$$\cosh^2 |AC| - \cosh^2 |AF| > 2 \cosh |AC| - 2 \iff$$

$$\cosh |AC| > \sinh |AF| + 1.$$

Using the Cosinus-theorem for hyperbolic functions: $\cosh |AC| = \cosh |AF| \cdot \cosh |CF|$, see E 6.16, we obtain

(3) $$\cosh |CF| > \frac{\sinh |AF| + 1}{\cosh |AF|} .$$

Now if (1) and (3) are to be satisfied simultaneously we must have

$$\cosh |AF| > \frac{\sinh |AF| + 1}{\cosh |AF|} ,$$

which is equivalent (see E 6.15) to $\log(3 + 2\sqrt{2}) < |AB| = 2|AF|$. The points C on the perpendicular bisector of AB which satisfy the conditions (1) and (3) constitute a finite open interval when $|AB| > \log(3 + 2\sqrt{2})$. When $|AB| < \log(3 + 2\sqrt{2})$, (1) and (3) are not simultaneously satisfiable.

Now the area of the commutator surface is determined by A, B and C, and for fixed AB it is a monotonic decreasing function of $|CF|$. At the same time one sees from the condition $\cos \gamma_2 > (\cosh \frac{|AC|}{2})^{-1}$ that the supremum of the areas of all unfree commutator surfaces constructible on AB grows with the length of the base side $|AB|$.

Now let AB be given with $|AB| > \log(3+2\sqrt{2})$. Also let C be fixed in the open interval determined by AB (and with it the area of the unfree commutator surface). How can the point E be chosen? The necessary and sufficient condition on E is that the area of the triangle ACE is equal to $2\gamma_2$: $|ACE| = 2\gamma_2$. If one demands that E lies on the altitude on AC erected at G then by Lemma 6.10.4 this is only possible when G lies in an open interval around the midpoint of AC determined by $|AC|$ and γ_2. But if this condition is satisfied, then by Lemmas 6.10.1,2 there is exactly one point E on the altitude for which $|ACE| = 2\gamma_2$. On continuity grounds the geometric locus of all these points is a curve which leaves the upper half plane at two different points on the real axis and intersects each altitude on AC at most once. The set of possible points is therefore homeomorphic to an $\mathbb{R}^1$ when one topologizes it in a natural way corresponding to the topology of the space of canonical polygons from 6.9.1.

The above results may be summarized in

6.11.6 Lemma. *The areas of unfree commutator surfaces constructible on a segment* AB *constitute an open interval* (π,s), $s < 2\pi$. *Here* s *is a monotonic increasing function of* $|AB|$ *which approaches the value* 2π *asymptotically and takes the value* π *for* $|AB| = \log(3+2\sqrt{2})$. $(s \leq \pi \Leftrightarrow (\pi,s) = \emptyset)$. *For each constant* k *the space of unfree commutator surfaces with area greater than* k *is either empty or homeomorphic to* $\mathbb{R}^2$ *in the topology 6.11.2. If the area is required to equal* k, *then the space is either empty or homeomorphic to* $\mathbb{R}^1$.

□

6.11.7 *The free commutator surface.*

Like the unfree commutator surface, the free commutator surface satisfies $|s_1| = |s_3|$, $|s_2| = |s_4|$, but a weaker angle condition is imposed: the sum of the 5 interior angles must be smaller than π, i.e. the area of the free commutator surface must be greater than 2π.

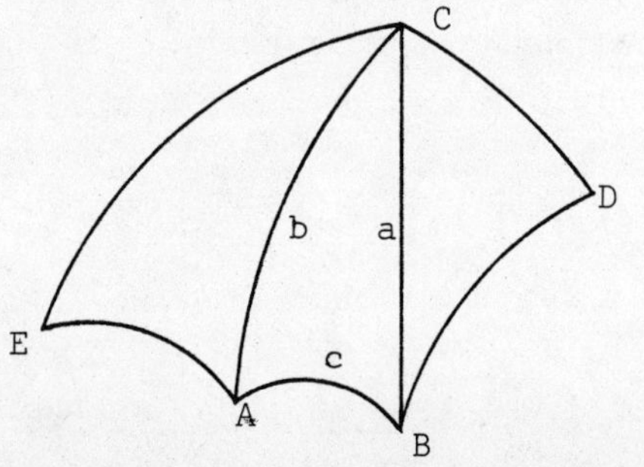

$|AE| = |CD|$

$|EC| = |DB|$

One easily sees that the supremum of the area of the commutator surface constructed on c also grows with $|c|$ here.

Now let AB = c be given. Then if C is also fixed the areas of the triangles ACE and BDC to be constructed on AC = b and BC = a satisfy

$$|ACE| + |BDC| < 2\text{arc cos}\frac{1}{\cosh\frac{|b|}{2}} + 2\text{arc cos}\frac{1}{\cosh\frac{|a|}{2}}.$$

On the other hand, with suitable choice of E (and D) one can bring $|ACE| + |BDC|$ arbitrarily close to this limit value, by e.g. letting E approach the boundary of the half-plane on the perpendicular bisector of AC. Thus if one moves C away from c on an altitude h_c then not only $|ABC|$, but also the supremum for $|ACE| + |BDC|$ increases.

Now we must clarify the relation between E and D. By the choice of E, a triangle CBD must be constructed on BC with side lengths $|CD| = |AE|$ and $|BD| = |CE|$. This is possible just in case $||AE| - |CE|| < |BC|$ holds as well as $|AE| + |CE| > |BC|$. Since in addition the orientation of the triangle is prescribed, there is exactly one point D satisfying the above conditions in this case.

We now fix A, B, C and move E away on an altitude from b. Then $|CE|$ and $|AE|$ increase while $||AE| - |CE|| = |\Delta|$ decreases (Lemma 6.10.2). Thus if there is a point E on this altitude for which it is possible to construct a triangle CBD on BC with $|BD| = |CE|$ and $|CD| = |AE|$, then it is also possible for all the points which follow, and $|ACE| + |CBD|$ grows monotonically as E moves away from AC (Lemmas 10.1,2). Now if one prescribes with A, B and C not an altitude, but rather the difference $\Delta = |AE| - |CE|$ between the side lengths, then the supremum of $|ACE| + |CBD|$ is a function F^* of $|a|$, $|b|$ and Δ and by Lemma 6.10.4:

$$F^*(|a|,|b|,\Delta) = 2\text{ arc cos}\frac{\cosh\frac{\Delta}{2}}{\cosh\frac{|a|}{2}} + 2\text{ arc cos}\frac{\cosh\frac{\Delta}{2}}{\cosh\frac{|b|}{2}}.$$

Since $\cosh\frac{\Delta}{2}$ is a strictly monotonically increasing function of $|\Delta|$ and arc cos is a strictly monotonically decreasing function on the above domain, F^* is a strictly monotonically decreasing function of $|\Delta|$.

Thus if one demands

$$|ACE| + |CBD| = K = \text{const} > \pi$$

for a given C, then the set of possible points E is either empty, or else constitutes a curve $\mathcal{E}$ which leaves the upper half-plane at two different places on the real axis and meets each altitude h_b at most once. (The proof is analogous to the proof of Lemma 6.11.6.) However, if one demands

$|ACE| + |CBD| > K = \text{const} > \pi$

then E can vary freely in the region between E and the real axis, so in the first case the space can be topologized as homeomorphic to $\mathbb{R}^1$, and in the second case as homeomorphic to $\mathbb{R}^2$.

Now suppose C is only required to lie on a fixed altitude h_c erected at F. If one moves C on h_c away from F then not only $|ABC|$, but also sup $(|ACE| + |CBD|)$, grows monotonically. The upper limits of both quantities depend on F and increase as F approaches the midpoint of AB. This allows C to vary in an $\mathbb{R}^2$ when $|ABC| + \sup(|ACE| + |CBD|)$ has to be greater than a fixed number and if such a C exists at all. For this reason the free commutator surfaces on AB with are $J > k$ (or $J = k$ respectively), where k is fixed and $3\pi > k > 2\pi$, constitute a space which is either empty or homeomorphic to $\mathbb{R}^4$ (or $\mathbb{R}^3$ respectively) in the topology from 6.11.2. Thus we have proved the following

<u>6.11.8 Lemma</u>. *When* k *is fixed and* $2\Pi < k < 3\pi$ *the space of free commutator surfaces with area* $J > k$ *(or* $J = k$ *respectively) constructible on a given side* AB *is either empty or homeomorphic to* $\mathbb{R}^4$ *(or* $\mathbb{R}^3$ *respectively) in the topology of 6.11.2. If the space is empty for some* k_o*, then it is also empty for any* $k > k_o$*. The infimum of all* k *for which the space is empty is a function which increases with* $|AB|$.

□

<u>6.11.9 *The commutator hexagon*</u>

The commutator hexagon consists of a triangle ABC and a free commutator surface placed on the side BC of the triangle.

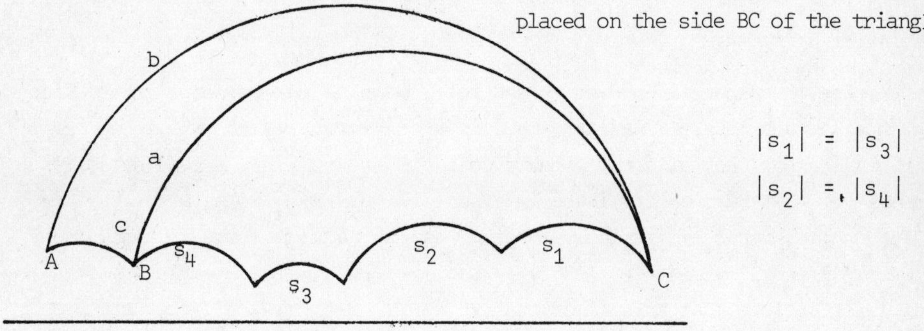

If one uses this polygon in order to construct the canonical polygon then the side $b = AC$ must be long enough to serve as base for a surface of sufficiently large area. If this is possible at all for the given AB then the induction yields the following situation: AB becomes the "free-edge" (cf. 6.11.10). The whole surface to be constructed on AB must have an area J_2 determined by the area J_1 of the surface previously constructed, namely by $J_1 + J_2 = 4(g-1)\pi$. If now the commutator hexagon has area $|H|$ then a surface of area $J_2 - |H|$ remains to be constructed on b. The lower limit for $|b|$ when AB is given proves to be a strictly monotonically decreasing function of $|H|$ and will be denoted by b_*, cf. 6.11.11 and Lemma 6.10.5.

Thus the problem regarding the commutator hexagon may be posed as follows: let AB be given, upon which a commutator hexagon H is to be constructed with $|AC| = |b| > b_*(|H|)$, where b_* is a continuous strictly monotonically decreasing function of $|H|$. Let h_c be an altitude on AB = c erected at D and C_o a point on h_c from which there is a commutator hexagon with base triangle ABC_o and $|b| > b_* = b_*(|H|)$. Then such commutator hexagons also exist for all points on h_c which are further removed from D than C_o. Namely, when C moves away from D on h_c, then $|a|$, $|b|$ and $|ABC|$ increase, and the upper limit of area of free commutator surfaces constructible on a also increases with $|a|$. Thus $|b|$ and the area supremum of all commutator hexagons with base triangle ABC increase with $|DC|$. In particular, for each C with $|DC| > |DC_o|$ there are commutator hexagons with area $|H|$ so large that $|b| > b_*(|H|)$.

How does this process depend on the point D at which the altitude is erected? By suitable choice of C one can make $|a|$ as well as $|b|$ arbitrarily large on each altitude. D therefore has no influence on the upper limit of area of all free commutator surfaces constructible on a, and imposes no bounds on $|a|$ or $|b|$. But the upper limit for the area of the base triangle ABC is dependent on D and grows monotonically as D approaches the midpoint of c, Lemmas 6.10.1,2 . Consequently, the set of all points C for which there is a commutator hexagon H with the base triangle ABC and $|b| > b_*(|H|)$ constitute a space which is either empty or homeomorphic to $\mathbb{R}^2$ in the topology of 6.11.2.

Now suppose this space is non-empty and let C be an arbitrary point in it. Then ABC and hence a and b(!) are fixed and there is a commutator hexagon H with $b_*(|H|) < |b|$. We now seek all free commutator surfaces on a whose area J satisfies the inequality $b_*(J+|ABC|) < |b|$. There is a positive number J_* with $b_*(J_*+|ABC|) = |b|$; for b_* is strictly monotone and continuous, there is a solution J of the above inequality and $b_*(J+|ABC|) \to \infty$ as $J \to 0$, because the angle sum in a free commutator surface must be smaller than π. We therefore seek all free commutator surfaces on a with area $J > J_*$. This space is homeomorphic to $\mathbb{R}^4$ by 6.11.8 since it is non-empty.

Since the possible C range over an $\mathbb{R}^2$, the commutator hexagons constitute a space which is either empty or homeomorphic to $\mathbb{R}^6$. This proves:

<u>6.11.10 Lemma</u>. *Let* $b_*: \{x \in \mathbb{R}: x > 0\} \to \{x \in \mathbb{R}: x > 0\} \cup \{\infty\}$ *be a continuous, strictly monotonically decreasing function and let* c = AB *be a given segment. Then the space of all commutator hexagons* H *on* c *with base triangle* ABC *and free commutator surface on* a = BC *whose area* $|H|$ *satisfies the condition*

$b_*(|H|) < |b| = |AC|$ *is either empty or else homeomorphic to* $\mathbb{R}^6$ *in the topology of 6.11.2.*

□

6.11.11 The space of canonical polygons.

Consider the figure in 6.11. 3 . Let the genus $g > 1$ be given. The vertices of the canonical polygon are denoted in order by $A_1,B_1,C_1,D_1,A_2,B_2,\dots,C_g,D_g,A_{g+1} = A_1$, where $A_1B_1C_1D_1A_2$ is the unfree commutator surface. Then for $1 < i \le g$, $A_iB_iC_iD_iA_{i+1}$ is a free commutator surface, while for $1 < i < g$, $A_1A_iB_iC_iD_iA_{i+1}$ is a commutator hexagon. We shall also denote the polygon $A_1A_iB_i \dots C_gD_g$ for $1 < i < g$ by Q_i. This is a (g-i+2)-gon with (g-i+1) free commutator surfaces attached. Let $Q_i^* = Q_i^*(|A_1A_i|)$ be the area supremum for all polygons of type Q_i on A_1A_i. Then it is easy to see that

$$Q_i^* = (4(g-i)+2)\pi + 2\operatorname{arc\,cos} \frac{1}{\cosh \frac{|A_1A_i|}{2}}$$

(cf. Lemma 6.10.4).

Q_i^* is a function which grows strictly monotonically with $|A_1A_i|$. The same holds for the *area supremum* $G^*(|A_1A_2|)$ of all unfree commutator surfaces constructed on A_1A_2 (Lemma 6.11.6). One begins the construction of the canonical polygon with the segment A_1A_2. Different positions of this segment in the non-euclidean plane must be regarded as equivalent, since they can be carried into each other by linear fractional transformations. However, segments of different lengths do not lead to congruent polygons.

The only condition imposed on A_1A_2 is that it permits the construction of a canonical polygon, which means that

$$G^* + Q_2^* = G^*(|A_1A_2|) + (4(g-2)+2)\pi + 2\operatorname{arc\,cos} \frac{1}{\cosh \frac{|A_1A_2|}{2}} > 4(g-1)\pi$$

i.e.
$$G^*(|A_1A_2|) + 2\operatorname{arc\,cos} \frac{1}{\cosh \frac{|A_1A_2|}{2}} > 2\pi \ .$$

The two terms on the left hand side of this inequality are strictly monotonically increasing functions of $|A_1A_2|$ whose sum becomes greater than 2π. (By Lemma 6.11.6 $\lim_{|A_1A_2|\to\infty} G^*(|A_1A_2|) = 2\pi$, moreover $\lim_{|A_1A_2|\to\infty} 2\operatorname{arc\,cos} \frac{1}{\cosh \frac{|A_1A_2|}{2}} = \pi$.)

On the other hand, we must have $|A_1A_2| > \log(3+2\sqrt{2})$ in order to be able to construct the unfree commutator surface on $|A_1A_2|$ (cf. 6.11.5). But now

$$\lim_{|A_1A_2|\downarrow \log(3+2\sqrt{2})} (G^*(|A_1A_2|) + 2\text{arc cos } (\cosh \frac{|A_1A_2|}{2})^{-1} = \frac{3\pi}{2} < 2\pi .$$

Because of the strict monotonicity of this function there is therefore a $k_* > \log(3+2\sqrt{2})$ such that a canonical polygon can be constructed on A_1A_2 just in case $|A_1A_2| > k_*$.[1] The set of all A_2 which permit the construction of a canonical polygon (with a fixed A_1, and a given half line on which A_2 is to lie) therefore constitute an $\mathbb{R}^1$.

The only condition on the unfree commutator surface G on the now fixed A_1A_2 is that it has area greater than

$$4(g-1)\pi - Q_2^*(|A_1A_2|) = 2 \quad - 2\text{arc cos } (\cosh \frac{|A_1A_2|}{2})^{-1}.$$

Since such an unfree commutator surface on A_1A_2 exists by construction of A_1A_2, the set of all these surfaces constitutes an $\mathbb{R}^2$ by Lemma 6.11.6.

The next step is the construction of a commutator hexagon $H_2 = A_1A_2B_2C_2D_2A_3$ on A_1A_2. (If g = 2 the last step already follows, see below). For this purpose $|A_1A_2|$ and $|G|$ must be such that Q_2 and thereby H_2 is constructible on A_1A_2, i.e. $Q_2^*(|A_1A_2|) + |G| > 2\pi$. But this step then corresponds exactly to the "induction step" in the construction of $H_i = A_1A_iB_iC_iD_iA_{i+1}$ $(1 < i < g-1)$ on A_1A_i where, by induction hypothesis, A_1A_i and the area $|P_i|$ of the surface $P_i = A_1B_1C_1D_1A_2 \ldots C_{i-1}D_{i-1}A_i$ already constructed satisfy the inequality $Q_i^*(|A_1A_i|) + |P_i| > 2\pi$.

By hypothesis it is possible to construct an H_i so that

$$Q_{i+1}^*(|A_1A_{i+1}|) + |H_i| + |P_i| > 2\pi .$$

This condition defines a function $b_{i*}(|H_i|)$ such that $|A_1A_{i+1}| > b_{i*}(|H_i|)$ is satisfies just in case

$$Q_{i+1}^*(|A_1A_{i+1}|) + |H_i| + |P_i| > 2\pi \quad \text{(cf. 6.11.9).}$$

As a result, b_{i*} is a continuous strictly monotonically decreasing function of $|H_i|$. Then by Lemma 6.11.10 the space of these hexagons H_i constitutes an $\mathbb{R}^6$, since it is not empty. The induction hypothesis is preserved by this construction, for $P_{i+1} = H_i \cup P_i$. After g-2 such steps the surface

[1] Thanks to a letter from Herr C.L. Siegel, k_* can be identified as $\log(9+4\sqrt{5})$.

$$P_g = A_1B_1C_1D_1A_2 \ldots A_g$$

is constructed and we have

$$Q_g^*(A_1A_g) + |P_g| > 2\pi.$$

We then seek all free commutator surfaces of fixed area $2\pi - |P_g|$ on A_1A_g. Since the existence of one such surface is secured by Lemma 6.11.8, these surfaces constitute an $\mathbb{R}^3$.

The space of canonical polygons is homeomorphic to

$$\mathbb{R}^1 \times \mathbb{R}^2 \times (\mathbb{R}^6)^{g-2} \times \mathbb{R}^3 = \mathbb{R}^{6g-6}$$

because one obtains successive fiberings over $\mathbb{R}^6$.

This finishes the proof of theorem 6.9.2.

□

6.12 AUTOMORPHISMS OF FINITE ORDER

In this section we prove a theorem about finite mapping classes.

6.12.1 Theorem. *Let G be a discontinuous group of the non-euclidean plane with compact fundamental domain, and let* $\alpha: G \to G$ *be an automorphism for which* α^n, $n \geq 2$ *is an inner automorphism. Then there is a homeomorphism* $\zeta: \mathbb{H} \to \mathbb{H}$ *such that* $\zeta g \zeta^{-1} = \alpha(g)$ *and* $\zeta^n \in G$. *In particular, for closed surfaces each mapping class of finite order contains a homeomorphism of finite order.*

The latter assertion is proved in [Nielsen 1942]. Here we follow the proofs in [Fenchel 1948, 1950] and [Macbeath 1962] and confine ourselves, as they do, to the case which deals with the fundamental group of a closed orientable surface and where a prime power p^m of α is an inner automorphism. From this the full theorem is obtained by induction, using the fact that the Teichmüller space of planar discontinous groups is also some euclidean space. For stronger results see 5.15.10.

A mapping of the space of marked Riemann surfaces of genus g onto itself is defined by $A: (R,\Sigma) \to (R,\alpha\Sigma)$. Since the generator system $\alpha\Sigma$ may be computed from Σ

independently of the Riemann surface, "$\alpha\Sigma$ varies continuously with Σ"; since α^{p^m} is an inner automorphism, $[R,\alpha^{p^m}\Sigma]$ equals $[R,\Sigma]$. As a result we obtain a mapping of $\mathbb{R}^{6g-6}$ onto itself of prime power order. Because of the fixed point theorem of [Smith 1934], A has a fixed point: $[R_o,\Sigma_o] = [R_o,\alpha\Sigma_o]$. The passage from the generator system Σ_o of R_o to $\alpha\Sigma_o$ may therefore be achieved by a conformal mapping $\varphi: R_o \to R_o$, and φ^{p^m} is isotopic to the identity. But there is only one conformal mapping in the isotopy class of the identity: the identity itself, cf. Corollary 6.7.18 to Theorem 6.7.16 or the following proof).

□

That a conformal mapping isotopic to the identity is the identity also follows like this: the fixed points of such a mapping different from the identity must lie discretely and must all have index 1. Therefore the number of fixed points is equal to the algebraic fixed point number, and hence given by the Lefschetz number

$\sum_{i=o}^{2} (-1)^i$ trace φ_i, where φ induces the mapping φ_i in the homology group $H_i(R, Z)$, see e.g. [Alexandroff-Hopf 1935, Ch. XIV, § 3, Theorem 1]. Since in our case all the φ_i are identity mappings we obtain the contradiction

$$0 \leq \text{number of fixed points} = 2 - 2g < 0.$$

□

Another more combinatorial group theoretical proof has been given in 4.16.

Additional literature to chapter 7:

[Appell-Goursat 1930], [Bailey 1961], [Ballmann 1978], [Cohen 1974], [Greenberg 1960], [Greenberg 1967], [Natanson 1972], [Herman 1973], [Ignatov 1978],[Jørgensen 1978], [Keen 1971], [Kneser 1928], [Larcher 1963], [Macbeath-Singerman 1975], [Maclachlan 1977], [Magnus 1972], [Magnus 1975], [Marden 1969], [Purzitsky 1974], [Purzitsky-Rosenberger 1972], [Rosenberger 1972], [Rosenberger 1973], [Scherrer 1929], [Stothers 1977], [Zarrow 1979], [Zieschang 1971], [Zieschang 1974], [Zimmermann 1977].

EXERCISES

E 6.1 (a) Each linear fractional transformation is a product of the transformations $w = z + b$, $w = -\frac{1}{z}$, $w = az$ $(a \neq 0)$. If the transformation has real ceofficients then a, b can be restricted to real numbers.
(b) Proof of 6.3.2.

E 6.2 Let $f, g \in SLF(2, \mathbb{R})$ be commuting elements different from the identity. Then both are of the same type and have the same fixed points (in $\bar{\mathbb{C}}$).

E 6.3 (a) Verify that the arc element from 6.3.7 and the distance from 6.3.3 correspond to each other.
(b) Prove that the non-euclidean segment PQ has the smallest (noneuclidean) length among all curves joining P and Q.

E 6.4 Let $f \in SLF(2, \mathbb{R})$ be hyperbolic and let A be the non-euclidean line connecting the two fixed points of f, the so-called *axis of* f. Prove that $\delta(P) = d(P,f(P))$ is a differentiable function of P which takes its minimal values on A. (In fact, $\delta(P)$ is a strictly monotonic increasing function of the (non-euclidean) distance of P to A.)

E 6.5 Let P_0, P_1 and Q_0, Q_1 be points of the upper halfplane $\mathbb{H}$ with $d(P_0,P_1) = d(Q_0,Q_1) \neq 0$. Prove that there is exactly one transformation in $SLF(2, \mathbb{R})$ mapping P_i to Q_i.

E 6.6 Proof of 6.3.8.

E 6.7 Proof of 6.4.3.

E 6.8 Prove the Riemann-Hurwitz formula 4.14.22 using the euclidean or non euclidean-geometry.

E 6.9 Determine the discontinuous groups of orientation preserving motions of the sphere and the euclidean plane, and give geometric descriptions of fundamental domains, generators and defining relations etc.

E 6.10 Give a detailed proof of 6.5.3 and 6.5.4.

E 6.11 Prove that the group SLF(2,$\mathbb{Z}$) of linear fractional transformations with integer coefficients is generated by

$$f,\ z \mapsto -\frac{1}{z},\ \text{and } g,\ z \mapsto \frac{1}{-z+1},$$

and that $SLF(2,\mathbb{Z}) = \langle f,g \mid f^2, g^3 \rangle \cong \mathbb{Z}_2 * \mathbb{Z}_3$.

Solve the analogous problem for the group SL(2,$\mathbb{Z}$) of integer 2-by-2 matrices with determinant 1. (Hint: Use the projection $SL(2,\mathbb{Z}) \to SLF(2,\mathbb{Z})$, $\begin{pmatrix} a & b \\ c & d \end{pmatrix} \mapsto w = \frac{az+b}{cz+d}$.)

E 6.12 Prove 6.6.11.

E 6.13 Let $f \in SLF(2,\mathbb{R})$ and $P \in \mathbb{H}$ be such that P, $f(P)$, $f^2(P) \neq P$ are on one (non-euclidean) line A. Then f is a hyperbolic transformation, maps A into itself and has the ends of A as fixed points. (A is the axis of f, see E 6.4.)

E 6.14 Prove $\sinh\frac{a-b}{2} \cdot \sinh\frac{a+b}{2} = \frac{1}{2}(\cosh a - \cosh b)$

$$\cosh\frac{a}{2} \cdot \cosh\frac{b}{2} = \frac{1}{2}(\cosh(a+b) + \cosh(a-b))$$

and describe the relation to formulas for the trigonometric functions sin and cos.

E 6.15 Prove that for $a > 0$ the inequality $(\cosh a)^2 > \sinh a + 1$ implies $\log(3 + 2\sqrt{2}) < 2a$.

E 6.16 Let ABC be a triangle in the hyperbolic plane with the angles α, β and $\frac{\pi}{2}$ at the vertices A, B and C, resp. The lengths of the sides AB, BC and CA are c, a and b, resp. Then:

(a) $\sinh a = \sinh c \cdot \sin\alpha$ (Sinus-theorem)

(b) $\cosh c = \cosh a \cdot \cosh b$ (Cosinus-theorem).

E 6.17 Determine the automorphisms of finite order of the fundamental group of the torus and prove the theorem corresponding to 6.12.1. (Hint: see E 6.11).

E 6.18 Determine the finite groups of mapping classes of the torus and the Klein bottle.

7. ON THE TOPOLOGICAL THEORY OF SURFACES

In this chapter we will prove that any topological surface is triangulable and that triangulations of homeomorphic surfaces are equivalent (a positive answer to the Hauptvermutung). Fundamental for our proof is the Schönflies theorem 7.4.1, a strengthening of the Jordan curve theorem which is well known from the books on algebraic topology. The proof of the Schönflies theorem is taken from [Newman 1951] with minor changes where we assumed some results of algebraic topology which Newman proves from scratch. (His book gives an excellent introduction to homology theory and its application in geometric topology.) The proof of the triangulation theorem is a copy of the original one in [Rado 1924].

7.1 HOMOLOGICAL PROPERTIES OF SUBSETS OF THE PLANE

We use singular homology with coefficients in $\mathbb{Z}_2$. The sphere S^2 is the Riemann sphere $\mathbb{R}^2 \cup \{\infty\}$. The following Jordan curve theorem is proved in all books on algebraic topology as an indicator of the strength of homology theory, see, for instance, [Spanier 1966, 4.8.15]. For a list of elementary proofs see [Dostal-Tindell 1978].

7.1.1 Theorem (Jordan Curve Theorem). *Let γ be a simple closed curve in S^2 (or $\mathbb{R}^2$). Then $S^2 \setminus |\gamma|$ (resp. $\mathbb{R}^2 \setminus |\gamma|$) consists of two path-connected open sets, which have γ as their common boundary.*

□

The following proposition is also well known:

7.1.2 Proposition. *Consider homology resp. cohomology with $\mathbb{Z}_2$ coefficients. Let X be a space. The points of X are identified with 0-simplexes, and 0-chains are then finite formal sums $x_1 + \ldots + x_n$, $x_i \in X$. Paths in X correspond to 1-chains.*

(a) Two points $x, y \in X$ lie in the same (path connected) component of X just in case $x + y \sim 0$.

(b) $H_o(X) := H_o(X;\mathbb{Z}_2)$ is isomorphic to $\mathbb{Z}_2^n$ $(n \geq 1)$ when X consists of n path-connected components.

(c) If X consists of n path-connected components, then $H^o(X)$ is isomorphic to $\mathbb{Z}_2^n$. The cohomology classes of $H^o(X)$ are represented in a unique way by the functions $X \to \mathbb{Z}_2$ which are constant on the components of X.

(d) *Let A be a subset of X, and $i: A \hookrightarrow X$ its injection. If A and X are path connected then $i^*: H^o(X) \to H^o(A)$ is an isomorphism.*

□

7.1.3 Lemma ([Alexander 1922]). *(a) Let* F_1, F_2 *be compact subsets of* S^2 *and* $x, y \in S^2 \setminus (F_1 \cup F_2)$. *Also suppose that for* $i = 1,2$ *there is a path* κ_i *in* $S^2 \setminus F_i$ *which connects* x *and* y, *and let* $\kappa_1 + \kappa_2 \sim 0$ *in* $S^2 \setminus (F_1 \cap F_2)$. *Then* $x \sim y$ *in* $S^2 \setminus (F_1 \cup F_2)$.

(b) The analogous holds when S^2 *is replaced by* $\mathbb{R}^2$. (Here the assumption of compactness for the F_i becomes necessary.)

Proof. (a) Without loss of generality we can assume that $\infty \notin F_1 \cup F_2$ and that κ_1 and κ_2 do not contain the point ∞, so that they lie in $\mathbb{R}^2$, and that they are polygonal paths. And in fact we can find a grid of lines parallel to the real and imaginary axes such that x and y are vertices of the grid and the paths κ_i run along its edges. This can always be achieved by replacing the original κ_i by a new curve $\bar{\kappa}_i$ which is homologous to κ_i in $S^2 \setminus F_i$ (only a "small" deformation is necessary for this, cf. [Spanier 1966, 3.3-4]). In particular we also have $\bar{\kappa}_1 + \bar{\kappa}_2 \sim 0$ in $S^2 \setminus (F_1 \cap F_2)$. We can therefore assume that the κ_i are already on the grid. Then $\kappa_1 + \kappa_2 \sim 0$ in $S^2 \setminus (F_1 \cap F_2)$ says that there is a 2-chain c_2 of rectangles of the grid with $\partial c_2 = \kappa_1 + \kappa_2$. (Here again we identify certain point sets with chains, as is usual and convenient in simplicial or cellular homology theory. Since we have $\mathbb{Z}_2$ coefficients, we need not take the trouble to give a more precise description.)

Let $|c_2|$ be the union of the rectangles of the grid which have coefficient 1 in c_2. Then

$$|c_2| \cap F_1 \cap F_2 = \emptyset.$$

Therefore there is a refinement of the grid, no cell of which meets both $|c_2| \cap F_1$ and $|c_2| \cap F_2$, because these two sets are disjoint and compact. We now consider c_2 as a sum of rectangles of this refined grid. Let $\tilde{c}_2$ be the sum of all rectangles which appear in c_2 and which meet F_1.

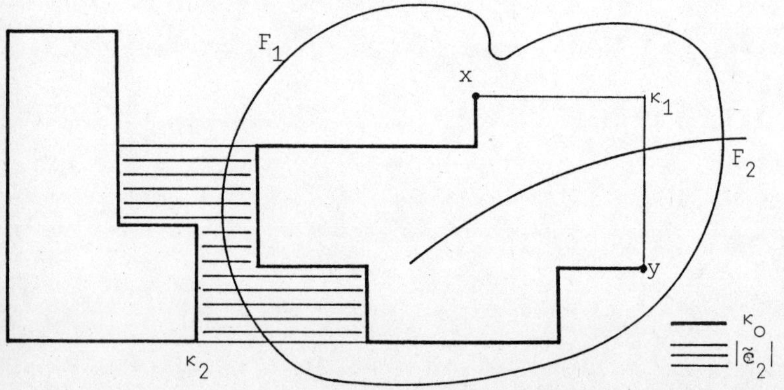

Let $\kappa_o = \kappa_2 + \partial\tilde{c}_2$. Then $\partial\kappa_o = \partial\kappa_2 = x+y$. We now show

(1) $$|\kappa_o| \cap (F_1 \cup F_2) = \emptyset$$

Altogether we then have $x+y \sim 0$ in $S^2 \setminus (F_1 \cup F_2)$, and hence assertion (a). By choice of κ_2,

$$|\kappa_2| \cap F_2 = \emptyset$$

Each rectangle of $\tilde{c}_2$ meets F_1 and hence, by choice of the refined grid, no rectangle of $\tilde{c}_2$ meets F_2, so

$$|\partial\tilde{c}_2| \cap F_2 \subset |\tilde{c}_2| \cap F_2 = \emptyset$$

and hence, since $|\kappa_2 + \partial\tilde{c}_2| \subset |\kappa_2| \cup |\partial\tilde{c}_2|$,

(2) $$|\kappa_o| \cap F_2 = |\kappa_2 + \partial\tilde{c}_2| \cap F_2 = \emptyset.$$

$|c_2 + \tilde{c}_2|$ does not meet F_1, because each retangle of c_2 which meets F_1 also appears in $\tilde{c}_2$, and hence not in $c_2 + \tilde{c}_2$. From this it follows that

$$\emptyset = F_1 \cap |\partial(c_2 + \tilde{c}_2)| = F_1 \cap |\kappa_1 + \kappa_2 + \partial\tilde{c}_2|.$$

By the hypothesis on κ_1 we have

$$\emptyset = F_1 \cap |\kappa_1|.$$

These two relations imply

(3) $$|\kappa_o| \cap F_1 = \emptyset;$$

because $|\kappa_o| = |\kappa_2 + \partial\tilde{c}_2| = |\kappa_1 + (\kappa_1 + \kappa_2 + \partial\tilde{c}_2)| \subset |\kappa_1| \cup |\kappa_1 + \kappa_2 + \partial\tilde{c}_2|$. Clearly, (2) and (3) imply (1).

(b) follows immediately from (a) when one completes $\mathbb{R}^2$ by ∞ to S^2. The κ_o constructed in the proof does not contain ∞.

□

<u>7.1.4 Lemma</u>. *Let* A *be a compact path connected subset of* S^2. *Then* $H_1(S^2 \setminus A) = 0$.

Proof. To the exact homology sequence of the pair $(S^2, S^2 \setminus A)$ we apply the Alexander duality theorem (in a slightly generalized form) cf. [Spanier 1966, 6.2.17]:

$$\begin{array}{ccccccccccc}
 & \mathbb{Z}_2 & & & & & & 0 & \\
\to H_2(S^2\setminus A) & \to & H_2(S^2) & \to & H_2(S^2, S^2\setminus A) & \to & H_1(S^2\setminus A) & \to & H_1(S^2) \to \\
\bar{\gamma}_U \downarrow \cong & & \bar{\gamma}_U \downarrow \cong & & \bar{\gamma}_U \downarrow \cong & & \bar{\gamma}_U \downarrow \cong & & \bar{\gamma}_U \downarrow \cong \\
\to \bar{H}^0(S^2,A) & \to & \bar{H}^0(S^2) & \xrightarrow{i^*} & \bar{H}^0(A) & \to & \bar{H}^1(S^2,A) & \to & \bar{H}^1(S^2) \to
\end{array}$$

Here we have $H_2(S^2) = \mathbb{Z}_2$, $H_1(S^2) = 0$. Moreover $\bar{H}^i(S^2) = H^i(S^2)$. (By definition $\bar{H}^i(X) = \varinjlim H^i(U)$ $(X \subset S^2)$, where U is a neighbourhood of X in S^2.) Similarly $\bar{H}^0(A) = H^0(A) = \mathbb{Z}_2$, because

$$\bar{H}^0(A) = \varinjlim \{H^0(U) \mid U \text{ a neighbourhood of } A\} .$$

Since A is compact and path connected, each neighbourhood U of A contains a path connected open neighbourhood U' such that $\bar{U}'$ is compact and $\bar{U}' \subset U$. Thus the neighbourhoods with the latter property constitute a cofinal subsystem, hence $\bar{H}^0(A) = \varinjlim \{H^0(\bar{U}') \mid \bar{U}' \text{ compact path connected neighbourhood of } A\}$. But here we have $H^0(\bar{U}') \cong \mathbb{Z}_2$ and for $U' \overset{j}{\hookrightarrow} U_1'$ we have $j^*: H^0(U_1') \to H^0(U')$ an isomorphism by 7.1.2 (d). Moreover $j^*: H^0(S^2) \to H^0(\bar{U}')$ is always an isomorphism. Thus $\bar{H}^0(A) = H^0(A) = \mathbb{Z}_2$ and i^* is an isomorphism.

The bottom row of the diagram therefore splits:

$$\bar{H}^0(S^2) \underset{\cong}{\to} \bar{H}^0(A) \to \bar{H}^1(S^2,A) \to 0$$
$$\searrow \; 0 \; \nearrow$$

It follows that $\bar{H}^1(S^2,A) = 0$, i.e. $H_1(S^2\setminus A) = 0$.

□

7.1.5 Proposition. (a) *Let* F_1 *and* F_2 *be compact subsets of* S^2. *Let* $x,y \in S^2\setminus(F_1 \cup F_2)$ *satisfy* $x \sim y$ *in* $S^2\setminus F_i$ *for* $i = 1,2$. *Moreover, let* $F_1 \cap F_2$ *be path connected. Then* $x \sim y$ *in* $S^2\setminus(F_1 \cup F_2)$.
(b) *The analogous statement holds when* S^2 *is replaced by* $\mathbb{R}^2$.

Proof. Follows from 7.1.4 by 7.1.3.

□

To this section see [Newman 1951, V. § 2].

Exercises: E 7.1-6

7.2 LOCAL PATH CONNECTIVITY

In the proof of the Schönflies theorem we must contract paths. Therefore we are concerned with the existence of paths in neighbourhoods of points.

7.2.1 Definition. Let E be a subset of a topological space X, a ∈ X a point. E is called *locally path connected at* a (lc) when each neighbourhood U of a contains a neighbourhood V of a such that any two points of V ∩ E are connected by a path in U ∩ E.

Trivially, a set E is locally path connected at all points of $X\setminus\bar{E}$. If X is a manifold and E is open, then E is obviously locally path connected at all points of E. For such sets, therefore, the question of local connectivity is interesting only for boundary points. The following sketch shows that local path connectivity need not hold even for "nice" sets: E is the open square minus the segment [a,b]. At each point c of the half-open segment [a,b[, E is not locally path connected.

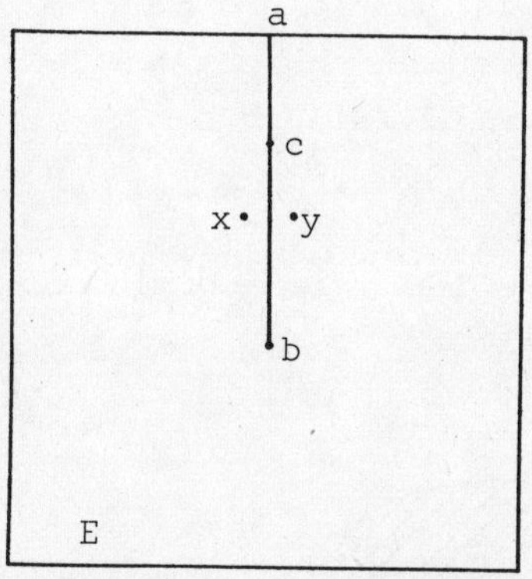

7.2.2 Definition. A metric space E is called *uniformly locally path connected* (ulc) when for each $\varepsilon > 0$ there is a $\delta > 0$ such that for any two points x, y ∈ E with $d(x,y) < \delta$ in E there is a path ω from x to y with $d(\omega) < \varepsilon$.

The set E in the above figure, under the metric induced by $\mathbb{R}^2$, is not ulc.

7.2.3 Proposition. *Let* E *be a subset of a metric space. Then:*
(a) If E *is* ulc, *then* E *is* lc *at all points of* $\bar{E}$.
(b) If E *is* lc *at all points of* $\bar{E}$ *and* $\bar{E}$ *is compact, then* E *is* ulc.

Proof. (a) Suppose that $E \neq \emptyset$ and that $a \in \bar{E}$. It suffices to test local path connectivity for the basis of neighbourhoods $U(a,\varepsilon) = \{x \in E \mid d(x,a) < \varepsilon\}$. Given $\varepsilon > 0$ we take the $\delta > 0$ which corresponds to $\varepsilon/2$ in the definition of ulc for E. Now let $x,y \in U(a, 1/2\ \delta)$. Then $d(x,y) < \delta$, and thus there is a path ω in E, connecting x and y, with $d(\omega) < \varepsilon/2$. Therefore ω lies in $U(a,\varepsilon) \cap E$.

(b) Assume that E is not ulc. Then there is an $\alpha > 0$ and a sequence of point pairs x_n, y_n with $d(x_n,y_n) \to 0$, such that x_n and y_n cannot be connected by a path of diameter $< \alpha$. Since $\bar{E}$ is compact, there is a subsequence x_{n_i} which converges to a point $a \in \bar{E}$. Then y_{n_i} likewise converges to a. Now since each set $E \cap U(a,\delta)$, $\delta > 0$, contains points x_{n_i}, y_{n_i}, E is not at a; namely, let U be the $\alpha/2$-neighbourhood of a.

□

7.2.4 Corollary. *Let* E *be a compact metric space. If* E *is* lc *at all points of* E, *then* E *is* ulc.

□

7.2.5 Corollary. *Let* X *be compact and* lc *at all its points. Let* $E \subset X$ *be open. Then:* E *is* ulc $\Leftrightarrow$ E *is* lc *at all points of* $\partial E = \bar{E} \setminus E$.

Proof. "$\Rightarrow$" follows from 7.2.3 (a). Now for "$\Leftarrow$". Since X is lc at all points, E is lc at all points of $\mathring{E}$. By hypothesis this also holds at the points of ∂E, and hence the assertion follows from 7.2.3 (b).

□

The lc-property is a topological property of the pair (X,E), however the ulc-property is not, as the figure shows:

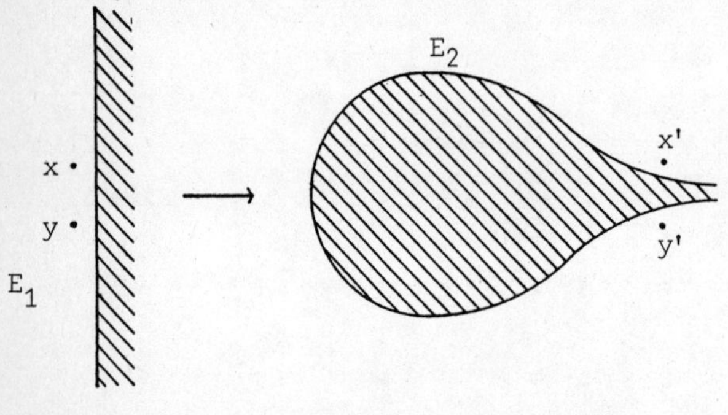

The $E_i \subset \mathbb{R}^2$ are the unshaded sets, with topologically equivalent embeddings. Of course, ulc is invariant under isometries.

7.2.6 Proposition. *Let* J *be a simple closed curve in* S^2 *or* $\mathbb{R}^2$ *and let* D *be one of the two path components of* $S^2 \setminus J$ *(resp.* $\mathbb{R}^2 \setminus J$*). (*D *is also called a Jordan domain.) Then:*

(a) D *is* lc *at all points of* $J = \partial D$.

(b) D *is* ulc.

Proof. If J lies in S^2 we can arrange that $\infty \notin J$. We now show that D is lc at all points of J. Let $a \in J$, $\varepsilon > 0$. Let L_1 be a closed arc on J, with $L_1 \subset U(a,\varepsilon)$, which has a as an interior point, see the figure. Let $\delta = d(a, J\setminus L_1)$. We now show:

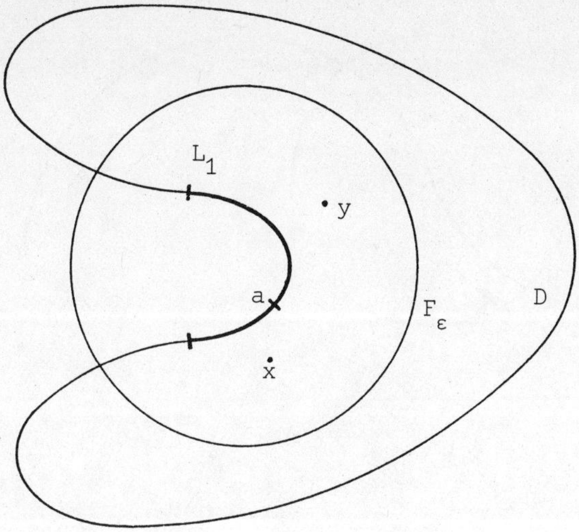

(1) *Two points* $x,y \in D \cap U(a,\delta)$ *may be connected by a path in* $D \cap U(a,\varepsilon)$.

To prove this let $F_\varepsilon = \partial U(a,\varepsilon) = \{x \in \mathbb{R}^2 \mid d(a,x) = \varepsilon\}$. Also, let ω be a path in $U(a,\delta)$ which connects x and y. Then

$$\omega \cap F_\varepsilon = \emptyset$$
$$\omega \cap \overline{J\setminus L_1} = \emptyset$$

i.e. $x \sim y$ in $\mathbb{R}^2\setminus(F_\varepsilon \cup \overline{J\setminus L_1})$.

Since $x,y \in D$, $x \sim y$ also holds in $\mathbb{R}^2\setminus J$. We have that $J \cap (F_{\varepsilon_2} \cup \overline{J\setminus L_1}) = \overline{J\setminus L_1}$ is path connected. By 7.1.5, $x \sim y$ in $\mathbb{R}^2\setminus(F_\varepsilon \cup \overline{J\setminus L_1} \cup J) = \mathbb{R}^2\setminus(F_\varepsilon \cup J)$. This implies that x and y can be connected by a path in $D \cap U(a,\varepsilon)$, i.e. (1) holds.

When $D \cup U = \bar{D}$ is compact, the proposition follows from (1) and 7.2.4. If D is the infinite component of $\mathbb{R}^2\setminus J$, we choose a circle K so large that J lies in the open disk determined by K. Let $\alpha = d(J,K)$; we have $K \subset D$. The region $\bar{D}_0$ between K and J is compact, so that $\bar{D}_0$ is ulc. Then, corresponding to the definition 7.2.2, let $\delta(\varepsilon) > 0$ belong to $\varepsilon > 0$. We extend this to D by taking $\delta' = \min\{\delta(\varepsilon),\alpha\}$ for the given $\varepsilon > 0$.

□

Analogously, one shows

7.2.7 Proposition. *The complement of a simple arc* L *in* S^2 *or* $\mathbb{R}^2$ *is* lc *at its endpoints*. □

This does not hold for interior points of L. as the following sketch shows:

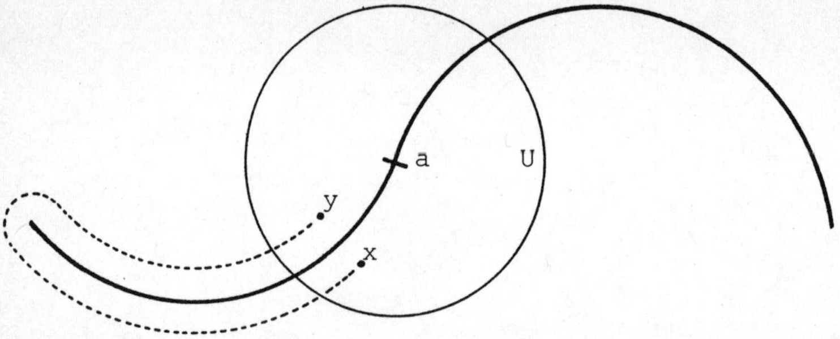

For this section see [Newman 1951, VI § 4].

7.3 CONSTRUCTION OF A CROSSCUT

In this section we find a way to reach a point x on the boundary of a Jordan domain by means of a curve which lies in the interior of the domain except at its endpoint x.

7.3.1 Definition. Let D be a domain in $\mathbb{R}^2$.
(a) A simple curve γ is called an *end cut* of D when an endpoint of γ lies on the boundary of D, but all other points lie inside D.
(b) A simple curve λ is called a *crosscut* of D when both endpoints of λ lie on the boundary of D, but all other points lie inside D.
(c) A point x on the boundary of D is called *accessible from* D when x is the endpoint of an end cut.

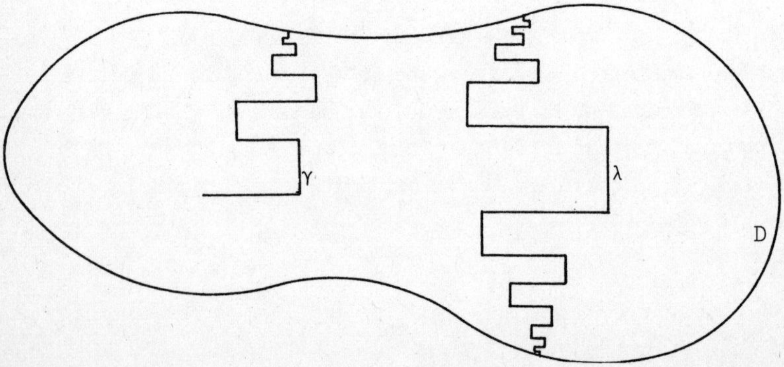

(d) An arc γ is called *quasilinear* when γ is the union of at most denumerably many segments and each point of γ with the possible exception of the endpoints is either in the interior of one of these segments or in the boundary of two of them. In other words, each closed proper subarc of γ consists of finitely many segments.

Accessibility of points is a topological property of the pair $(D,\partial D)$. The following gives an example of an inaccessible point:

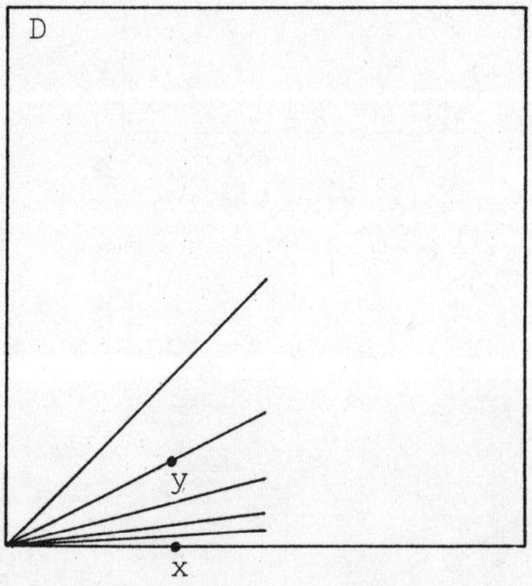

y accessible,
x not accessible

<u>7.3.2 Lemma</u>. *Let* D *be a domain in* $\mathbb{R}^2$ *or* S^2 .

(a) The points of ∂D *accessible from* D *are dense in* ∂D.

(b) If a point $a \in \partial D$ *is accessible, then it is accessible by quasilinear arcs. One can arrange that the segments of the arcs are alternately parallel to the real and imaginary axes.*

(c) If $a \in \partial D$ *is accessible, and* $b \in D$*, then* D *contains a quasilinear end cut from* b *to* a.

(d) If $a,b \in \partial D$ *are accessible, then* D *contains a quasilinear crosscut from* a *to* b.

In (c) and (d) one can also arrange that all segments are parallel to the real or imaginary axes.

Proof. (a) Suppose we have $\varepsilon>0$ and $a \in \partial D$, $x \in U(a,\varepsilon) \cap D$. Let γ be either the segment $[x,a]$ or an ${}^{x}\mathsf{L}_{a}$ consisting of one segment parallel to the real axis and one parallel to the imaginary axis. The first point of γ, after x, which lies in ∂D is accessible and lies in $U(a,\varepsilon)$.

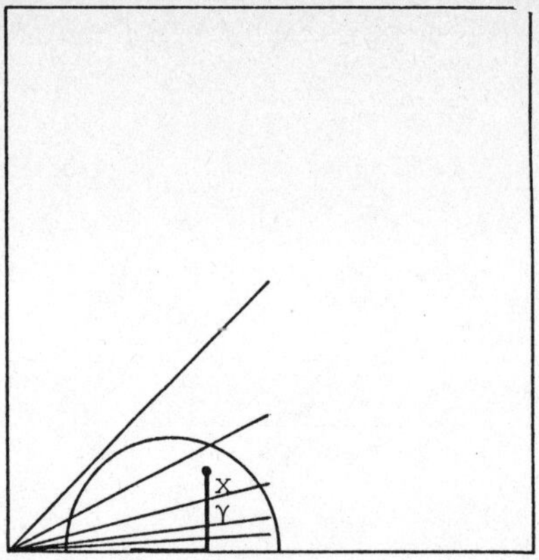

(b) If γ is an endcut to a, then $\gamma\setminus\{a\}$ may be covered by a system of circular disks which lie in D. Inside them, one can alter γ into the desired form.

(c) Let γ be an endcut of D to the point a. Let the other endpoint of γ be x. Then $x \in D$ and x may be connected to b by a simple arc λ. Now let y be the first point of γ, after a, which lies on λ. Let γ_1 be the subarc of γ between a and y, λ_1 the subarc of λ between y and b. Then $\gamma_1 \cup \lambda_1$ is an endcut from b to a.

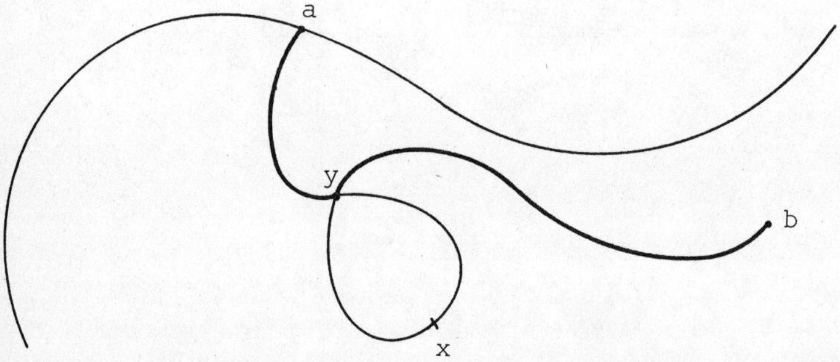

(d) follows analogously.

□

7.3.3 Proposition. *Let* D *be a domain in* S^2 *or* $\mathbb{R}^2$. *If* D *is* lc *at* $a \in \partial D$, *then* a *is accessible from* D.

The proof uses

7.3.4 Lemma. *Let* $a \in \mathbb{R}^2$ *and let* A *be the union of a sequence of segments* $[x_n, y_n]$ *with* $\lim_{n\to\infty} x_n = \lim_{n\to\infty} y_n = a$. *Suppose that* a *lies in the same path component of* $\bar{A}$ *as the point* $b \in A, b \neq a$. *Then there is a simple arc in* $\bar{A}$ *with endpoints* a *and* b.

Proof. Let $C_n = \{x \in \mathbb{R}^2 \mid d(x,a) = \frac{1}{2n} d(b,a)\}$ be the circle with radius $\frac{1}{2n} d(b,a)$ and center a. Since only finitely many of the segments $[x_n, y_n]$ meet the region outside C_n, we use subdivision to arrange that outside of any C_n two segments have at most an endpoint in common, and a common point of a segment and a circle C_n is an endpoint of the segment. Now let A_n be the part of A which lies outside or on C_n. Since a and b lie in the same path component of $\bar{A}$, there is a simple arc J_n in A_n which leads from b to a point on C_n. Thus $J_n \cap C_n$ is a point. Since only finitely many segments of A lie outside C_1, there is a simple arc J_1^* which is the initial arc of infinitely many of the J_n. Let the endpoint of J_1^* equal b_1, so that $J_1^* \cap C_1 = b_1$. Infinitely many of these arcs have the same beginning J_2^* from b to a point $b_2 \in C_2$. We have that $J_2^* \setminus \{b_2\}$ is disjoint from C_2. Thus for each n we find a simple arc J_n^* in A with endpoints b and $b_n = C_n \cap J_n^*$; moreover $J_n^* \subset J_{n+1}^*$. Let $b_0 = b$.

There is a homeomorphism φ of the subarc of J_n^* between b_n and b_{n-1} onto the interval $[\frac{1}{2^n}, \frac{1}{2^{n-1}}]$ with $\varphi(b_n) = \frac{1}{2^n}$ These homeomorphisms can be combined into a homeomorphism $\varphi\colon \bigcup_{n=1}^{\infty} J_n^* \to]0,1]$. If one sets $\varphi(a) = 0$, the result is a homeomorphism $\varphi\colon \{a\} \cup \bigcup_{n=1}^{\infty} J_n^* \to [0,1]$, which is the desired arc.

□

Proof of 7.3.3. We choose $\delta_n > 0$ so that any two points $x,y \in D \cap U(a,\delta_n)$ can be connected by a polygonal path in $D \cap U(a,\frac{1}{n})$. Let $x_n \in U(a,\delta_n) \cap D$ and let L_n be a polygonal path from x_n to x_{n+1} in $D \cap U(a,\frac{1}{n})$. Let $A = \bigcup_{n=1}^{\infty} L_n$. We can apply 7.3.4 to A. (To do this, we decompose the polygonal paths into segments.) Thus there is a polygonal path in $\bar{A} = A \cup \{a\}$ which connects $x_1 \in A$ with a and which, with the exception of its endpoint a, lies entirely in $A \subset D$. This is an endcut of D to a.

□

From 7.2.6 (a) we have

<u>7.3.5 Corollary.</u> *The points of a Jordan curve are accessible from both sides.*

□

From 7.2.7, 7.3.3 and 7.3.2 (d) we have

<u>7.3.6 Corollary.</u> *Each simple path in $\mathbb{R}^2$ or S^2 is a subarc of a simple closed curve in $\mathbb{R}^2$ or S^2.*

□

Now we come to the main result of this section.

<u>7.3.7 Theorem.</u> *Let* J *be a simple closed curve in* $\mathbb{R}^2$ *or* S^2, d *a metric on* $\mathbb{R}^2$ *resp.* S^2. *Let* L *be an arc on* J *with endpoints* a *and* b, *and let* D *be one of the path components of* $\mathbb{R}^2 \setminus J$ *(resp.* $S^2 \setminus J$*). Then:*

(a) For each $\varepsilon > 0$ *there is a crosscut* K *in* D *from* a *to* b *with* $d(K) < d(L) + \varepsilon$.

(b) The crosscut K *may be chosen to be quasilinear; moreover, the segments can be assumed parallel to the real or imaginary axes.*

Proof. By 7.3.5 there are endcuts γ_a from a to a point $p \in D$ and γ_b from b to $q \in D$. We may assume that $d(\gamma_a)$, $d(\gamma_b) < \varepsilon/2$. We now show that p and q can be connected by a path in $D \cap U(L, \varepsilon/2)$.

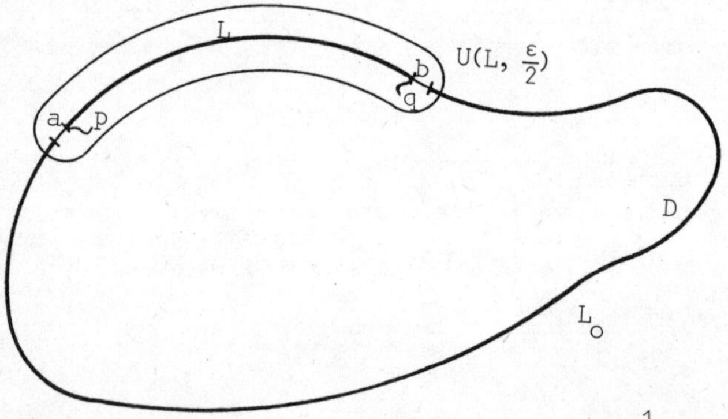

To prove this let L_o be an arc in $J \setminus L$ such that $J \setminus L_o \subset U(L, \frac{1}{2}\varepsilon)$. We will now apply 7.1.5 to the points p,q and the compact sets $L_o \cup \partial U(L, \frac{\varepsilon}{2})$ and J.

$L_o \cup \partial U(L, \frac{\varepsilon}{2})$ does not meet the curve $\gamma_a \cup L \cup \gamma_b$, and likewise J does not separate a and b. Moreover, $J \cap \{L_o \cup \partial U(L, \frac{\varepsilon}{2})\} = J \cap L_o = L_o$ is path connected. Thus by 7.1.5 there is a quasilinear simple path γ_1 from p to q which does not meet $J \cup (L_o \cup \partial U(L, \frac{\varepsilon}{2}))$. Since $p, q \in D \cap U(L, \frac{\varepsilon}{2})$ it follows that $\gamma_1 \subset D \cap U(L, \frac{\varepsilon}{2})$. Then the desired path K may be put together from γ_a, γ_b and γ_1 as in the proof of 7.3.2 (c).

For this section see [Newman 1951, VI, § 4].

Exercises: E 7.8-10 □

7.4 THE SCHÖNFLIES THEOREM.

Using the construction of 7.3 we will reduce the Schönflies theorem 7.4.1 to the polygonal case.

7.4.1 (Schönflies theorem). *Let γ be a simple closed curve in S^2 (resp. $\mathbb{R}^2$). Then there is a homeomorphism* f *of S^2 onto itself (resp. $\mathbb{R}^2$ onto itself) which maps γ onto the unit circle. (We will even show that the mapping* f *can arbitrarily be prescribed on γ.)*

Here S^2 is to be viewed as $\mathbb{R}^2 \cup \{\infty\}$ and the unit circle corresponds to the equator. The proof will be finished in 7.4.8.

7.4.2 A decomposition of the unit square.

We decompose the unit square $Q = \{(x,y) \mid 0 \le x,\ y \le 1\}$ inductively into Z_n's consisting of rectangles parallel to the axes as follows:

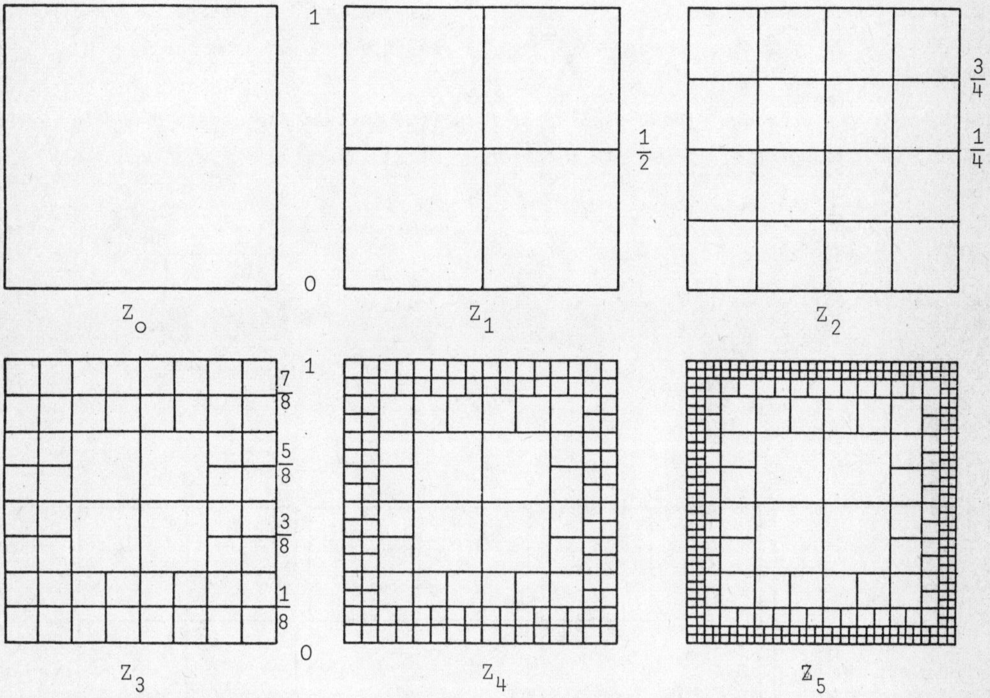

Z_0 is the whole Q. Z_1 and Z_2 are obtained by subdividing each square by the perpendicular bisectors of the sides. Z_{n+1} is obtained from Z_n by subdividing the squares of Z_n which meet the boundary of Q by their perpendicular bisectors. The squares of Z_n at the boundary therefore have sides of length $1/2^n$, and hence diameter $< 2/2^n$.

7.4.3 Definition. Let $\tilde{Z}$ be the grid on Q which is the limit of Z_n, and let $Z = |\tilde{Z}| \cup \partial Q$. Moreover, let $\tilde{Z}^{(o)}$ be the set of vertices of $\tilde{Z}$, and $\tilde{Z}^{(1)}$ the set of its *closed 1-cells*. The latter are segments parallel to the real or imaginary axes which contain no points of $\tilde{Z}^{(o)}$ in their interior, but which are bounded by two points of $\tilde{Z}^{(o)}$. The *2-cells* of Z are the closed squares in Q whose boundaries equal their intersections with $\tilde{Z}^{(1)}$. (In the interior of Q, Z induces a cell decomposition in the usual sense, however this is not the case for the boundary.)

7.4.4 Proposition. *Let* γ *be a simple closed curve in* $\mathbb{R}^2$ *and* D *the bounded path component of the complement of* γ. *Let* $g\colon \partial Q \to |\gamma|$ *be a homeomorphism. Then* D *contains a 1-dimensional complex* $\tilde{Y}$ *whose 1-cells are compact polygonal paths, and if* $Y = \tilde{Y} \cup |\gamma|$ *we have:*

(a) There is an isomorphism $\tilde{f}\colon \tilde{Z}^{(1)} \to \tilde{Y}$. *To each 1-cell of* $\tilde{Z}^{(1)}$ *there is an associated cell of* $\tilde{Y}$. *The isomorphism* $\tilde{f}$ *may be realized by a homeomorphism* $f\colon |\tilde{Z}^{(1)}| \to |\tilde{Y}|$.

(b) If ω *is the boundary of a 2-cell of* Z, *then* $f(\omega)$ *is a simple closed curve in* $D \subset \mathbb{R}^2$, *with no point of* $|\tilde{Y}|$ *in its interior.*

(c)

$$x \mapsto \begin{cases} f(x) & x \in |\tilde{Z}^{(1)}| \\ g(x) & x \in \partial Q \end{cases}$$

defines a homeomorphism $F^{(1)}\colon Z \to Y$

Proof. We take g as a parametrization of γ. Let d be the euclidean metric on $\mathbb{R}^2$. Quartering of a square may be accomplished by the images of three crosscuts:

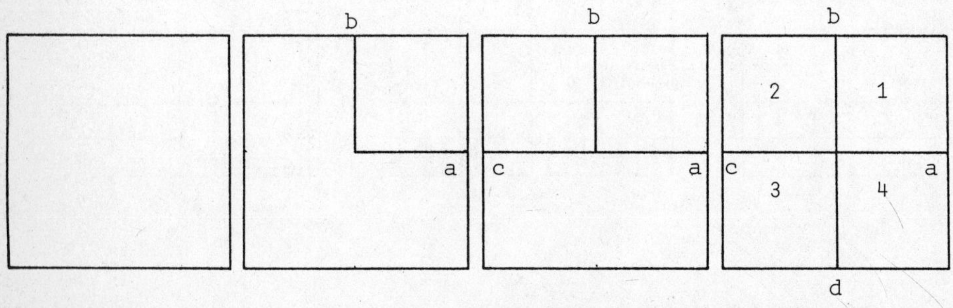

Now let δ be a simple closed curve and let a', b', c', d' be four of its points, which occur in that order. By 7.3.7 we can find crosscuts in the interior of δ which connect the points a', b', c', d' according to the same schema:

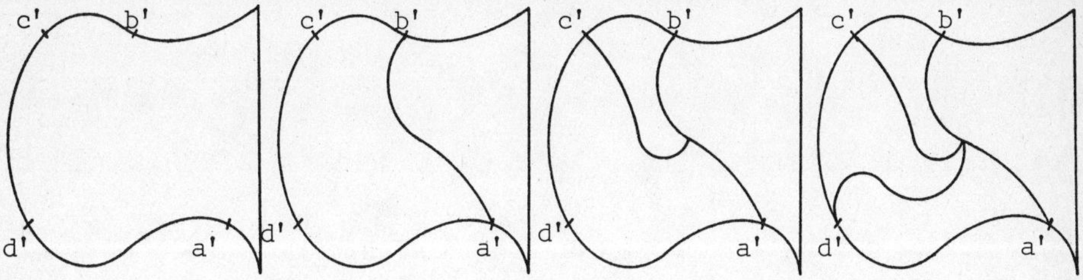

With γ as Y_o we can construct a Y_1 which is isomorphic to Z_1. The endpoints of the paths on γ are the images under g of the midpoints to the sides of Q. Obviously we can iterate to find complexes Y_1, Y_2, ... which are isomorphic to Z_1, Z_2, ... Also, each of these isomorphisms may be realized by a homeomorphism, with the mapping for Y_o in particular being g.

This however does not guarantee continuity in passing to the limit, as the following sketch shows.

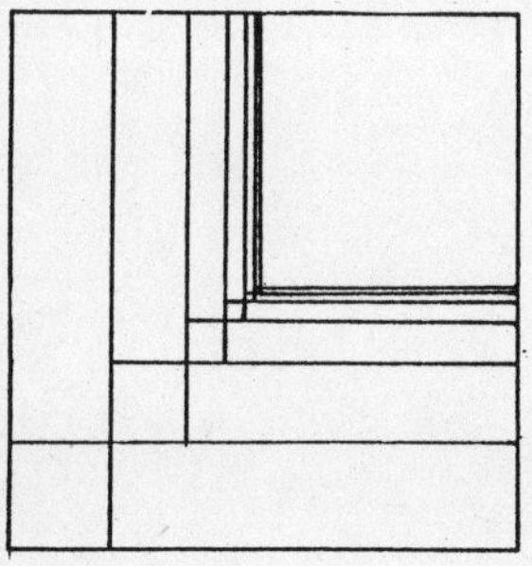

Not converging to the corner point.

To achieve this we must sharpen the conditions for quartering the domains in passing from Y_n to Y_{n+1}. We consider a domain which does not lie at one of the four corners, and use the following notation:

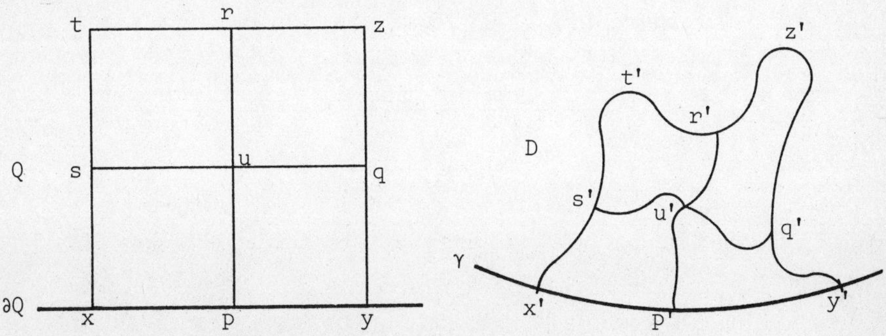

We let $\overline{x'p'}$ denote the arc from x' to p' on γ, and similarly for the other arcs.

Let $p' = g(p)$, where p is the midpoint of the segment $[x,y]$. The points q' and s' are chosen so that

$$d(\overline{x's'}) \leq d(\overline{x'p'})$$
$$d(\overline{y'q'}) \leq d(\overline{p'y'}).$$

By 7.3.7 we can then find the arc $\overline{s'p'}$ so that

$$d(\overline{s'p'}) \leq 3d(\overline{x'p'}).$$

Now choose $u' \in \overline{s'p'}$ so that

$$d(\overline{u'p'}) \leq d(\overline{p'y'})$$

Then by 7.3.7 it is possible to find an arc $\overline{u'q'}$ such that

$$d(\overline{u'q'}) \leq 4d(\overline{p'y'})$$

Thus the boundaries of the "curvilinear squares" satisfy

$$d(\overline{x'p'u's'}) \leq 5d(\overline{x'p'}) + d(\overline{p'y'}) \leq 6d(\overline{x'y'})$$
$$d(\overline{y'q'u'p'y'}) \leq 7d(\overline{p'y'}) \leq 7d(\overline{x'y'}).$$

The diameter of the domain enclosed by these curves therefore does not exceed $7d(\overline{x'y'})$.

The quartering of the "curvilinear squares" at the boundary used to pass from Y_n to Y_{n+1} now takes place as just described. Since we choose the points s' and q' at the same time and independently of each other, the choices do not interfere with each other. (There is a similar rule for the 4 squares at the corners of Q.)

We now realize the isomorphisms $Z_n \to Y_n$ successively by homeomorphisms $f_n\colon |Z_n| \to |Y_n|$ such that $f_{n+1}\,|\,|Z_n| = f_n$. In particular, $f_n|\partial Q = g$. Let $f(x) = f_n(x)$ for $x \in |Z_n|$. This defines an injective and surjective mapping $Z \to Y$ with $f|\partial Q = g$.

If $d(x,\partial Q) > \frac{1}{2^n}, \frac{1}{2^m}$, $x \in Q$, then $f(x) = f_n(x) = f_m(x)$. This implies the continuity of f in the interior of Q. Now let $x \in \partial Q$ be a point of Q which is not a corner and suppose $\varepsilon > 0$. Then by the continuity of γ there is a $\delta > 0$ such that $y_1, y_2 \in \partial Q$ and $d(x_i,y_i) < \delta$ imply $d(g(y_1),g(y_2)) < \frac{\varepsilon}{14}$.

Now let n be chosen so that $\frac{1}{2^n} < \delta$. As neighbourhood U of x we take the (resp. both) square (s) of Z_n with x in the boundary. By construction of Z, the image of the boundary ∂U has diameter $< \varepsilon$. The images of $U \cap Z_{n+1}$, $U \cap Z_{n+2}$, ... lie in the interior of the image of $U \cap Z_n$, hence $d(f(U \cap Z), f(x)) < \varepsilon$.

The proof of continuity of the four corners of Q is analogous.

□

Next we will show that the mapping $f: Z \to Y$ may be extended to a homeomorphism $F: Q \to D$. To do this we recall the following (well-known) theorems:

7.4.5 Theorem. *Let* E_1 *and* E_2 *be disks* $(E_i \cong Q)$ *and* $h: \partial E_1 \to \partial E_2$ *a homeomorphism. Then there is a homeomorphism* $H: E_1 \to E_2$ *with* $H|\partial E_1 = h$.

□

7.4.6 Theorem (Schönflies theorem for polygonal Jordan curves). *Let* γ *be a simple closed polygonal path in* $\mathbb{R}^2$ *and* D *the bounded path component of the complement. Then* $\bar{D}$ *is homeomorphic to* Q.

Proof. See 4.1.4.

□

7.4.7 Theorem. *With the hypotheses and notation of 7.4.4;* $g: \partial Q \to |\gamma|$ *may be extended to a homeomorphism* $F: Q \to \bar{D}$. *More precisely,* $F^{(1)}: Z \to Y$ *may be extended to the homeomorphism* F.

Proof. We need only show that $F^{(1)}$ may be extended to the interior on each 2-cell ζ of Z. But $F^{(1)}(\partial\zeta)$ is a simple closed polygonal path. By 7.4.6 $F^{(1)}(\partial\zeta)$ bounds a disk E_ζ and by 7.4.5 $F^{(1)}|\partial\zeta$ may be extended to a homeomorphism $F_\zeta: \zeta \to E_\zeta$. By 7.4.4 (b), the images E_ζ and $E_{\zeta'}$ have no common interior point for two different 2-cells ζ, ζ' of Z. Therefore the homeomorphisms $F^{(1)}$ and $\{F_\zeta | \zeta \text{ a 2-cell of } Z\}$ can be combined into the desired topological mapping $F: Q \to \bar{D}$.

□

7.4.8 *Proof of the Schönflies theorem* 7.4.1.

If γ lies on S^2, then one can apply theorem 7.4.7 to both path components of $S^2 \setminus \gamma$. A prescribed homeomorphism of γ onto the equator may therefore be extended to a homeomorphism which maps one component onto the upper hemisphere and the other onto the lower. Their combination gives the desired homeomorphism of S^2 onto itself.

If we are dealing with a curve in $\mathbb{R}^2$, then we close $\mathbb{R}^2$ to S^2 by ∞ and apply the previous argument. The homeomorphism obtained, $F: S^2 \to S^2$, may be taken to map the component containing ∞ to the hemisphere containing ∞. Now for any two points in the interior of the unit disk there is a homeomorphism which maps one to the other and which is the identity on the boundary. We can therefore assume that our homeomorphism $S^2 \to S^2$ leaves the point ∞ fixed. Then $F|\mathbb{R}^2$ is the desired topological mapping.

For this section see [Newman 1951, VI, § 5].
Exercises: E 7.11-14 □

7.5 TRIANGULATION AND HAUPTVERMUTUNG FOR SURFACES

This section contains the following two theorems:

7.5.1 (Triangulation of surfaces). *Each topological 2-manifold* F *with a denumerable basis is triangulable.*

7.5.2 (Positive answer to the Hauptvermutung for dimension 2). *Two simplicial complexes on a surface have isomorphic subdivisions.*

For 7.5.1 we give a proof which follows [Rado 1925], which was the first proof of this theorem. Other proofs are found in [Ahlfors - Sario 1960] and [Doyle - Moran 1968], [Moise 1977].

The proof of 7.5.2 assumes the topological invariance of the combinatorial surface invariants of 3.2, see e.g. [Massey 1967], [Seifert - Threlfall 1934].

The proof of 7.5.1 is based on

7.5.3 Lemma. *There is a sequence of closed domains* $B_1, B_2, \ldots$ *on* F *with the following properties:*

(a) $B_i \cong \mathbb{D}^2 = \{z \in \mathbb{C} \mid |z| \le 1\}$.

(b) $\bigcup_{i=1}^{\infty} \mathring{B}_i = F$.

(c) $\partial B_i \cap \partial B_j$ *is finite if* $i \ne j$.

Proof. There is a covering of F by sets B_1^*, B_2^*, ... homeomorphic to $\mathbb{D}^2$ and a system E_1, E_2, ... of sets with $B_i^* \subset E_i$. Here $E_i \cong \mathbb{R}^2$ if $B_i^* \cap \partial F = \emptyset$, and E_i is homeomorphic to the right half-plane $\mathbb{R}_+^2$ when $B_i^* \cap \partial F \neq \emptyset$. In the latter case $B_i^* \cap \partial F = B_i^* \cap \partial E_i$. We now seek a system B_1, B_2, ... with $B_i^* \subset B_i$ which has the properties (a), (b), (c).

Let $B_1 := B_1^*$. Now assume that $B_1, \ldots, B_{n-1}$ ($n \geq 2$) have already been obtained with the properties (a) and (c), and also that $B_i^* \subset B_i$ ($i = 1, \ldots, n-1$). Let $R_{n-1} := \bigcup_{i=1}^{n-1} \partial B_i$. We now consider the case $E_n \cong \mathbb{R}^2$. Let $\varphi: E_n \to \mathbb{D}^2$ be a homeomorphism onto the interior of the unit disk. There is a concentric circle K such that $\varphi B_n^* =: B_n'$ lies in the interior of K and so that the annulus G between K and $\partial\mathbb{D}^2$ contains no double point of φR_{n-1}, i.e. no point which lies on more than one of the curves $\varphi(\partial B_i)$ ($i = 1, \ldots, n-1$).

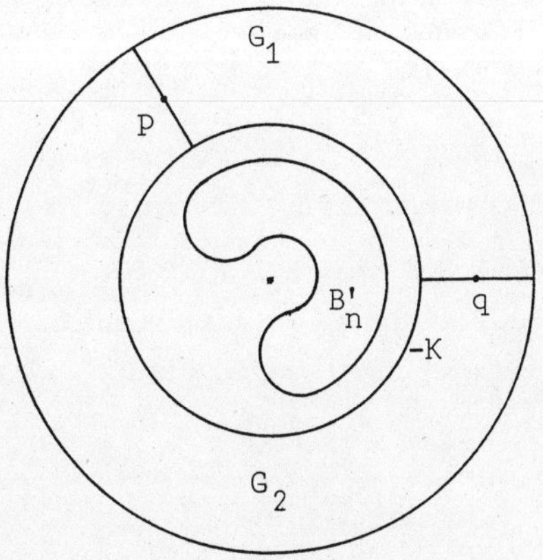

Since $R_{n-1} \cap E_n$ is nowhere dense, there are two points p and q in G, having neighbourhoods disjoint from R_{n-1}, and which lie on different radii. The radii through p and q divide G into two domains G_1 and G_2. The curves of $R_{n-1} \cap G$ possess disjoint regular neighbourhoods, since G contains no double points of R_{n-1}. The curves of $R_{n-1} \cap G_i$ divide G_i into several parts. If we extend each part by the intersection of the regular neighbourhoods of its boundary curves with G_i, then we obtain a covering of G_i by sets open in G_i. Two points which lie in one of these sets may be connected by a simple path which has at most two points in common with R_{n-1}. We now connect p to q by a simple path ω_i in G_i. The covering of G_i by sets open in G_i yields an open covering of ω_i. Then there is a finite subcovering of the path. It follows that ω_i may be replaced by a path ν_i which meets R_{n-1} only finitely often. The two paths ν_1 and ν_2 have only the points p and q in common, hence $\nu_1 \cdot \nu_2$ is a Jordan curve which has only finitely many points of intersection with R_{n-1}. By 7.4.1 it bounds a disk B_n''. Then $B_n := \varphi^{-1} B_n''$ has the desired properties.

The proof runs analogously for the case $E_n \cong \mathbb{R}^2_+$.

□

7.5.4 *Proof of* 7.5.1. We construct successive decompositions of B_n into finitely many closed domains $A_j^{(n)}$ such that

(a) $B_n \setminus \bigcup_{i<n} B_i = \bigcup_j A_j^{(n)}$

(b) $\partial A_j^{(n)} \subset R_n$

(c) $A_i^{(m)} \cap A_j^{(n)} = \partial A_i^{(m)} \cap \partial A_j^{(n)}$ for $(m,i) \neq (n,j)$

(d) Each $A_j^{(n)}$ is bounded by a finite number of disjoint simple closed curves.

Let $A_1^{(1)} := B_1$. Suppose that the decomposition of $\bigcup_{i<n} B_i$ has already been obtained. We use induction on i to decompose B_n by the boundary curves ∂B_i into closed domains for which the assertions (b) and (c) hold. If B_n lies inside $\bigcup_{i<n} B_i$ then all the $A_j^{(n)}$ are empty. If $B_1 \cap B_n = \emptyset$ we do not change B_n. If $\partial B_1 \cap \partial B_n = \emptyset$ and $B_1 \subset B_n$ then B_n is divided into B_1 and the closed complement of B_1. If $\partial B_1 \cap \partial B_n \neq \emptyset$ then $\partial B_1 \cap B_n$ decomposes into several simple arcs whose endpoints lie on the boundary of B_n. These arcs decompose B_n into several closed domains which have the properties (b) and (c). Next we carry out this decomposition for ∂B_2 and each domain obtained at the first step. Finally we obtain a decomposition of B_n with the properties (b), (c) and (d). Then we leave out all the domains already contained in $\bigcup_{i<n} B_i$.

Altogether we obtain a decomposition of F into closed domains with the properties (a-d). Now if $\partial A_j^{(n)}$ consists of k components (k > 1) we can introduce k arcs to decompose $A_j^{(n)}$ into two domains, each bounded by a simple closed curve. Since $A_j^{(n)}$ lies in B_n and $B_n \cong D_n \subset \mathbb{R}^2$ it follows from the Schönflies theorem 7.4.1 that each domain finally obtained is a disk. A point of F belongs to only finitely many domains $A_j^{(n)}$; therefore we have obtained a cell decomposition of F, and this yields a triangulation.

□

7.5.5 *Proof of* 7.5.2. We confine ourselves to the case of surfaces of finite type, i.e. surfaces with finitely generated homology group. Then the topological invariants are; see E 7.15:

(1) Orientability character,

(2) Number of boundary components,

(3) Number of "ends",

(4) Genus.

Here the number of ends is determined as follows: one takes a compact connected subset K of the surface F and considers the number of components of F \ K which do not have compact closure. The supremum over all K is the number of ends. The orientability character and genus may be determined from the number of ends and the first homology group. The combinatorial classification of surfaces yields the same invariants, see 3.2, and can be carried ont similarly for the infinite case, see [Kérékjartó 1923], [Richards 1963].

□

The above theorem proves the Hauptvermutung for dimension 2, i.e. gives a positive answer to the following question:

Are any two triangulations of a space related? I.e., do they possess isomorphic subdivisions?

In particular, one can ask whether two homeomorphic triangulable manifolds carry related complexes. This is obviously correct for 1-dimensional manifolds, it has also been known for a long time in dimension 2 although the proof - as we have seen - requires some effort. In dimension 3 it was proved in [Moise 1952] and also [Bing 1959]. In higher dimensions the Hauptvermutung is false in general, [Milnor 1961]. The answer is not known for manifolds of dimension 4. The Hauptvermutung is correct for most manifolds of dimension ≥ 5, but it is false for a few in each dimension, as Kirby, Siebenmann and Wall have shown.

Exercise: E 7.15.

Additional literature to chapter 7:
[Doyle-Moran 1968], [Narens 1971], [Wagner 1974], [Young 1947].

EXERCISES

E 7.1 Repeat proofs of the theorems 7.1.1, 7.1.2 and the Alexander duality theorem.

E 7.2 If D_1 and D_2 are domains in S^2 (i.e. open connected subsets) and if $\partial D_1 \cap \partial D_2$ is path connected then $D_1 \cap D_2$ is connected. (Hint: Use 7.1.3.)

E 7.3* If D_1 and D_2 are domains in S^2 and $\partial D_1 \cap \partial D_2$ has not more than two components then $D_1 \cap D_2$ has not more than two components. (Apply an argument as for the proof that a simple arc does not separate S^2.)

E 7.4 Let the boundary of a domain $D \subset S^2$ consist of two non-intersecting simple closed curves γ_1 and γ_2 and let λ_1 and λ_2 be disjoint simple arcs each of which has one endpoint on γ_1, the other on γ_2 and all other points in D. Then $D \setminus (\lambda_1 \cup \lambda_2)$ has two components, each of which is a domain bounded by one simple closed curve.

E 7.5 Let $D \subset S^2$ be a domain and λ an arc such that the endpoints of λ are on different components of ∂D and all other points of λ are in D. Then $D \setminus \lambda$ is connected.

E 7.6 Let the notation be as in 7.1.3. Assume that $\kappa_1 + \kappa_2 \sim 0$ in $S^2 \setminus (F_1 \cap C)$ for each component C of F_2. Then $x \sim y$ in $S^2 \setminus (F_1 \cup F_2)$.

E 7.7 Let $x, y \in S^2$, F_1 a compact connected subset of S^2 which contains at least two points (F_1 is called a continuum) and F_2 any closed subset of S^2. Assume that no set $F_1 \cap C$ where C is a component of F_2 separates x and y. Then $F_1 \cup F_2$ does not separate x and y.

E 7.8 Prove: In the domain D from the figure (the same as in 7.3.1) the points $\{(t,0) \mid 0 < t < \frac{1}{2}\}$ are not accessible from D . All other points of ∂D are accessible.

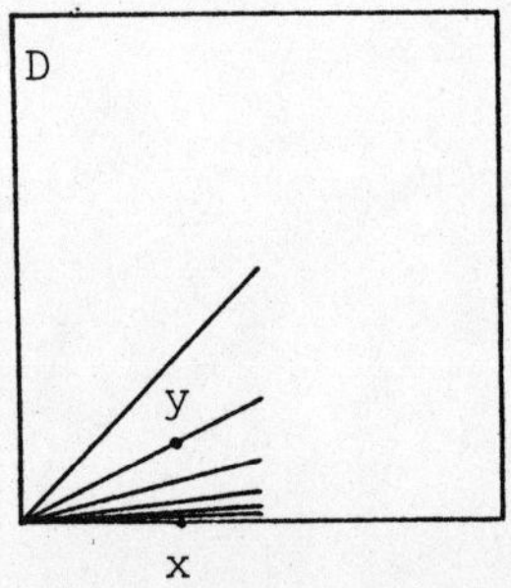

E 7.9 Construct domains whose boundaries have components (a) containing no accessible points, (b) containing just one accessible point and at least one other point.

E 7.10 If A is a connected set then a cut point of A is a point x such that $A \setminus \{x\}$ is not connected. Prove: If D is a simple connected domain in S^2 then every cut point of $S^2 \setminus D$ is accessible from D.

E 7.11 If α is a simple arc in S^2 then there is a topological mapping of S^2 onto itself which maps α onto a straight segment.

E 7.12 Let α be a simple arc in $\mathbb{R}^2$. Then there is a homeomorphism of $\mathbb{R}^2$ onto itself that maps α to exactly one of the following subsets of the real axis: closed segment, half-open segment, open segment, reals axis, closed halfline,

open halfline

E 7.13 If λ is a cross cut in the Jordan domain D then $\bar{D}$ can be mapped topologically onto the unit disk $\mathbb{D}^2$ such that λ is mapped onto a diameter.

E 7.14 Let $\alpha_1, \dots, \alpha_n$ be simple arcs in S^2 which have a common endpoint but no further intersection. Then S^2 can be mapped topologically onto itself so that the arcs are mapped onto segments radiating from a point. Prove that for any two such mappings the cyclical orders of the images are the same or inverse to one another.

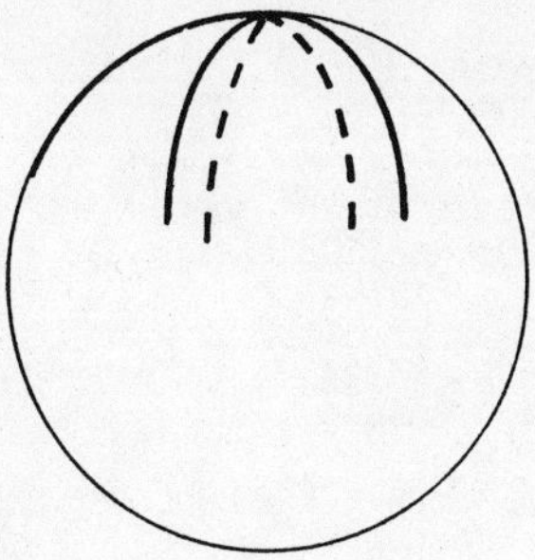

E 7.15 Characterize topologically for a surface of finite type: orientability, number of boundary components and 'ends', genus.

Books on surface theory and related subjects

Ahlfors, L.V., Sario, L., 1960: Riemann surfaces. Princeton University Press, Princeton, N.J. 1960

Aleksandrov, P.S., 1956: Combinatorial topology. Vol. 1. Graylock Press, Rochester 1956

Birman, J., 1974: Braids, links, and mapping class groups. Ann. Math. Studies 82, Princeton University Press, Princeton 1974

Cohn, H., 1967: Conformal mapping on Riemann surfaces. McGraw-Hill, New York 1967

Giblin, P.J., 1977: Graphs, Surfaces and Homology. Chapman & Hall, London 1977

Gramain, A., 1971: Topologie des surfaces. Presses Univ. France, Paris 1971

Griffiths, H.B., 1976: Surfaces. Cambridge University Press, Cambridge 1976

Kerékjártó von, B., 1923: Vorlesungen über Topologie. Springer, Berlin 1923

Kurosh, A.G., 1967: The Theory of Groups I, II. Chelsea, London 1960

Lehner, J., 1964: Discontinuous groups and automorphic functions. Math. Surveys VIII, Amer. Math. Soc., Providence, R.I., 1964

Levi, F., 1929: Geometrische Konfigurationen. Hirzel Verlag, Leipzig 1929

Liebmann, H., 1912: Nichteuklidische Geometrie. 2. Aufl., Berlin und Leipzig: Göschen'sche Verlagshandlung 1912

Lyndon, R.C., Schupp, P.E., 1977: Combinatorial group theory. Ergebn. Math. Grenzgebiete 89, Springer Verlag, Berlin 1977

Magnus, W., 1974: Noneuclidean tesselations and their groups. Academic Press, New York 1974

Massey, W.S., 1967: Algebraic Topology: An Introduction. Harcourt, Brace & World, Inc., New York 1967

Moise, E.E., 1977: Geometric topology in dimension 2 and 3. Graduate Texts in Math. 47, Springer, Berlin 1977

Newman, M.H.A., 1951: Elements of the topology of plane sets of points. Cambridge University Press, Cambridge 1951

Reidemeister, K., 1932: Einführung in die kombinatorische Topologie. Friedr. Vieweg und Sohn, Braunschweig 1932

Seifert, H., Threlfall, W.: 1934: Lehrbuch der Topologie. Teubner, Leipzig 1934

Bibliography

Accola, R.D.M., 1967: Automorphisms of Riemann surfaces. J. d'Anal. Math. 18, 1-5 (1967)

Accola, R.D.M., 1968: On the number of automorphisms of a closed Riemann surface. Trans. Amer. Math. Soc. 131, 398-408 (1968)

Accola, R.D.M., 1969: Riemann surfaces with automorphism groups admitting partitions. Proc. Amer. Math. Soc. 21, 477-482 (1969)

Accola, R.D.M., 1970: Two theorems on Riemann surfaces with nonyclic automorphism groups. Proc. Amer. Math. Soc. 25, 598-602 (1970)

Ahlfors, L.V., 1953: On quasiconformal mappings. J. d'Analyse Math. 3, 1-58 (1953/54)

Ahlfors, L.V., 1960: The complex analytic structure of the space of closed Riemann surfaces. Analytic functions, p. 45-66, Princeton University Press, Princeton 1960

Ahlfors, L.V., Sario, L., 1960: Riemann surfaces. Princeton University Press, Princeton, N.J., 1960

Aleksandrov, P.S., 1956: Combinatorial topology. Vol. 1. Rochester, N.Y. 1956

Alexander, J.W., 1922: A proof and extension of the Jordan-Brouwer separation theorem. Trans. Amers. Math. Soc. 23, 333-349 (1922)

Alexander, J.W., 1923: Invariant points of a surface transformation of a given class. Trans. Amer. Math. Soc. 25, 173-184 (1923)

Alexandroff, P., Hopf, H., 1935: Topologie I. Springer Verlag, Berlin 1935

Andrea, S.A., 1967: On homeomorphism of the plane which have no fixed points. Abhandl. math. Sem. Univ. Hamburg 30, 61-74 (1967)

Appell, P., Goursat, E., 1930: Theorie des Fonctions algébriques. pp. 125-133, Gauthier-Villars, Paris 1930

Armstrong, M.A., 1968: The fundamental group of the orbit space of a discontinuous group. Proc. Cambridge Phil. Soc. 64, 299-301 (1968)

Арнольд, В.И., 1969: Кольцо когомологий группы крашеных кос. Мат. заметки 5, 227-231 (1969). Engl. transl.:
Arnold, V.I., 1969: The cohomology ring of the colored braid groups. Math. Notes Acad. Sci. USSR 5, 227-232 (1969)

Арнольд, В.И., 1970: Топологические инварианты алгебраических функций, II. Функц. анализ прил. 4:2, 1-9 (1970). Engl. transl.:
Arnold, V.I., 1970: Topological invariants of algebraic functions, II. Functional Analysis Appl. 4, 91-98 (1970)

Арнольд, В.И., 1971: О некоторых топологических инвариантах алгебраических функций. Труды Моск. Мат. общ. 21, 27-46 (1970). Engl. transl.:
Arnold, V.I., 1971: On some topological invariants of algebraic functions. Trans. Moscow Math. Soc. 21, 30-52 (1971)

Artin, E., 1926: Theorie der Zöpfe. Abh. Math. Sem. Univ. Hamburg 4, 47-72 (1926)

Artin, E., 1947: Theory of braids. Ann. of Math. (2) 48, 101-126 (1947)

Artin, E., 1947': Braids and permutations. Ann. of Math. (2) 48, 643-649 (194)

Baer, R., 1927: Kurventypen auf Flächen. J. reine angew. Math. 156, 231-246 (1927)

Baer, R., 1928: Isotopien auf Kurven auf orientierbaren, geschlossenen Flächen. J. reine angew. Math. 159, 101-116 (1928)

Baer, R., 1928': Die Abbildungstypen der orientierbaren, geschlossenen Flächen vom Geschlecht 2. J. reine angew. Math. 160, 1-25 (1928)

Baer, R., Levi, F., 1936: Freie Produkte und ihre Untergruppen. Compositio Math. 3, 391-398 (1936)

Bailey, W.L., 1961: On the automorphism group of a generic curve of genus > 2. J. Math. Kyoto Univ. 1, 101-108 (1961)

Ballmann, W., 1978: Doppelpunktfreie geschlossene Geodätische auf kompakten Flächen. Math. Z. 161, 41-46 (1978)

Baumslag, G., 1962: On generalized free products. Math. Z. 78, 423-438 (1962)

Behnke, H., Sommer, F., 1955: Theorie der analytischen Funktionen einer komplexen Veränderlichen. Springer Verlag, Berlin 1955

Bell, H., 1976: A fixed point theorem for plane homeomorphisms. Bull. Amer. Math. Soc. 82, 778-780 (1976)

Bergau, P., Mennicke, J., 1960: Über topologische Abbildungen der Brezelfläche vom Geschlecht 2. Math. Z. 74, 414-435 (1960)

Bers, L., 1958: Spaces of Riemann surfaces. Proc. Int. Kongr. Math. Edinburg p. 349-361, 1958

Bers, L., 1960: Quasiconformal mappings and Teichmüller's theorem. Analytic functions, p. 89-120, Princeton Univ. Press, Princeton, N.J., 1960

Bers, L., 1973: Uniformization, moduli, and Kleinian groups. Bull. London Math. Soc. 4, 257-300 (1973)

Best, L.A., 1973: Subgroups of one-relator Fuchsian groups. Canad. J. Math. 25, 888-891 (1973)

Betten, D., 1975: Sperner-Homöomorphismen auf Ebene, Zylinder und Möbiusband. Abhandl. Math. Sem. Univ. Hamburg 44, 263-272 (1975)

Birman, J., 1969: On braid groups. Comm. Pure and Appl. Math. 22, 41-72 (1969)

Birman, J., 1969': Mapping class groups and their relationship to braid groups. Commun. Pure Appl. Math. 22, 213-238 (1969)

Birman, J., 1969": Automorphisms of the fundamental group of a closed, orientable 2-manifold. Proc. Amer. Math. Soc. 21, 351-354 (1969)

Birman,J.S.,1970: Abelian quotients of the mapping class group of a 2-manifold Bull. Amer. Math. Soc. 76, 147-150 (1970)

Birman,J.S.,1971: On Siegel's modular group. Math. Ann. 191, 59-68 (1971)

Birman,J.S.,1974: Braids, links, and mapping class groups. Ann. Math. Studies 82, Princeton University Press, Princeton, N.J., 1974

Birman, J.S., Chillingworth, D.R.J., 1972: On the homeotopy group of a non-orientable surface. Proc. Cambridge Phil. Soc. 71, 437-448 (1972)

Birman, J.S., Hilden, H., 1971: On the mapping class groups of closed surfaces as covering spaces. Advances in the theory of Riemann surfaces. Annals of Math. Studies No. 66, 81-115, Princeton University Press, Princeton 1971

Birman, J.S., Hilden, H.M., 1972: Lifting and projecting homeomorphisms. Archiv Math. 23, 428-434 (1972)

Birman, J.S., Hilden, H.M., 1973: On isotopies of homeomorphisms of Riemann surfaces. Ann. of Math. 97, 424-439 (1973)

Bourgin, D.G., 1968: Homeomorphism of the open disk. Studia math. 31, 433-438 (1968)

Britton, J.L., 1963: The word problem. Ann. of Math. 77, 16-32 (1963)

Brödel, W., 1935: Über die Deformationsklassen zweidimensionaler Mannigfaltigkeiten. Ber. Verhandl. Sächs. Akad. Leipzig 87, 85-120 (1935)

Brown, E.M., Messer, R., 1979: The classification of two-dimensional manifolds. Trans. Amer. Math. Soc. 255, 377-402 (1979)

Brunner, A.M., 1976: A group with an infinite number of Nielsen inequivalent one-relator presentations. J. Algebra 42, 81-84 (1976)

Bundgaard, S., Nielsen, J., 1946: Forenklede Beviser for nogle Saetninger i Flachtopologien. Mat. Tidsskrift B 1946, 1-16

Burde, G., 1963: Zur Theorie der Zöpfe. Math. Ann. 151, 101-107 (1963)

Burkhardt, H., 1890: Grundzüge einer allgemeinen Systematik der hyperelliptischen Functionen I. Ordnung. Math. Ann. 35, 198-296 (1890)

Călugăreanu, G., 1966: Sur les courbes fermées simples tracées sur une surface fermée orientable. Mathematica (Cluj) 8, 29-38 (1966)

Călugăreanu, G., 1967: Considérations directes sur la génération des noeuds (II). Studia Univ. Babes-Bolyai, ser. Math.-Phys. 2, 25-30 (1967)

Călugăreanu, G., 1967': Courbes fermées simples sur une surface fermée orientable. Mathematica (Cluj) 9, 225-231 (1967)

Călugăreanu, G., 1968: Sur les générateurs de certains groupes d'automorphismes. Mathematica (Cluj) 10, 245-251 (1968)

Călugăreanu, G., 1975: Sur certains invariants attachés aux groupes dénombrables. Mathematica (Cluj) 17, 11-58 (1975)

Călugăreanu, G., 1975': Sur un théorème de H. Zieschang. L'Enseign. math. 21, 15-30 (1975)

Chalk, J.H.H., 1976: Generators of Fuchsian groups. Tôhuku Math. J. 28, 89-94 (1976)

Chang, B., 1960: The automorphism group of a free group with two generators. Mich. Math. J. 7, 79-81 (1960)

Chillingworth, D.R.J., 1969: Simple closed curves on surfaces. Bull. London Math. Soc. 1, 310-314 (1969)

Chillingworth, D.R.J., 1971: An algorithm for families of disjoint simple closed curves on surfaces. Bull. London Math. Soc. 3, 23-26 (1971)

Chillingworth, D.R.J., 1972: Winding numbers on surfaces, I. Math. Ann. 196, 218-249 (1972)

Chillingworth, D.R.J., 1972': Winding numbers on surfaces, II. Applications. Math. Ann. 199, 131-153 (1972)

Chiswell, I.M., 1976: The Grushko-Neumann theorem. Proc. London Math. Soc. 33, 385-400 (1976)

Chiswell, I.M., 1976': Euler characteristics of groups. Math. Z. 148, 1-11 (1976)

Chiswell, I.M., 1976": Abstract length functions in groups. Math. Proc. Cambridge Phil. Soc. 80, 451-463 (1976)

Clebsch, A., Gordan, P., 1866: Theorie der Abelschen Functionen. Teubner, Leipzig 1866. Neudruck: Physica-Verlag, Würzburg 1967

Cohen, D.B., 1974: The Hurwitz monodromy group. J. Algebra 32, 501-517 (1974)

Cohen, D.E., Lyndon, R.C., 1963: Free bases for normal subgroups of free groups. Trans. Amer. Math. Soc. 108, 526-537 (1963)

Cohn, H., 1967: Conformal mapping on Riemann surfaces. McGraw-Hill, New York 1967

Coldewey, H.-D., 1971: Kanonische Polygone endlich erzeugter Fuchsscher Gruppen. Dissertation Ruhr-Universität Bochum 1971

Collins, D.J., 1978: Presentations of the amalgamated free product of two infinite cycles. Math. Ann. 237, 233-241 (1978)

Coxeter, H.S.M., Moser, H.W.O.J., 1957: Generators and relations for discrete groups. Ergebn. Math. N.F. 14, Springer Verlag, Berlin 1957

Dehn, M., 1910: Über die Topologie des dreidimensionalen Raumes. Math. Ann. 69, 137-168 (1910)

Dehn, M., 1912: Über unendliche diskontinuierliche Gruppen. Math. Ann. 71, 116-144 (1912)

Dehn, M., 1912': Transformation der Kurven auf zweiseitigen Flächen. Math. Ann. 72, 413-421 (1912)

Dehn, M., 1922: Über Kurvensysteme auf zweiseitigen Flächen mit Anwendungen auf das Abbildungsproblem. Autogr. Vortrag im Math. Kolloquium, Breslau, 11. Febr. 1922

Dehn, M., 1938: Die Gruppe der Abbildungsklassen. Acta Math. 69, 135-206 (1938)

Dehn, M., 1939: Über Abbildungen. Mat. Tidskript B 1939, 25-48

Dehn, M., 1950: Über Abbildungen geschlossener Flächen auf sich. Mat. Tidsskript B 1950, 146-151

Dostal, M., Tindell, R., 1978: The Jordan Curve Theorem Revisited. Jahresber. Deutsche Math. Ver. 80, 111-128 (1978)

Doyle, P.H., Moran, D.A., 1968: A short proof that compact 2-manifolds can be triangulated. Inventiones math. 5, 160-162 (1968)

v. Dyck, W., 1882: Gruppentheoretische Studien. Math. Ann. 20, 1-44 (1882)

v. Dyck, W., 1888: Beiträge zur Analysis situs, I. Ein- und zweidimensionale Mannigfaltigkeiten. Math. Ann. 32, 457-512 (1888)

Earle, C.J., Eells, J., 1967: The diffeomorphism group of a compact Riemann surface. Bull. Amer. Math. Soc. 73, 557-559 (1967)

Edmonds, A.L., 1979: Deformation of maps to branched coverings in dimension 2. Ann. of Math. 110, 113-125 (1979)

Elvin, J.D., Short, D.R., 1975: Branched immersions between 2-manifolds of higher topological type. Pacific J. Math. 58, 361-370 (1975)

Epstein, D.B.A., 1966: Curves on 2-manifolds and isotopies. Acta Math. 115, 83-107 (1966)

Epstein, D.B.A., Zieschang, H., 1966: Curves on 2-manifolds: A counterexample. Acta Math. 115, 109-110 (1966)

Ezell, C.L., 1978: Branch point structure of covering maps onto orientable surfaces. Trans. Amer. Math. Soc. 243, 123-133 (1978)

Fadell, E., 1962: Homotopy groups of configuration spaces and the string problem of Dirac. Duke Math. J. 29, 231-242 (1962)

Fadell, E., Neuwirth, L., 1962: Configuration spaces. Math. Scand. 10, 111-118 (1962)

Fadell, E., van Buskirk, J.: 1962: The braid groups of E^2 and S^2. Duke Math. J. 29, 243-258 (1962)

Fenchel, W., 1948: Estensioni di gruppi descontinui e transformazioni periodiche delle superficie. Rend. Acc. Naz. Lincei (Sc. fis. mat. e nat.) 5, 326-329 (1948)

Fenchel, W., 1950: Bemaerkingen om endelige gruppen af abbildungsklasser. Mat. Tidsschrift B 1950, 90-95

Feuer, R.D., 1971: Torsion-free subgroups of triangle groups. Proc. Amer. Math. Soc. 30, 235-250 (1971)

Fox, R.H., 1952: On Fenchel's conjecture about F-groups. Mat. Tidsschrift B 1952, 61-65

Fox, R.H., Neuwirth, L.P., 1962: The braid groups. Math. Scand. 10, 119-126 (1962)

Fricke, R., Klein, F., 1897: Vorlesungen über die Theorie der automorphen Funktionen 1. Teubner, Leipzig 1897. Reprinted by Johnson Reprint Corp. New York. Teubner, Stuttgart 1965

Fricke, R., Klein, F., 1912: Vorlesungen über die Theorie der automorphen Funktionen 2. Teubner, Leipzig 1912. Reprinted by Johnson Reprint Corp. New York. Teubner, Stutgart 1965

Funcke, K., 1975: Gegenbeispiele zu einer Vermutung von Magnus. Math. Z. 141, 205-217 (1975)

Garside, F.A., 1969: The braid group and other groups. Quart. J. Math. Oxford 20, 235-254 (1969)

Gerstenhaber, M., 1953: On the algebraic structure of discontinuous groups. Proc. Amer. Math. Soc. 4, 745-750 (1953)

Giblin, P.J., 1977: Graphs, Surfaces and Homology. Chapman & Hall, London 1977

Gillette, R., van Buskirk, J., 1968: The word problem and consequences for braid groups and mapping class groups of the 2-sphere. Trans. Amer. Math. Soc. 131, 277-296 (1968)

Gilman, J., 1976: On conjugacy classes in the Teichmüller modular group. Mich. Math. J. 23, 53-63 (1976)

Gilman, J., 1977: A matrix representation for automorphisms of compact Riemann surfaces. Linear Algebra Appl. 17, 139-147 (1977)

Gilman, J., 1980: On the Nielsen Type and the classification for the mapping-class group. Preprint 1980

Gilman, J., Patterson, D., 1979: Intersection matrices for bases adapted to automorphisms of a compact Riemann surface. Preprint 1979

Goeritz, L., 1933: Normalformen der Systeme einfacher Kurven auf orientierbaren Flächen. Abh. Math. Sem. Univ. Hamburg 9, 223-243 (1933)

Goeritz, L., 1933': Die Abbildungen der Brezelflächen und Vollbrezel vom Geschlecht 2. Abh. Math. Sem. Univ. Hamburg 9, 244-259 (1933)

Goldman, M.E., 1971: An algebraic classification of non-compact manifolds. Trans. Amer. Math. Soc. 156, 241-258 (1971)

Graeub, W., 1950: Die semilinearen Abbildungen. Sitz. Ber. Heidelberger Akad. Wiss. Math. Nat. Kl. 1950, 205-272

Gramain, A., 1971: Topologie des surfaces. Presses Univ. France, Paris 1971

Gramberg, E., Zieschang, H., 1979: Order reduced Reidemeister-Schreier subgroup presentations and applications. Math. Z. 168, 53-70 (1979)

Greenberg, L., 1960: Discrete Groups of motions. Canad. J. Math. 12, 415-426 (1960)

Greenberg, L., 1960': Conformal transformations of Riemann surfaces. Amer. J. Math. 82, 749-760 (1960)

Greenberg, L., 1963: Maximal Fuchsian groups. Bull. Amer. Math. Soc. 69, 569-573 (1963)

Greenberg, L., 1967: Fundamental polygons for Fuchsian groups. J. d'Analyse 18, 99-105 (1967)

Griffiths, H.B. 1963: The fundamental group of a surface, and a theorem of Schreier. Acta Math. 110, 1-17 (1963)

Griffiths, H.B., 1967: A covering-space approach to theorems of Greenberg in Fuchsian, Kleinian and other groups. Commun. Pure Appl. Math. 20, 365-399 (1967)

Griffiths, H.B., 1976: Surfaces. Cambridge Univ. Press, Cambridge 1976

Grossman, E.K., 1974: On the residual finiteness of certain mapping class groups. J. London Math. Soc. 9, 160-164 (1974)

Grushko, I.A., 1940: Über die Basen eines freien Produktes von Gruppen. Mat. Sb. 8, 169-182 (1940)

Hamstrom, M.E., 1962: Some global properties of the space of homeomorphisms of a disk with holes. Duke Math. J. 29, 675-662 (1962)

Hamstrom, M.E., 1965: Homotopy properties of the space of homeomorphisms on P^2 and the Klein bottle. Trans. Amer. Math. Soc. 110, 37-45 (1965)

Hamstrom, M.E., 1965': The space of homeomorphisms on a torus. Illinois J. Math. 9, 59-65 (1965)

Hamstrom, M.E., 1966: Homotopy groups of the space of homeomorphims on a 2-manifold. Illinois Journ. Math. 10, 563-573 (1966)

Hamstrom, M.E., Dyer, E., 1958: Regular mappings and the space of homeomorphisms. Duke Math. J. 25, 521-531 (1958)

Hansen, V.L., 1974: On the space of maps of a closed surface into the 2-sphere. Math. Scand. 35, 149-158 (1974)

Harvey, W.J., 1966: Cyclic groups of automorphisms of a compact Riemann surface. Quart. J. Math. 17, 86-97 (1966)

Harvey, W.J., 1977: Discrete groups and automorphic functions. Academic Press London 1977

Hatcher, A., Thurston, W., 1980: A presentation for the mapping class group of closed orientable surfaces. Topology 19, 221-237 (1980)

Heimes, R., Stöcker, R., 1978: Coverings of surfaces. Archiv Math. 30, 181-187 (1978)

Heineken, H., Strambach, K., 1974: Gruppen auf ebenen nichtkompakten Netzen. Abhandl. math. Sem. Univ. Hamburg 42, 255-265 (1974)

Hemion, G., 1979: On the classification of homeomorphisms of 2-manifolds and the classification of 3-manifolds. Acta Math. 142, 123-155 (1979)

Hempel, J., 1972: Residual finiteness of surface groups. Proc. Amer. Math. Soc. 32, 323 (1972)

Hendriks, H., Shastri, A.R., 1978: A splitting theorem for surfaces. Preprint 1978

Herman, M.R., 1973: Le groupe des difféomorphismes des tores. Ann. L'Institut Fourier, Grenoble, 23, 75-86 (1973)

Higgins, P.J., Lyndon, R.C., 1974: Equivalence of elements under automorphisms of a free group. J. London Math. Soc. 8, 254-258 (1974)

Hirsch, U., 1976: Offene Abbildungen von Flächen auf die 2-Sphäre mit minimalem Defekt. Archiv Math. 27, 649-656 (1976)

Hoare, A.H.M., 1976: On length functions and Nielsen methods in free groups. J. London Math. Soc. (2) 14, 188-192 (1976)

Hoare, A.H.M., 1979: Nielsen methods in groups with a length function. Preprint 1979

Hoare, A.H.M., 1979': On length functions and Nielsen methods in free groups II. Preprint 1979

Hoare, A.H.M., 1980: On lifting Fuchsian groups. Math. Proc. Cambridge Phil. Soc. 87, 61-67 (1980)

Hoare, A.H.M., Karrass, A., Solitar, D., 1971: Subgroups of finite index of Fuchsian groups. Math. Z. 120, 289-298 (1971)

Hoare, A.H.M., Karrass, A., Solitar, D., 1972: Subgroups of infinite index in Fuchsian groups. Math. Z. 125, 59-69 (1972)

Hoare, A.H.M., Karrass, A., Solitar, D., 1973: Subgroups of NEC groups. Commun. Pure Appl. Math. 26, 731-744 (1973)

Hopf, H., 1931: Beiträge zur Klassifikation von Flächenbildungen. J. reine angew. Math. 165, 225-236 (1931)

Hua, L.K., Reiner, L., 1949: On the generators of the symplectic modular group. Trans. Amer. Math. Soc. 65, 415-426 (1949)

Humphries, S.P., 1979: Generators for the mapping class group Topology of low-dimensional manifolds. Lecture Notes in Math. 722, 44-47, Springer Berlin 1979

Hurewicz, W., 1931: Zu einer Arbeit von O. Schreier. Abh. Math. Seminar Univ. Ham. burg 8, 307-314 (1931)

Игнатов, Ю.А., 1978: Свободные и несвободные подгруппы $PSL_2(C)$ порожденные двумя параболическими элементами. Мат. сб. 106, 372-379 (1978). Engl. transl.: Ignatov, Ju.A., 1978: Free and non-free subgroups of $PSL_2(C)$ generated by two parabolic elements. Math. USSR Sbornik 35 (1979)

Jaco, W., Shalen, P.B., 1977: Surface homeomorphisms and periodicity. Topology 16, 347-367 (1977)

Johansson, I., 1931: Topologische Untersuchungen über unverzweigte Überlagerungsflächen. Norsk Vid. Ak. Skr. Mat. Naturw. Kl. 1, 1-69 (1931)

Johnson, D.L., 1979: Homeomorphisms of a surface which act trivially on homology. Proc. Amer. Math. Soc. 75, 119-125 (1979)

Jones, G.A., Singerman, D., 1978: Theory of maps on orientable surfaces. Proc. London Math. Soc. (3) 37, 273-307 (1978)

Jonsson, W., 1970: On a result of A.M. Macbeath on normal subgroups of a Fuchsian group. Canad. Math. Bull. 13, 15-16 (1970)

Jordan, C., 1866: Sur la déformation des surfaces. J. Mathém. Pures Appl. (2) 11, 105-109 (1866)

Jordan, C., 1866': Des contours tracés sur les surfaces. J. Mathém. Pures Appl. (2) 11, 110-130 (1866)

Jucovič, E., Trenkler, E., 1973: A theorem on the structure of cell-decompositions of orientable 2-manifolds. Mathematika 20, 63-82 (1973)

Jørgensen, T., 1978: Closed geodesics on Riemann surfaces. Proc. Amer. Math. Soc. 72, 140-142 (1978)

Karrass, A., Solitar, D., 1957: A note on a theorem of Schreier. Proc. Amer. Math. Soc. 8, 696-697 (1957)

Keen, L., 1966: Canonical polygons for finitely generate Fuchsian groups. Acta Math. 115, 1-16 (1966)

Keen, L., 1971: On Fricke moduli. Advances theory of Riemann surfaces. Annals of Math. Studies 66, 205-224. Princeton University Press, Princeton N.J. 1971

Keen, L., 1971': On infinitely generated Fuchsian groups. J. Indian Math. Soc. 35, 67-85 (1971)

Keen, L., 1973: A correction to "On Fricke moduli". Proc. Amer. Math. Soc. 40, 60-62 (1973)

Keller, R., 1973: Zur Klassifikation ebener nichteuklidischer krystallographischer Gruppen mit kompaktem Fundamentalbereich. Diplomarbeit Ruhr-Universität Bochum 1973

von Kerékjártó, B., 1923: Vorlesungen über Topologie. Springer Verlag, Berlin 1923

von Kerékjártó, B., 1934: Über reguläre Abbildungen von Flächen auf sich. Acta Sci. Math. Szeged 7, 65-75 (1934)

von Kerékjártó, B., 1934': Bemerkung über reguläre Abbildungen von Flächen. Acta Sci. Math. Szeged 7, 206 (1935)

Kerckhoff, S.P., 1980: The Nielsen realization problem. Bull. Amer. Math. Soc. 2, 452-454 (1980)

Kiley, W.T., 1970: Automorphism groups on compact Riemann surfaces. Trans. Amer. Math. Soc. 150, 557-563 (1970)

Klein, F., 1879: Über die Transformationen siebenter Ordnung der elliptischen Funktionen. Math. Ann. 14, 428-471 (1879)

Klingen, H., 1956: Über die Erzeugenden gewisser Modulgruppen. Nachr. Akad. Wiss. Göttingen, Math.-Phys. Klasse Nr. 8, 1956, 173-185

Klingen, 1961: Charakterisierung der Siegelschen Modulgruppe durch ein endliches System definierender Relationen. Math. Ann. 144, 64-82 (1961)

Knapp, W., 1968: Doubly generated Fuchsian groups. Mich. Math. J. 15, 289-304 (1968)

Kneser, H., 1926: Die Deformationssätze der einfach zusammenhängenden Flächen. Math. Z. 25, 362-372 (1926)

Kneser, H., 1928: Glättung von Flächenabbildungen. Math. Ann. 100, 609-617 (1928)

Kneser, H., 1930: Die kleinste Bedeckungszahl innerhalb einer Klasse von Flächenabbildungen. Math. Ann. 103, 347-358 (1930)

Kra, I., 1972: Automorphic forms and Kleinian groups. Benjamin, Reading, Mass., 1972

Kravetz, S., 1959: On the geometry of Teichmüller spaces and the structure of their modular group. Ann. Acad. Sci. Fenn. Ser. A VI 278 (1959)

Kroll, M., 1974: Fundamentalpolygone und Fricke Moduli Riemannscher Flächen. Staatsexamensarbeit Ruhr-Universität Bochum 1974

Kuhn, H.W., 1952: Subgroup theorems for groups presented by generators and relations. Ann. of Math. 58, 22-46 (1952)

Kurosch, A.G., 1934: Die Untergruppen der freien Produkte von beliebigen Gruppen. Math. Ann. 109, 647-660 (1934)

Курош, А.Г., 1967: Теория групп. Изд. Наука, Москва 1967
Kurosh, A.G., 1967: The Theory of Groups, I, II. Chelsea, London 1960

Ladegaillerie, Y., 1974: Classification topologique des plongements des 1-complexes compactes dans les surfaces. I: Cas des surfaces orientées. Comptes Rend. Ac. Sci. Ser. A 278, 1401-1404 (1974)
II: surfaces non orientées, plongements minimaux. Comptes Rend. Ac. Sci. Ser. A 279, 129-132 (1974)

Ladegaillerie, Y., 1975: Un théorème d'isotopie sur les surfaces. Comptes Rend. Ac. Sci. Ser. A 281, 195-197 (1975)

Ladegaillerie, Y., 1976: Groupes de tresses et problème des mots dans les groupes de tresses. Bull. Sc. math. 100, 255-267 (1976)

Larcher, H., 1963: A necessary and sufficient condition for a discrete group of linear fractional transformations to be discontinuous. Duke Math. J. 30, 433-436 (1963)

Lauritzen, S., 1946: Om Afbildning af ikke orientierbare Flader paa sic selv. Mat. Tidsskript B 1946, 92-96

Lee, J.P., 1972: Homeotopy groups of orientable 2-manifolds. Fund. Math. 77, 115-124 (1972)

Lehner, J., 1964: Discontinuous groups and automorphic functions. Math. Surveys VIII, Amer. Math. Soc. Providence, Rhode Island 1964

Levi, F., 1929: Geometrische Konfigurationen. Hirzel Verlag, Leipzig 1929

Levine, H.J., 1963: Homotopic curves on surfaces. Proc. Amer. Math. Soc. 14, 986-990 (1963)

Lickorish, W.B.R., 1963: Homeomorphisms of non-orientable 2-manifolds. Proc. Cambridge Phil. Soc. 59, 307-317 (1963)

Lickorish, W.B.R., 1964: A finite set of generators for the homeotopy group of a 2-manifold. Proc. Cambridge Phil. Soc. 60, 769-778 (1964)

Lickorish, W.B.R., 1966: A finite set of generators for the homeotopy group of a 2-manifold (corrigendum). Proc. Cambridge Phil. Soc. 62, 679-681 (1966)

Liebmann, H., 1912: Nichteuklidische Geometrie. 2. Aufl. Berlin und Leipzig: Göschen'sche Verlagshandlung, Berlin und Leipzig 1912

Luft, E., 1971: Covering of 2-dimensional manifolds with open cells. Archiv Math. 22, 536-544 (1971)

Luke, R., Mason, W.K., 1972: The space of homeomorphisms on a compact two-manifold is an absolute neighborhood retract. Trans. Amer. Math. Soc. 164, 275-285 (1972)

Lyndon, R.C., 1959: The equation $a^2b^2 = c^2$ in free groups. Mich. Math. J. 6, 89-94 (1959)

Lyndon, R.C., 1962: Depedence and independence in free groups. J. reine angew. Math. 210, 148-174 (1962)

Lyndon, R.C., 1966: On Dehn's algorithm. Math. Ann. 166, 208-238 (1966)

Lyndon, R.C., 1976: On the combinatorial Riemann-Hurwitz formula. In: Symposia Math. 17, 435-439. Istituto Naz. Alta Math. Academic Press, London-New York 1976

Lyndon, R.C., 1978: Quadratic words in free products with amalgamation. Houston Math. J. 4, 91-103 (1978)

Lyndon, R.C., Schupp, P.E., 1977: Combinatorial group theory. Ergebn. Math. Grenzgebiete 89, Springer Verlag, Berlin 1977

Macbeath, A.M., 1961: Fuchsian groups. Proc. summer school, Queen's College, Dundee (Scotland), 1961

Macbeath, A.M., 1961': On a theorem of Hurwitz. Proc. Glasgow Math. Assoc. 5, 90-96 (1961)

Macbeath, A.M., 1962: On a theorem by J. Nielsen. Quart. J. Math., Ser. 2, 13, 235-236 (1962)

Macbeath, A.M., 1965: Geometrical realisations of isomorphisms between plane groups. Bull. Amer. Math. Soc. 71, 629-630 (1965)

Macbeath, A.M., 1965': On a curve of genus 7. Proc. London Math. Soc. 15, 527-542 (1965)

Macbeath, A.M., 1967: The classification of non euclidean plane crystallographic groups. Canad. J. Math. 6, 1192-1205 (1967)

Macbeath, A.M., 1973: Action of automorphisms of a compact Riemann surface on the first homology group. Bull. London Math. Soc. 5, 103-108 (1973)

Macbeath, A.M., Hoare, A.H.M., 1976: Groups of hyperbolic crystallography. Math. Proc. Cambridge Phil. Soc. 79, 235-249 (1976)

Macbeath, A.M., Singerman, D., 1975: Spaces of subgroups and Teichmüller space. Proc. London Math. Soc. 31, 211-256 (1975)

Maclachlan, C., 1965: Abelian groups of automorphisms of compact Riemann surfaces. Proc. London Math. Soc. 15, 699-712 (1965)

Maclachlan, C., 1969: A bound on the number of automorphisms of a compact Riemann surface. J. London Math. Soc. 44, 265-272 (1969)

Maclachlan, C., 1971: Maximal normal Fuchsian groups. Illinois J. Math. 15, 104-113 (1971)

Maclachlan, C., 1977: Note on the Hurwitz-Nielsen realization problem. Proc. Amer. Math. Soc. 64, 87-90 (1977)

Maclachlan, C., 1978: On Representations of Artin's braid groups. Mich. Math. J. 25, 235-244 (1978)

Maclachlan, C., Harvey, W.J., 1975: On mapping-class groups and Teichmüller spaces. Proc. London Math. Soc. (3) 30, 496-512 (1975)

MacLane, S., 1958: A proof of the subgroup theorem for free products. Mathematica 5, 13-18 (1958)

Magnus, W., 1930: Über diskontinuierliche Gruppen mit einer definierenden Relation. J. reine angew. Math. 163, 141-165 (1930)

Magnus, W., 1932: Das Identitätsproblem für Gruppen mit einer definierenden Relation. Math. Ann. 106, 295-307 (1932)

Magnus, W., 1934: Über Automorphismen von Fundamentalgruppen berandeter Flächen. Math. Ann. 109, 617-648 (1934)

Magnus, W., 1972: Braids and Riemann surfaces. Commun. Pure Appl. Math. 25, 151-161 (1972)

Magnus, W., 1975: Two generator subgroups of PSL(2,C). Nachr. Akad. Wiss. Göttingen 1975 Nr. 7, 1-14

Magnus, W., 1974: Noneuclidean tesselations and their groups. Academic Press, New York 1974

Magnus, W., 1975: Braid groups: a survey. Proc. Second Intern. Conf. Theory of Groups, Canberra 1973. Lect. Notes in Math. 372, 463-487, Springer Verlag 1975

Magnus, W., Karrass, A., Solitar, D., 1966: Combinatorial Group Theory: Presentations of groups in terms of generators and relations. Interscience Publishers, John Wiley + Sons, Inc. New York, London, Sidney, 1966

Mal'zev, A.J., 1962: Über Gleichungen $zxyx^{-1}y^{-1}z^{-1} = aba^{-1}b^{-1}$ in freien Gruppen. Algebra i Logica 1, 45-50 (1962)

Mangler, W., 1939: Die Klassen topologischer Abbildungen einer geschlossenen Fläche auf sich. Math. Z. 44, 541-554 (1939)

Marden, A., 1967: On finitely generated Fuchsian Groups. Comment. Math. Helvetici 42, 81-85 (1967)

Marden, A., 1969: On homotopic mappings of Riemann surfaces. Ann. of Math. 90, 1-8 (1969)

Markov, A.A., 1958: Unsolvability of the problem of homeomorphy. Proc. International Congr. of Mathematicians Edinburgh 1958, pp. 300-306

Maskit, B., 1965: A theorem of planar covering surfaces with applications to 3-manifolds. Ann. of Math. 81, 341-355 (1965)

Massey, W.S., 1974: Finite covering spaces of 2-manifolds with boundary. Duke Math. J. 41, 875-887 (1974)

Massey, W.S., 1967: Algebraic Topology: An Introduction. Harcourt, Brace & World, Inc., New York 1967

May, C.L., 1977: Large automorphism groups of compact Klein surfaces with boundary, I. Glasgow Math. J. 18, 1-10 (1977)

May, C.L., 1975: Automorphisms of compact Klein surfaces with boundary. Pacific J. Math. 59, 199-210 (1975)

May, C.L, 1979: Cyclic automorphism groups of compact bordered Klein surfaces. Houston J. Math., Preprint 1979

McCool, J., 1975: Some finitely presented subgroups of the automorphism group of a free group. J. Algebra 35, 205-213 (1975)

McCool, J., Pietrowski, A., 1971: On free products with amalgamation of two infinite cyclic groups. J. Algebra 18, 337-383 (1971)

Медньк, А.Д., 1979: О неразветвленных накрытиях компактных римановых поверхностей. Доклады АН СССР 244, 529-532 (1979). Engl. Transl.:
Mednyk, A.D., 1979: On unramified coverings of compact Riemann surfaces. Soviet Math., Doklady, 20, 85-88 (1979)

Mellis, W., Schadowski, U., Zieschang, H., 1980: Zu einem Satz von Hurwitz über Abbildungen von Flächen. Preprint 1980

Mennicke, J., 1961: A note on regular coverings of closed orientable surfaces. Proc. Glasgow Math. Ass. 5, 49-66 (1961)

Mennicke, J., 1967: Eine Bemerkung über Fuchs'sche Gruppen. Invent. math. 2, 301-305 (1967)

Mennicke, J., 1968: Corrigendum zu "Eine Bemerkung über Fuchs'sche Gruppen". Invent. math. 6, 106 (1968)

Möbius, A.F., 1863: Theorie der elementaren Verwandtschaft. Ber. Verhandl. Königl. Sächs. Gesellsch. Wiss., math.-phys. Kl. 15, 18-57 (1863). Gesammelte Werke 2. Band, 433-471, S. Hirzel Verlag, Leipzig 1886, Neudruck: Dr. Martin Sändig oHG, Wiesbaden 1967

Möbius, A.F., 1886: Zur Theorie der Polyeder und der Elementarverwandtschaften. Gesammelte Werke 2. Band, 511-559, S. Hirzel Verlag, Leipzig 1886. Neudruck: Dr. Martin Sändig oHG, Wiesbaden 1967

Moise, E.E., 1977: Geometric topology in dimension 2 and 3. Springer Verlag, Berlin 1977

Moore, M.J., 1970: Fixed points of automorphisms of compact Riemann surfaces. Canad. J. Math. 22, 922-932 (1970)

Moore, M.J., 1972: Riemann surfaces as orbit spaces of Fuchsian groups. Canad. J. Math. 24, 612-616 (1972)

Morton, H.R., 1967: The space of homeomorphisms of a disc with n holes. Illinois, J. Math. 11, 40-48 (1967)

Narens, L., 1971: A nonstandard proof of the Jordan curve theorem. Pacific J. Math. 36, 219-229 (1971)

Натанзон, С.М., 1972: Инвариантные прямые фуксовых групп. Успехи Мат. Наук 27:4, 145-160 (1972). Engl. transl.:
Natanzon, S.M., 1972: Invariant lines of Fuchsian groups. Russian Math. Surveys 27:4, 161-177 (1972)

Натанзон, С.М., 1978: О порядке конечной группы диффеоморфизмов поверхности на себя и числе вещественных форм комплексной алгебраической кривой. Доклады АН СССР 242, 765-768 (1978). Engl. transl.:
Natanzon, S.M., 1978: On the order of a finite group of homeomorphisms of a surface onto itself and the number of real forms of a complex algebraic curve. Soviet Math., Doklady, 19, 1195-1199 (1978)

Neumann, B.H., 1943: On the number of generators of a free product. J. London Math. Soc. 18, 12-20 (1943)

Neumann, B.H., 1954: An essay on free products of groups with amalgamation. Phil. Trans. Roy. Soc. London Ser. A 246, 503-554 (1954)

Newman, M.H.A., 1951: Elements of the topology of plane sets of points. Cambridge University Press, Cambridge 1951

Nielsen, J., 1918: Die Isomorphismen der allgemeinen, unendlichen Gruppe mit zwei Erzeugenden. Math. Ann. 78, 385-397 (1918)

Nielsen, J., 1919: Über die Isomorphismen unendlicher Gruppen ohne Relation. Math. Ann. 79, 269-272 (1919)

Nielsen, J., 1924: Die Isomorphismengruppen der freien Gruppen. Math. Ann. 91, 169-209 (1924)

Nielsen, J., 1927: Untersuchungen zur Topologie der geschlossenen zweiseitigen Flächen I. Acta Math. 50, 189-358 (1927)

Nielsen, J., 1929: Untersuchungen zur Theorie der geschlossenen zweiseitigen Flächen II. Acta Math. 53, 1-76 (1929)

Nielsen, J., 1932: Untersuchungen zur Theorie der geschlossenen zweiseitigen Flächen, III. Acta Math. 58, 87-167 (1932)

Nielsen, J., 1935: Einige Sätze über topologische Flächenabbildungen. Acta Sci. Math. Szeged 7, 200-205 (1935)

Nielsen, J., 1936: Topologischer Beweis eines Satzes von Wiman. Mat. Tidsskrift B 1936, 11-24

Nielsen, J., 1937: Die Struktur periodischer Transformationen von Flächen. Det Kgl. Dansk Videnskabernes Selskab Mat.-fys. Meddelerer 15, 1-77 (1937)

Nielsen, J., 1937': Aekvivalenzproblemet for periodike Transformationer. Mat. Tidsskrift B 1937, 33-41

Nielsen, J., 1940: Über Gruppen linearer Transformationen. Mitteilungen Math. Gesellsch. Hamburg 8, 82-104 (1940)

Nielsen, J., 1942: Abbildungsklassen endlicher Ordnung. Acta Math. 75, 23-115 (1942)

Nielsen, J., 1955: A basis for subgroups of free groups. Math. Scand. 3, 31-43 (1955)

Jakob Nielsen in Memoriam (by W. Fenchel) 1960. Acta Math. 103, VII-XIV (1960)

Novikov, P.S., 1955: On the algorithmic unsolvability of the word problem in group theory (Russian). Trudy Mat. Inst. Steklov 44, 143 pp. (1955)

Peczynski, N., 1972: Eine Kennzeichnung der Relation der Fundamentalgruppe einer nicht-orientierbaren geschlossenen Fläche. Diplomarbeit Ruhr-Universität Bochum 1972

Peczynski, N., Reiwer, W., 1978: On cancellations in HNN-groups. Math. Z. 158, 79-86 (1978)

Peczynski, N., Rosenberger, G., Zieschang, H., 1975: Über Erzeugende ebener diskontinuierlicher Gruppen. Invent. math. 29, 161-180 (1975)

Pettey, D.H., 1972: Mappings onto 2-dimensional spaces. Bull. Amer. Math. Soc. 78, 53-54 (1972)

Poincaré, H., 1882: Theorie des groupes fuchsiens. Acta Math. 1, 1-62 (1882)

Powell, J., 1978: Two theorems on the mapping class group of a surface. Proc. Amer. Math. Soc. 68, 347-350 (1978)

Purzitsky, N., 1974: Canonical generators of fuchsian groups. Illinois J. Math. 18, 484-490 (1974)

Purzitsky, N., Rosenberger, G., 1972: Two generator Fuchsian groups of genus one. Math. Z. 128, 245-251 (1972)

Quine, J.R., 1977: Tangent winding numbers and branched mappings. Pacific J. Math. 73, 161-167 (1977)

Quintas, L.V., 1965: The homotopy groups of the space of homeomorphisms of a multiply punctured sphere. Illinois J. Math. 9, 721-725 (1965)

Quintas, L.V., 1968: Solved and unsolved problems in the computation of the homeotopy groups of 2-manifolds. Trans. N.Y. Acad. Sci. 30, 919-938 (1968)

Rabin, M.O., 1958: Recursive unsolvability of group theoretic problems. Ann. of Math. 67, 172-194 (1958)

Rado, T., 1924: Über den Begriff der Riemannschen Fläche. Acta Univ. Szeged 2, 101-121 (1924-26)

Rapaport, E.S., 1958: On free groups and their automorphisms. Acta Math. 99, 139-163 (1958)

Reidemeister, K., 1927: Über unendliche diskrete Gruppen. Abh. math. Sem. Univ. Hamburg 5, 33-39 (1927)

Reidemeister, K., 1932: Einführung in die kombinatorische Topologie. Friedr. Vieweg und Sohn, Braunschweig 1932

Reidemeister, K., Brandis, A., 1959: Über freie Erzeugendensysteme der Wegegruppen eines zusammenhängenden Graphen. Sammelband zu Ehren des 250. Geburtstages Leonhard Eulers, pp. 284-292. Akademie Verlag, Berlin 1959

Reinhart, B.L., 1960: Simple curves on compact surfaces. Proc. Nat. Acad. Sci. 46, 1242-1243 (1960)

Reinhart, B.L., 1962: Algorithms for Jordan curves on compact surfaces. Ann. of Math. 75, 209-222 (1962)

Reinhart, B.L., 1963: Further remarks on the winding number. Ann. Inst. Fourier (Grenoble) 13, 155-160 (1963)

Richards, J., 1963: On the classification of noncompact surfaces. Trans. Amer. Math. Soc. 106, 259-269 (1963)

Riemann, B., 1851: Grundlagen für eine allgemeine Theorie der Functionen einer veränderlichen complexen Größe. Inauguraldissertation, Göttingen 1951. Gesammelte Mathematische Werke 1892. Dover Publ. Inc., New York 1953

Riemann, B., 1857: Theorie der Abel'schen Functionen. J. reine angew. Math. 54, 115-155 (1857). Gesammelte Mathematische Werke 1892. Dover Publ. Inc. New York, 1953

Ritter, G.X., 1978: A characterization of almost periodic homeomorphisms on the 2-sphere and the annulus. General Topology Appl. 9, 185-191 (1978)

Rosenberger, G., 1972: Fuchssche Gruppen, die freies Produkt zweier zyklischer Gruppen sind, und die Gleichung $x^2+ y^2+ z^2 = xyz$. Math. Ann. 199, 213-227 (1972)

Rosenberger, G., 1973: Eine Bemerkung zu den Triangel-Gruppen. Math. Z. 132, 239-244 (1973)

Rosenberger, G., 1974: Zum Rang- und Isomorphieproblem für freie Produkte mit Amalgam. Habilitationsschrift, Hamburg 1974

Rosenberger, G., 1978: Alternierende Produkte in freien Gruppen. Pacific J. Math. 78, 243-250 (1978)

Rosenberger, G., 1978': Produkte von Potenzen und Kommutatoren in freien Gruppen. J. Algebra 53, 416-422 (1978)

Sah, C.-H., 1969: Groups related to compact Riemann surfaces. Acta Math. 123, 13-42 (1969)

Sanatani, S., 1967: On planar group diagrams. Math. Ann. 172, 203-208 (1967)

Schafer, J.A., 1976: Representing homology classes on surfaces. Canad. Math. Bull. 19, 373-374 (1976)

Schattschneider, D., 1978: The plane symmetry groups: their recognition and notation. Amer. Math. Monthly 85, 439-450 (1978)

Scherrer, W., 1929: Zur Theorie der endlichen Gruppen topologischer Abbildungen von geschlossenen Flächen in sich. Comment. Math. Helv. 1, 69-119 (1929)

Schläfli, L., 1872: Quand'è che dalla superficie generale di terzo ordine si stacca una patre che non sia realmente segata da ogni piano reale? Annali Mat. pua appl. 2 5, 289-295 (1872) Gesamm. Math. Abhandlungen, Birkhäuser, Basel 1956

Schläfli, L., 1873: Über die linearen Relationen zwischen den 2p Kreiswegen erster Art und den 2p Kreiswegen zweiter Art in der Theorie der Abelschen Funktionen der Herren Clebsch und Gordan. J. reine angew. Math. 76, 149-155 (1973). Gesamm. Math. Abhandl., Birkhäuser, Basel 1956

Schreier, O., 1924: Über die Gruppen $A^aB^b = 1$. Abh. math. Skm. Univ. Hamburg 3, 167-169 (1924)

Schreier, O., 1927: Die Untergruppen der freien Gruppen. Abh. math. Sem. Univ. Hamburg 5, 161-183 (1927)

Schubert, H., 1953: Knoten und Vollringe. Acta Math. 90, 131-286 (1953)

Schupp, P.E., 1968: On Dehn's algorithm and the conjugacy problem. Math. Ann. 178, 119-130 (1968)

Scott, G.P., 1970: The space of homeomorphisms of a 2-manifold. Topology 9, 97-109 (1970)

Scott, G.P., 1970': Braid groups and the group of homeomorphisms of a surface. Proc. Cambridge Phil. Soc. 68, 605-617 (1970)

Scott, P., 1978: Subgroups of surface groups are almost geometric. J. London Math. Soc. (2) 17, 555-565 (1978)

Seifert, H., 1937: Bemerkungen über stetigen Abbildungen von Flächen. Abh. math. Sem. Univ. Hamburg 12, 23-37 (1937)

Seifert, H., Threlfall, W., 1934: Lehrbuch der Topologie. Teubner, Leipzig 1934

Selberg, A., 1960: On discontinuous groups in higher dimensional symmetric spaces. Colloquium Function Theory pg. 147-164, Bombay 1960

Serre, J.-P., 1960/61: Rigidité du fonctour de Jacobi d'échelon $n \geq 3$. Séminaire Henri Cartan 1960/61, pg. 17-18 - 17-20, Paris 1960/1961

Shepardson, C.B., 1973: Generalized Hurwitz-Riemann formulas. Indiana Univ. Math. J. 23, 277-285 (1973)

Siegel, C.L., 1945: Some Remarks on Discontinuous groups. Ann. of Math. 46, 708-718 (1945)

Siegel, C.L., 1950: Bemerkungen zu einem Satz von Jakob Nielsen. Mat. Tidsskrift B 1950, 66-70

Siegel, C.L., 1964: Vorlesungen über ausgewählte Kapitel der Funktionentheorie I. Math. Institut Göttingen 1964. Topics in complex function theory. Vol. 1 Wiley, New York 1969

Siegel, C.L., 1964': Vorlesungen über ausgewählte Kapitel der Funktionentheorie II. Math. Institut Göttingen 1964. Topics in complex function theory. Vol. 2 Wiley, New York 1971

Singerman, D., 1971: Automorphisms of non-orientable Riemann surfaces. Glasgow Math. J. 12, 50-59 (1971)

Singerman, D., 1974: On the structure of non-Euclidean crystallographic groups. Proc. Cambridge Phil. Soc. 76, 233-240 (1974)

Singerman, D., 1974': Symmetries of Riemann surfaces with large automorphism groups. Math. Ann. 210, 17-32 (1974)

Singerman, D., 1976: Automorphisms of maps, permutation groups and Riemann surfaces. Bull. London Math. Soc. 8, 65-68 (1976)

Smith, P.A., 1934: A theorem on fixed points of periodic transformations. Ann. of Math. 35, 572-578 (1934)

Smith, P.A., 1967: Abelian action on 2-manifolds. Michigan Math. J. 14, 257-279 (1967)

Spanier, E.H., 1966: Algebraic topology. New York: McGraw-Hill, New York 1966

Sperner, E., 1934: Über fixpunktfreie Abbildungen der Ebene. Abhandl. Math. Sem. Univ. Hamburg 10, 1-48 (1934)

Sprows, D.J., 1975: Homeotopy groups of compact 2-manifolds. Fund. Math. 90, 99-103 (1975)

Stallings, J.R., 1965: A topological proof of Grushko's theorem on free products. Math. Z. 90, 1-8 (1965)

Stillwell, J.C., 1979: Isotopy in surface complexes from the computational viewpoint. Bull. Austr. Math. Soc. 20, 1-6 (1979)

Stillwell, J.C., 1979': Unsolvability of the knot problem for surface complexes. Bull. Austr. Math. Soc. 20, 131-137 (1979)

Stothers, W.W., 1977: Subgroups of the (2,3,7) triangle group. manuscripta math. 20, 323-334 (1977)

Teichmüller, O., 1940: Extremale quasikonforme Abbildungen und quadratische Differentiale. Preuß. Akad. Ber. 22, 1-197 (1940)

Teichmüller, O., 1943: Bestimmung der extremalen quasikonformen Abbildung bei geschlossenen orientierbaren Riemannschen Flächen. Preuß. Akad. Ber. 4 (1943)

Threlfall, W., 1932: Gruppenbilder. Abh. sächs. Akad. Wiss. math.-phys. Kl. 41, Nr. 6, 1-59 (1932)

Tietze, H., 1908: Über die topologischen Invarianten mehrdimensionaler Mannigfaltigkeiten. Monatsh. Math. Phys. 19, 1-118 (1908)

Tietze, H., 1914: Über stetige Abbildungen einer Quadratfläche auf sich selbst. Rend. Circ. Mat. Palermo 38, 1-58 (1914)

Timmann, S., 1976: A bound for the number of automorphisms of a finite Riemann surface. Kodai Math. Sem. Rep. 28, 104-109 (1978)

Tukia, P., 1972: On discrete groups of the unit disk and their isomorphisms. Annales Acad. Sci. Fennicae A I. 504, 5-44 (1972)

Тураев, В.Г., 1978: Пересечения петель в двумерных многообразиях. Мат. сб. 106, 566-588 (1978). Engl. transl.:
Turaev, V.G., 1978: Intersection of loops in two-dimensional manifolds. Math. USSR Sbornik 35, 229-250 (1979)

Van Buskirk, J., 1966: Braid groups of compact 2-manifolds with elements of finite order. Trans. Amer. Math. Soc. 122, 81-97 (1966)

van der Waerden, B.L., 1948: Free products of groups. Amer. J. Math. 70, 527-528 (1948)

van der Waerden, B.L., 1955: Algebra 1. 4. Aufl. Springer Verlag, Berlin 1955

Vollmer, C., Bekemeier, B., 1978: Die Moduln markierter Fuchsscher Gruppen. Diplomarbeit, Ruhr-Universität Bochum 1978

Vollmerhaus, W., 1963: Über die Automorphismen ebener Gruppen. Dissertation, Göttingen 1963

Wagner, N.R., 1974: A continuity property with applications to the topology 2-manifolds. Trans. Amer. Math. Soc. 200, 369-393 (1974)

Weir, A.J., 1956: The Reidemeister-Schreier and Kuroš subgroup theorem. Mathematika 3, 47-55 (1956)

Whitehead, J.H.C., 1936: On certain sets of elements in a free group. Proc. Lond. Math. Soc. 41, 48-56 (1936)

Whitehead, J.H.C., 1936': On equivalent sets of elements in a free group. Ann. of Math. 37, 782-800 (1936)

Wilkie, M.C., 1966: On noneuclidean cristallographic groups. Math. Z. 91, 87-102 (1966)

Wiman, A., 1895/96: Über die hyperelliptischen Curven und diejenigen vom Geschlecht p = 3, welche eindeutige Transformationen auf sich zulassen. Bihang Til Kongl. Svenska Veienskaps-Akademiens Hadlingar, Stockholm 1985-86

Witt, E., 1941: Eine Identität zwischen Modulformen zweiten Grades. Abhandl. Math. Sem. Univ. Hamburg 14, 323-337 (1941)

Young, G.S., 1947: A characterization of 2-manifolds. Duke Math. J. 14, 979-990 (1947)

Zarrow, R., 1979: Orientation reversing maps of surfaces. Illinois J. Math. 23, 82-92 (1979)

Zarrow, R., 1979': Orientation reversing square roots of involutions. Illinois J. Math. 23, 71-81 (1979)

Zieschang, H., 1962: Über Worte $S_1^{a_1}S_2^{a_2}\ldots S_q^{a_q}$ in freien Gruppen mit p freien Erzeugenden. Math. Ann. 147, 143-153 (1962)

Цишанг, Х., 1963': Теорема Нильсена, некоторые ее приложения и обобщения. Труды IV Всесоюзн. топол. конф. Ташкент 1963, Изд. ФАН Узб. ССР, Ташкент 1967
Zieschang, H., 1963': A theorem of Nielsen, some of its applications and generalizations. Trudy IV Allunion conf. on topology, Taschkent 1963. FAN-publ. Usbecist. SSR, Taschkent 1967

Zieschang, H., 1964: Alternierende Produkte in freien Gruppen. Abh. math. Sem. Univ. Hamburg 27, 13-31 (1964)

Цишанг, Х., 1964': Об автоморфизмах плоских групп. Доклады АН СССР 155, 57-60 (1964). Engl. transl.:
Cisang, H., (H. Zieschang) 1964': Automorphisms of planar groups. Soviet Math., Doklady, 5, 364-367 (1964)

Zieschang, H., 1965: Alternierende Produkte in freien Gruppen II. Abh. math. Sem. Univ. Hamburg 28, 219-233 (1965)

Zieschang, H., 1965': Algorithmen für einfache Kurven auf Flächen. Math. Scand. 17, 17-40 (1965)

Цишанг, Х., 1966: Дискрвтые группы движений плоскости и плоские групповые образы. Успехи Мат. Наук 21:3, 195-212 (1966)
Zieschang, H., 1966: Discrete groups of motions of the plane and planar group diagrams. Uspechi Mat. Nauk 21:3, 195-212 (1966)

Zieschang, H., 1966': Über Automorphismen ebener diskontinuierlicher Gruppen. Math. Ann. 166, 148-167 (1966)

Zieschang, H., 1969: Algorithmen für einfache Kurven auf Flächen II. Math. Scand. 25, 49-58 (1969)

Zieschang, H., 1970: Über die Nielsensche Kürzungsmethode in freien Produkten mit Amalgam. Invent. math. 10, 4-37 (1970)

Zieschang, H., 1971: On extensions of fundamental groups of surfaces and related groups. Bull. Amer. Math. Soc. 77, 1116-1119 (1971)

Zieschang, H., 1973: On the homeotopy groups of surfaces. Math. Ann. 206, 1-21 (1973)

Zieschang, H., 1973': Lifting and projecting homeomorphisms. Archiv Math. 24, 416-421 (1973)

Zieschang, H., 1974: Addendum to "On extensions of fundamental groups of surfaces and related groups". Bull. Amer. Math. Soc. 80, 366-367 (1974)

Цишанг, Х., 1976': О треугольных группах. Успехи Мат. Наук 31:5, 177-183 (1976). Engl. transl.:
Zieschang, H., 1976': On triangle groups. Russian Math. Surveys 31:5, 226-233 (1976)

Zieschang, H., 1977: Generators of the free product with amalgamation of two infinite cyclic groups. Math. Ann. 227, 195-221 (1977)

Zieschang, H., 1980: On finite groups of mapping classes of surfaces. Lecture Notes in Math., Preprint 1980

Цишанг, Х., 1980: О разложениях дискпетных групп движении плоскости. Успехи Мат. Наук
Zieschang, H., 1980: On decompositions of groups of motions of the plane. Uspechi Mat. Nauk. Preprint 1980

Zieschang, H., Vogt, E., Coldewey, H.-D., 1970: Flächen und ebene diskontinuierliche Gruppen. Lecture Notes in Math. 122, Springer Verlag, Berlin 1970

Zimmermann, B., 1977: Endliche Erweiterungen nichteuklidischer kristallographischer Gruppen. Math. Ann. 231, 187-192 (1977)

Zimmermann, B., 1977': Eine Verallgemeinerung der Formel von Riemann-Hurwitz. Math. Ann. 229, 279-288 (1977)

LIST OF NOTATION

INDEX

A

B

C

D

E

U

V

W

Z

Vol. 670: Fonctions de Plusieurs Variables Complexes III, Proceedings, 1977. Edité par F. Norguet. XII, 394 pages. 1978.

Vol. 671: R. T. Smythe and J. C. Wierman, First-Passage Perculation on the Square Lattice. VIII, 196 pages. 1978.

Vol. 672: R. L. Taylor, Stochastic Convergence of Weighted Sums of Random Elements in Linear Spaces. VII, 216 pages. 1978.

Vol. 673: Algebraic Topology, Proceedings 1977. Edited by P. Hoffman, R. Piccinini and D. Sjerve. VI, 278 pages. 1978.

Vol. 674: Z. Fiedorowicz and S. Priddy, Homology of Classical Groups Over Finite Fields and Their Associated Infinite Loop Spaces. VI, 434 pages. 1978.

Vol. 675: J. Galambos and S. Kotz, Characterizations of Probability Distributions. VIII, 169 pages. 1978.

Vol. 676: Differential Geometrical Methods in Mathematical Physics II, Proceedings, 1977. Edited by K. Bleuler, H. R. Petry and A. Reetz. VI, 626 pages. 1978.

Vol. 677: Séminaire Bourbaki, vol. 1976/77, Exposés 489–506. IV, 264 pages. 1978.

Vol. 678: D. Dacunha-Castelle, H. Heyer et B. Roynette. Ecole d'Eté de Probabilités de Saint-Flour. VII-1977. Edité par P. L. Hennequin. IX, 379 pages. 1978.

Vol. 679: Numerical Treatment of Differential Equations in Applications, Proceedings, 1977. Edited by R. Ansorge and W. Törnig. IX, 163 pages. 1978.

Vol. 680: Mathematical Control Theory, Proceedings, 1977. Edited by W. A. Coppel. IX, 257 pages. 1978.

Vol. 681: Séminaire de Théorie du Potentiel Paris, No. 3, Directeurs: M. Brelot, G. Choquet et J. Deny. Rédacteurs: F. Hirsch et G. Mokobodzki. VII, 294 pages. 1978.

Vol. 682: G. D. James, The Representation Theory of the Symmetric Groups. V, 156 pages. 1978.

Vol. 683: Variétés Analytiques Compactes, Proceedings, 1977. Edité par Y. Hervier et A. Hirschowitz. V, 248 pages. 1978.

Vol. 684: E. E. Rosinger, Distributions and Nonlinear Partial Differential Equations. XI, 146 pages. 1978.

Vol. 685: Knot Theory, Proceedings, 1977. Edited by J. C. Hausmann. VII, 311 pages. 1978.

Vol. 686: Combinatorial Mathematics, Proceedings, 1977. Edited by D. A. Holton and J. Seberry. IX, 353 pages. 1978.

Vol. 687: Algebraic Geometry, Proceedings, 1977. Edited by L. D. Olson. V, 244 pages. 1978.

Vol. 688: J. Dydak and J. Segal, Shape Theory. VI, 150 pages. 1978.

Vol. 689: Cabal Seminar 76–77, Proceedings, 1976–77. Edited by A.S. Kechris and Y. N. Moschovakis. V, 282 pages. 1978.

Vol. 690: W. J. J. Rey, Robust Statistical Methods. VI, 128 pages. 1978.

Vol. 691: G. Viennot, Algèbres de Lie Libres et Monoïdes Libres. III, 124 pages. 1978.

Vol. 692: T. Husain and S. M. Khaleelulla, Barrelledness in Topological and Ordered Vector Spaces. IX, 258 pages. 1978.

Vol. 693: Hilbert Space Operators, Proceedings, 1977. Edited by J. M. Bachar Jr. and D. W. Hadwin. VIII, 184 pages. 1978.

Vol. 694: Séminaire Pierre Lelong – Henri Skoda (Analyse) Année 1976/77. VII, 334 pages. 1978.

Vol. 695: Measure Theory Applications to Stochastic Analysis, Proceedings, 1977. Edited by G. Kallianpur and D. Kölzow. XII, 261 pages. 1978.

Vol. 696: P. J. Feinsilver, Special Functions, Probability Semigroups, and Hamiltonian Flows. VI, 112 pages. 1978.

Vol. 697: Topics in Algebra, Proceedings, 1978. Edited by M. F. Newman. XI, 229 pages. 1978.

Vol. 698: E. Grosswald, Bessel Polynomials. XIV, 182 pages. 1978.

Vol. 699: R. E. Greene and H.-H. Wu, Function Theory on Manifolds Which Possess a Pole. III, 215 pages. 1979.

Vol. 700: Module Theory, Proceedings, 1977. Edited by C. Faith and S. Wiegand. X, 239 pages. 1979.

Vol. 701: Functional Analysis Methods in Numerical Analysis, Proceedings, 1977. Edited by M. Zuhair Nashed. VII, 333 pages. 1979.

Vol. 702: Yuri N. Bibikov, Local Theory of Nonlinear Analytic Ordinary Differential Equations. IX, 147 pages. 1979.

Vol. 703: Equadiff IV, Proceedings, 1977. Edited by J. Fábera. XIX, 441 pages. 1979.

Vol. 704: Computing Methods in Applied Sciences and Engineering, 1977, I. Proceedings, 1977. Edited by R. Glowinski and J. L. Lions. VI, 391 pages. 1979.

Vol. 705: O. Forster und K. Knorr, Konstruktion verseller Familien kompakter komplexer Räume. VII, 141 Seiten. 1979.

Vol. 706: Probability Measures on Groups, Proceedings, 1978. Edited by H. Heyer. XIII, 348 pages. 1979.

Vol. 707: R. Zielke, Discontinuous Čebyšev Systems. VI, 111 pages. 1979.

Vol. 708: J. P. Jouanolou, Equations de Pfaff algébriques. V, 255 pages. 1979.

Vol. 709: Probability in Banach Spaces II. Proceedings, 1978. Edited by A. Beck. V, 205 pages. 1979.

Vol. 710: Séminaire Bourbaki vol. 1977/78, Exposés 507–524. IV, 328 pages. 1979.

Vol. 711: Asymptotic Analysis. Edited by F. Verhulst. V, 240 pages. 1979.

Vol. 712: Equations Différentielles et Systèmes de Pfaff dans le Champ Complexe. Edité par R. Gérard et J.-P. Ramis. V, 364 pages. 1979.

Vol. 713: Séminaire de Théorie du Potentiel, Paris No. 4. Edité par F. Hirsch et G. Mokobodzki. VII, 281 pages. 1979.

Vol. 714: J. Jacod, Calcul Stochastique et Problèmes de Martingales. X, 539 pages. 1979.

Vol. 715: Inder Bir S. Passi, Group Rings and Their Augmentation Ideals. VI, 137 pages. 1979.

Vol. 716: M. A. Scheunert, The Theory of Lie Superalgebras. X, 271 pages. 1979.

Vol. 717: Grosser, Bidualräume und Vervollständigungen von Banachmoduln. III, 209 pages. 1979.

Vol. 718: J. Ferrante and C. W. Rackoff, The Computational Complexity of Logical Theories. X, 243 pages. 1979.

Vol. 719: Categorial Topology, Proceedings, 1978. Edited by H. Herrlich and G. Preuß. XII, 420 pages. 1979.

Vol. 720: E. Dubinsky, The Structure of Nuclear Fréchet Spaces. V, 187 pages. 1979.

Vol. 721: Séminaire de Probabilités XIII. Proceedings, Strasbourg, 1977/78. Edité par C. Dellacherie, P. A. Meyer et M. Weil. VII, 647 pages. 1979.

Vol. 722: Topology of Low-Dimensional Manifolds. Proceedings, 1977. Edited by R. Fenn. VI, 154 pages. 1979.

Vol. 723: W. Brandal, Commutative Rings whose Finitely Generated Modules Decompose. II, 116 pages. 1979.

Vol. 724: D. Griffeath, Additive and Cancellative Interacting Particle Systems. V, 108 pages. 1979.

Vol. 725: Algèbres d'Opérateurs. Proceedings, 1978. Edité par P. de la Harpe. VII, 309 pages. 1979.

Vol. 726: Y.-C. Wong, Schwartz Spaces, Nuclear Spaces and Tensor Products. VI, 418 pages. 1979.

Vol. 727: Y. Saito, Spectral Representations for Schrödinger Operators With Long-Range Potentials. V, 149 pages. 1979.

Vol. 728: Non-Commutative Harmonic Analysis. Proceedings, 1978. Edited by J. Carmona and M. Vergne. V, 244 pages. 1979.

Vol. 729: Ergodic Theory. Proceedings, 1978. Edited by M. Denker and K. Jacobs. XII, 209 pages. 1979.

Vol. 730: Functional Differential Equations and Approximation of Fixed Points. Proceedings, 1978. Edited by H.-O. Peitgen and H.-O. Walther. XV, 503 pages. 1979.

Vol. 731: Y. Nakagami and M. Takesaki, Duality for Crossed Products of von Neumann Algebras. IX, 139 pages. 1979.

Vol. 732: Algebraic Geometry. Proceedings, 1978. Edited by K. Lønsted. IV, 658 pages. 1979.

Vol. 733: F. Bloom, Modern Differential Geometric Techniques in the Theory of Continuous Distributions of Dislocations. XII, 206 pages. 1979.

Vol. 734: Ring Theory, Waterloo, 1978. Proceedings, 1978. Edited by D. Handelman and J. Lawrence. XI, 352 pages. 1979.

Vol. 735: B. Aupetit, Propriétés Spectrales des Algèbres de Banach. XII, 192 pages. 1979.

Vol. 736: E. Behrends, M-Structure and the Banach-Stone Theorem. X, 217 pages. 1979.

Vol. 737: Volterra Equations. Proceedings 1978. Edited by S.-O. Londen and O. J. Staffans. VIII, 314 pages. 1979.

Vol. 738: P. E. Conner, Differentiable Periodic Maps. 2nd edition, IV, 181 pages. 1979.

Vol. 739: Analyse Harmonique sur les Groupes de Lie II. Proceedings, 1976-78. Edited by P. Eymard et al. VI, 646 pages. 1979.

Vol. 740: Séminaire d'Algèbre Paul Dubreil. Proceedings, 1977-78. Edited by M.-P. Malliavin. V, 456 pages. 1979.

Vol. 741: Algebraic Topology, Waterloo 1978. Proceedings. Edited by P. Hoffman and V. Snaith. XI, 655 pages. 1979.

Vol. 742: K. Clancey, Seminormal Operators. VII, 125 pages. 1979.

Vol. 743: Romanian-Finnish Seminar on Complex Analysis. Proceedings, 1976. Edited by C. Andreian Cazacu et al. XVI, 713 pages. 1979.

Vol. 744: I. Reiner and K. W. Roggenkamp, Integral Representations. VIII, 275 pages. 1979.

Vol. 745: D. K. Haley, Equational Compactness in Rings. III, 167 pages. 1979.

Vol. 746: P. Hoffman, τ-Rings and Wreath Product Representations. V, 148 pages. 1979.

Vol. 747: Complex Analysis, Joensuu 1978. Proceedings, 1978. Edited by I. Laine, O. Lehto and T. Sorvali. XV, 450 pages. 1979.

Vol. 748: Combinatorial Mathematics VI. Proceedings, 1978. Edited by A. F. Horadam and W. D. Wallis. IX, 206 pages. 1979.

Vol. 749: V. Girault and P.-A. Raviart, Finite Element Approximation of the Navier-Stokes Equations. VII, 200 pages. 1979.

Vol. 750: J. C. Jantzen, Moduln mit einem höchsten Gewicht. III, 195 Seiten. 1979.

Vol. 751: Number Theory, Carbondale 1979. Proceedings. Edited by M. B. Nathanson. V, 342 pages. 1979.

Vol. 752: M. Barr, *-Autonomous Categories. VI, 140 pages. 1979.

Vol. 753: Applications of Sheaves. Proceedings, 1977. Edited by M. Fourman, C. Mulvey and D. Scott. XIV, 779 pages. 1979.

Vol. 754: O. A. Laudal, Formal Moduli of Algebraic Structures. III, 161 pages. 1979.

Vol. 755: Global Analysis. Proceedings, 1978. Edited by M. Grmela and J. E. Marsden. VII, 377 pages. 1979.

Vol. 756: H. O. Cordes, Elliptic Pseudo-Differential Operators – An Abstract Theory. IX, 331 pages. 1979.

Vol. 757: Smoothing Techniques for Curve Estimation. Proceedings, 1979. Edited by Th. Gasser and M. Rosenblatt. V, 245 pages. 1979.

Vol. 758: C. Nǎstǎsescu and F. Van Oystaeyen; Graded and Filtered Rings and Modules. X, 148 pages. 1979.

Vol. 759: R. L. Epstein, Degrees of Unsolvability: Structure and Theory. XIV, 216 pages. 1979.

Vol. 760: H.-O. Georgii, Canonical Gibbs Measures. VIII, 190 pages. 1979.

Vol. 761: K. Johannson, Homotopy Equivalences of 3-Manifolds with Boundaries. 2, 303 pages. 1979.

Vol. 762: D. H. Sattinger, Group Theoretic Methods in Bifurcation Theory. V, 241 pages. 1979.

Vol. 763: Algebraic Topology, Aarhus 1978. Proceedings, 1978. Edited by J. L. Dupont and H. Madsen. VI, 695 pages. 1979.

Vol. 764: B. Srinivasan, Representations of Finite Chevalley Groups. XI, 177 pages. 1979.

Vol. 765: Padé Approximation and its Applications. Proceedings, 1979. Edited by L. Wuytack. VI, 392 pages. 1979.

Vol. 766: T. tom Dieck, Transformation Groups and Representation Theory. VIII, 309 pages. 1979.

Vol. 767: M. Namba, Families of Meromorphic Functions on Compact Riemann Surfaces. XII, 284 pages. 1979.

Vol. 768: R. S. Doran and J. Wichmann, Approximate Identities and Factorization in Banach Modules. X, 305 pages. 1979.

Vol. 769: J. Flum, M. Ziegler, Topological Model Theory. X, 151 pages. 1980.

Vol. 770: Séminaire Bourbaki vol. 1978/79 Exposés 525-542. IV, 341 pages. 1980.

Vol. 771: Approximation Methods for Navier-Stokes Problems. Proceedings, 1979. Edited by R. Rautmann. XVI, 581 pages. 1980.

Vol. 772: J. P. Levine, Algebraic Structure of Knot Modules. XI, 104 pages. 1980.

Vol. 773: Numerical Analysis. Proceedings, 1979. Edited by G. A. Watson. X, 184 pages. 1980.

Vol. 774: R. Azencott, Y. Guivarc'h, R. F. Gundy, Ecole d'Eté de Probabilités de Saint-Flour VIII-1978. Edited by P. L. Hennequin. XIII, 334 pages. 1980.

Vol. 775: Geometric Methods in Mathematical Physics. Proceedings, 1979. Edited by G. Kaiser and J. E. Marsden. VII, 257 pages. 1980.

Vol. 776: B. Gross, Arithmetic on Elliptic Curves with Complex Multiplication. V, 95 pages. 1980.

Vol. 777: Séminaire sur les Singularités des Surfaces. Proceedings, 1976-1977. Edited by M. Demazure, H. Pinkham and B. Teissier. IX, 339 pages. 1980.

Vol. 778: SK1 von Schiefkörpern. Proceedings, 1976. Edited by P. Draxl and M. Kneser. II, 124 pages. 1980.

Vol. 779: Euclidean Harmonic Analysis. Proceedings, 1979. Edited by J. J. Benedetto. III, 177 pages. 1980.

Vol. 780: L. Schwartz, Semi-Martingales sur des Variétés, et Martingales Conformes sur des Variétés Analytiques Complexes. XV, 132 pages. 1980.

Vol. 781: Harmonic Analysis Iraklion 1978. Proceedings 1978. Edited by N. Petridis, S. K. Pichorides and N. Varopoulos. V, 213 pages. 1980.

Vol. 782: Bifurcation and Nonlinear Eigenvalue Problems. Proceedings, 1978. Edited by C. Bardos, J. M. Lasry and M. Schatzman. VIII, 296 pages. 1980.

Vol. 783: A. Dinghas, Wertverteilung meromorpher Funktionen in ein- und mehrfach zusammenhängenden Gebieten. Edited by R. Nevanlinna and C. Andreian Cazacu. XIII, 145 pages. 1980.

Vol. 784: Séminaire de Probabilités XIV. Proceedings, 1978/79. Edited by J. Azéma and M. Yor. VIII, 546 pages. 1980.

Vol. 785: W. M. Schmidt, Diophantine Approximation. X, 299 pages. 1980.

Vol. 786: I. J. Maddox, Infinite Matrices of Operators. V, 122 pages. 1980.